XII International Conference on Solar-Terrestrial Relations and Physics of Earthquake Precursors 2021

EPJ Web of Conferences Volume 254 (2021)

Paratunka, Russia
27 September - 1 October 2021

Editors:

Alexei Dmitriev
Gleb Vodinchar
Zafar Rakhmonov

ISBN: 978-1-7138-3893-7

Printed from e-media with permission by:

Curran Associates, Inc.
57 Morehouse Lane
Red Hook, NY 12571

Some format issues inherent in the e-media version may also appear in this print version.

For additional information, please contact EDP Sciences – Web of Conferences at the address below.

EDP Sciences – Web of Conferences
17, Avenue du Hoggar
Parc d'Activité de Courtabœuf
BP 112
F-91944 Les Ulis Cedex A
France

Phone: +33 (0) 1 69 18 75 75
Fax: +33 (0) 1 69 28 84 91

contact-edps@webofconferences.org

Additional copies of this publication are available from:

Curran Associates, Inc.
57 Morehouse Lane
Red Hook, NY 12571 USA
Phone: 845-758-0400
Fax: 845-758-2633
Email: curran@proceedings.com
Web: www.proceedings.com

TABLE OF CONTENTS

INSTITUTE OF COSMOPHYSICAL RESEARCH AND RADIO WAVE PROPAGATION FEB RAS

SOLAR-TERRESTRIAL RELATIONS AND PHYSICS OF EARTHQUAKE PRECURSORS

12th International Conference

September 27 - October 1, 2021, Paratunka, Russia

http://www.ikir.ru/en/Events/Conferences/2021-XII/index.html

The following topical areas in solar-terrestrial physics and in geosphere processes dynamics have been discussed at the conference:

- solar activity impact on geosphere processes;
- geophysical fields and their interaction;
- dynamic processes in the atmosphere;
- radiophysics and acoustics of energy-active zones;
- simulation of geophysical processes;
- technology of data acquisition and analysis in helio-geophysical monitoring systems;
- seismic activity impact on geosphere processes and earthquake precursor physics.

Common Use Center
North-Eastern
Heliogeophysical
Center
CKP 558279

Conference Organizers

Institute of Cosmophysical Research and Radio Wave Propagation FEB RAS

Common Use Center
«North-Eastern Heliogeophysical Center»

Scientific Committee

Alexei Dmitriev, Taiwan
Boris Shevtsov, Russia
János Lichtenberger, Hungary
Kazuo Shiokawa, Japan
Oksana Mandrikova, Russia
Sergey Khomutov, Russia
Vadim Bogdanov, Russia
Veneta Guineva, Bulgaria
Vladimir Sivokon', Russia
Wojciech Miloch, Norway
Yuriy Senkevich, Russia
Zafar Rakhmonov, Uzbekistan

Statement of Peer review

In submitting conference proceedings to *Web of Conferences*, the editors of the proceedings certify to the Publisher that

1. They adhere to its **Policy on Publishing Integrity** in order to safeguard good scientific practice in publishing.
2. All articles have been subjected to peer review administered by the proceedings editors.
3. Reviews have been conducted by expert referees, who have been requested to provide unbiased and constructive comments aimed, whenever possible, at improving the work.
4. Proceedings editors have taken all reasonable steps to ensure the quality of the materials they publish and their decision to accept or reject a paper for publication has been based only on the merits of the work and the relevance to the journal.

Title, date and place of the conference

XII International Conference "Solar-Terrestrial Relations and Physics of Earthquake Precursors", September 27 - October 1, 2021, Paratunka, Kamchata, Russia

Proceedings editor(s):

Dr. Alexei Dmitriev
Dr. Gleb Vodinchar
Dr. Zafar Rakhmonov

Date and editor's signature

Annual variation of atmospheric electricity diurnal variation maximum in Kamchatka

Sergey Smirnov[1,*]

[1]Institute of Cosmophysical Researches and Radio Wave Propagation FEB RAS, Paratunka, Kamchatsky Kray, Russia

Abstract. Atmospheric electric field diurnal variation measured in fair weather conditions over the ocean surface has a typical form which is called a unitary variation. It is associated with the global time and occurs simultaneously all over the planet. However, the diurnal variation, measured over the ground, depends on many local factors. The diurnal variation maximum of the electric field potential gradient, measured at Paratunka observatory, has the maximum close in time to the unitary variation maximum. In the paper we show that this maximum is determined by local conditions and is associated in time with the sunrise. The diurnal variation maximum of the electric field potential gradient, measured at Paratunka observatory in fair weather conditions, has annual variation coinciding with the annual variation of local sunrise.

1 Introduction

In fair weather conditions, a characteristic diurnal variation of the electric field potential gradient (PG) is observed over the oceans. This diurnal variation does not depend on local time and is observed simultaneously throughout the planet. It was called the unitary variation (UT variation). UT-variation is determined by thunderstorm activity all over the globe. The maximum variation is observed around 19 hours of universal time (UT) [1, 2]. Local time (LT) at Paratunka observatory is 12 hours ahead of UT. Thus, the UT variation maximum occurs durning the morning hours for the observation point. But in the morning time, besides the UT-variation, other mechanisms work as well.

At the observatories located on the continents, the effect of morning maximum is observed in PG diurnal variations [3–5]. Brown [3] explained this effect as follows. During the night, positively charged condensation nuclei are accumulated in the exchange layer of the atmosphere, and in the morning, as air temperature increases, these nuclei are transported upward as a result of turbulence and convection processes in the atmosphere. In the paper [5] the author investigated the statistical characteristics of the sunrise effect at Paratunka observatory. The work [6] gave physical explanation of this phenomenon.

Paratunka observatory is located 30 km from the Pacific coast. PG measurements can be influenced both by local and global factors. The authors set the task to find out if the morning maximum of the PG daily variation at the observatory are caused by the UT variation or by the sunrise effect.

*e-mail: sergey@ikir.ru

2 Measurement methods

Measurements of atmospheric electricity at Paratunka observatory are carried out with the Pole-2 fluxmeter. It was developed at the Voeikov Main Geophysical Observatory, Russia. Since 2016, simultaneously with the Pole-2 sensor, the measurements have been carried out with the CS110 sensor developed by Campbell Scientific, Inc., USA. The article [7] shows the identity of the measurements carried out by the both sensors. Double control of the measurements carried out by the instruments from different manufacturers increases the quality and liquidity of the data.

The electrical parameters of surface atmosphere are highly dependent on meteorological conditions at the observation site. Measurements of meteorological parameters (temperature, air pressure, wind speed and direction, precipitation level) are carried out by Davis digital weather station with 10-minute interval -. When processing the data of PG diurnal variations , the meteorological situation on the selected days is taken into account.

In Fig. 1 shows the shape-stable diurnal variations in PG under fair weather conditions. The data were processed from January 1, 2004 to November 1, 2020. The statistics included variations in fair weather conditions with a maximum after 17h local time. During the period under the study there were 300 such daily variations.

3 Main results and discussion

We made a Table with the data on the time of astronomical sunrise for the geographical point, where the Paratunka observatory is located (52.97° N; 158.25° E) , and the time of PG daily cycle maximum. Then the data were sorted in ascending order. Then approximation was carried out by the formula

$$Y = AX + B, \tag{1}$$

where X is the sunrise time in hours, Y is the PG maximum time in hours. If the coefficient A in this formula is close to zero, then the PG maximum at Paratunka observatory is determined by the UT variation. If the coefficient A is close to 1, then the PG maximum is determined by local factors. In Fig. 2, the abscissa shows the sunrise time for the observation point, and the ordinate shows the time of the PG maximum. The dots indicate the observed maxima of the PG daily variation in fair weather conditions, and the straight line shows their approximation by formula 1. As a result of the approximation, we obtained A = 0.92, B = 2.5, root-mean-square error = 0.03. The correlation coefficient is 0.9. Thus, the time of the PG daily cycle maximum shifts during the year following the time of sunrise in the region.

The physical mechanism of this phenomenon is as follows. The density of electric current in the absence of lightning sources is written as:

$$j = \lambda E + \rho V + D_t \nabla \rho, \tag{2}$$

where λ is the atmospheric electrical conductivity, the main contribution to it is made by light ions; E is the electric field strength; ρ – is the electric charge density; V is the medium hydrodynamic velocity; D_t is the turbulent diffusion coefficient. In the quasistationary case, the current density is determined by the first term and is attributed to the action of thunderstorm generators. The first term in formula 2 is responsible for the conduction current, which is determined by to the discharge current in the areas of fair weather in the global electrical circuit. After sunrise, the main contribution to the electric current density is made by the second and third terms of formula 2. As a result of turbulent heat transfer, turbulent mixing processes ($D_t \nabla \rho$) come into action and the positive space charge, accumulated near the earth's surface

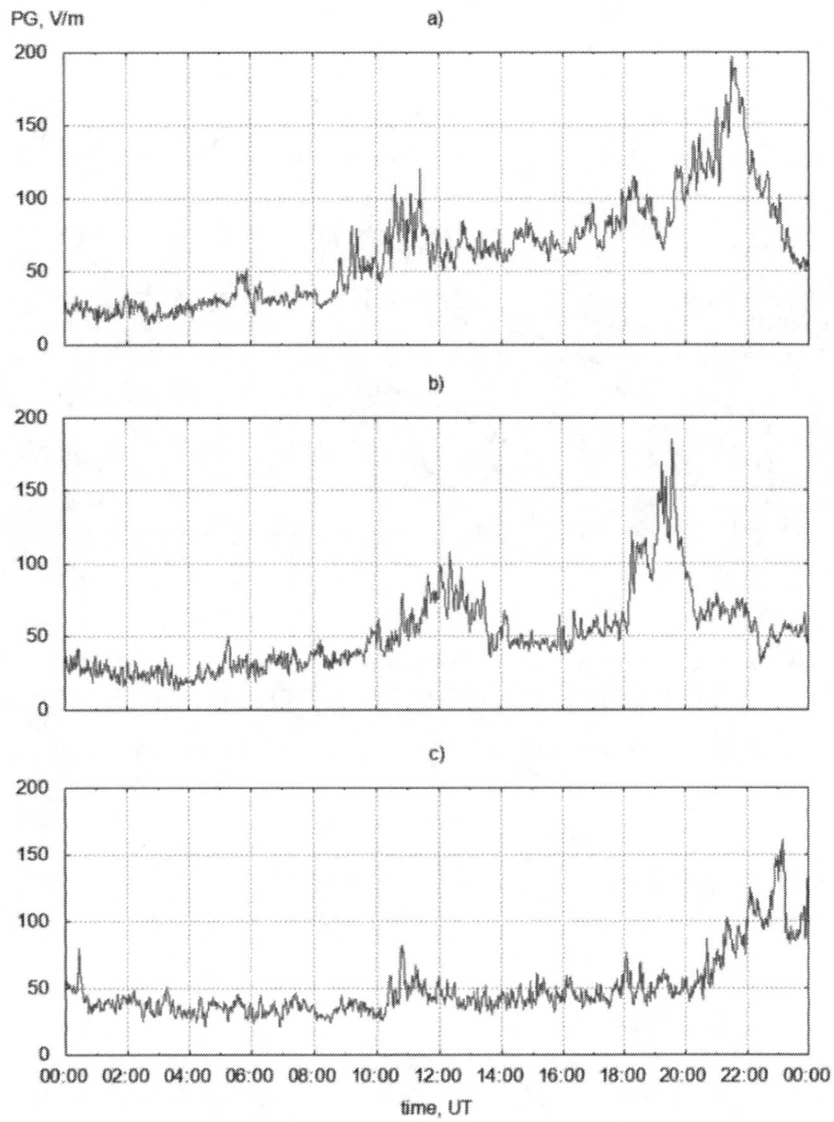

Figure 1. PG daily variations (a) for October 26, 2012, (b) for August 25, 2016, (c) for February 12, 2020.

at night, is transfered upwards (ρV) by a convective air flow . That, in its turn, leads to PG increase near the earth's surface and electric conduction current amplification. This mechanism is confirmed by experimental data. The PG daily variation with a maximum in the morning correlates with the difference in air temperatures measured at different heights [6].

To clarify the question of when a stable PG daily variation with a maximum in the morning appears, a histogram of the distribution of such days over months was made (Fig. 3). It

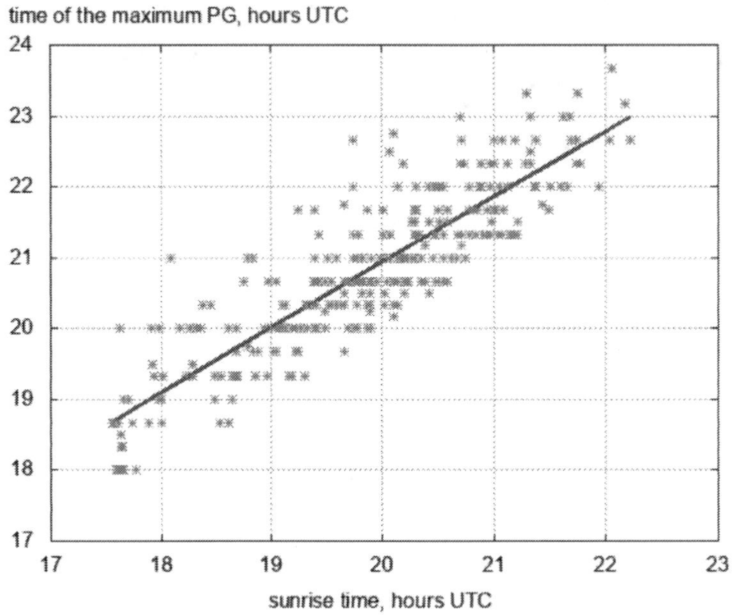

Figure 2. The stars indicate the PG daily cycle maxima. Straight line is the value linear approximation. The abscissa axis is the sunrise time, the ordinate axis is the time of the PG daily cycle maximum.

turned out that the most often the PG daily variations with maximum values in the morning hours in fair weather conditions appear in March and October.

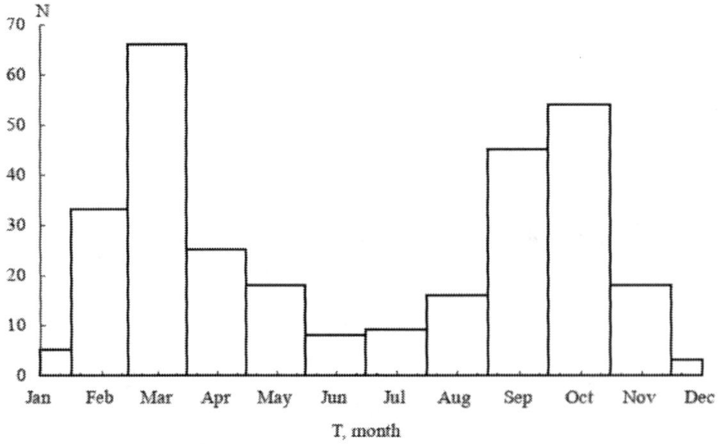

Figure 3. Histogram of the distribution of the number of days with the PG morning maximum over months.

4 Conclusions

1. In fair weather conditions, the maximum daily variation of atmospheric electricity in the surface layer is determined by the sunrise effects.

2. Most often, characteristic days with a maximum in the atmospheric electricity daily variation are observed in March and October.

Funding

The work was carried out as part of the realization of the State Task AAAA-A21-121011290003-0.

Acknowledgements

The authors are appreciative of «Paratunka» observatory FEB RAS.

References

[1] H. Kasemir, Pure Appl. Geophys. **100**, 70–80 (1972)
[2] J. Tacza, J.-P. Raulin, L. Macotela, E. Norabuena, G. Fernandez, E. Correia, M. Rycroft, R. Harrison, J. Atmos. Sol. Terr. Phys. **120**, 70–79 (2014)
[3] J. Brown, Terr. Magn. Atmos. Elect. **41**, 279–285 (1936)
[4] S. Gurmani, N. Ahmad, J. Tacza, T. Iqbal, J. Atmos. Sol. Terr. Phys. **179**, 441–449 (2018)
[5] S. Smirnov, G. Mikhailova, O. Kapustina, Geomagnetism and Aeronomy **52**, 507–512 (2012)
[6] S. Smirnov, Geomagnetism and Aeronomy **53**, 515–521 (2013)
[7] S. Khomutov, S. Smirnov, S. Butin, I. Babakhanov, E3S Web of Conferences **11**, 00008 (2016)

Development of a substorms catalog including the MPB observed at Panagjurishte station, Bulgaria

Veneta Guineva[1,*], *Rolf* Werner[1], *Andris* Lubchich[2], *Atanas* Atanassov[1], *Rumiana* Bojilova[3], *Lyubomira* Raykova[1], *Dimitar* Valev[1], and *Irina* Despirak[2]
[1]Space Research and Technology Institute (SRTI), Bulgarian Academy of Sciences, Stara Zagora Department, Bulgaria
[2]Polar Geophysical Institute, Apatity, Russia
[3]National Institute of Geophysics, Geodesy and Geography (NIGGG), Bulgarian Academy of Sciences, Sofia, Bulgaria

Abstract. In this work the first results of the creation of a substorms catalog including the mid-latitude positive bays (MPB) registered at the Bulgarian station Panagjurishte (~37° GMLat, ~97° GMLon) are presented. MPB index characterizes the mid-latitude effect of substorm, which developed at auroral latitudes, and it is associated with the substorm current wedge.
The work went in several lines. First, the catalog design and content were taken. Second, tools of data processing have been developed. Third, substorms during two months, namely January and February 2013 were identified. And finally, data processing was implemented and the characteristics of the positive bays were determined.
The obtained data and their visualizations were put in a convenient folder and file structure, ftp positioned.

1. Introduction

Substorms are a typical event at auroral latitudes and a substantial source of magnetic activity. It is known, that the magnetic disturbances during substorms are connected to the auroral electrojets development. The dynamics, intensity and location of the electrojets have been investigated still since the 70s of the past century [e.g. 1-3]. Depending on the conditions in the solar wind and the geomagnetic activity, substorm disturbances, characteristic for auroral latitudes (~60° to ~71° GMlat), could reach as very high latitudes (latitudes of the polar cap, >70° GMlat) [2, 4, 5], as well as middle (~50° GMlat) and even low latitudes (<20° GMlat). After McPherron et al. [1] midlatitudes are from 20° to 52°. It is to be noted, that substorms at middle and low latitudes are usually observed during magnetic storms, when the equatorial boundary of the auroral oval is shifted as far as ~50° GMlat [6].

In contrast to the auroral latitudes, where the magnetic substorms are expressed by sharp negative bays in the X magnetic component, at midlatitudes the magnetic substorms are observed as positive bays (maxima) in the X component [1]. First, it was assumed, that the

*Corresponding author: v_guineva@yahoo.com

low latitude return currents from the westward electrojet have given rise to the positive bays [7], later the positive bays were explained as a result of the upwards field-alligned currents [8, 9]. Subsequently it was found out, that the midlatitude positive bays were connected to the substorm current wedge (SCW) – a current system including the deviation of the magnetospheric tail current along the magnetic field lines to the ionosphere in the dawn side and its return to the magnetosphere through the dusk side [10-12]. The structure of the SCW was studied in order to verify whether this current system is formed by an ensemble of wedglets [13] and it was suggested, that the SCW is basically a globally coherent system.

A special index was developed, the midlatitude positive bay index (MPB) [14-16], as an indicator of the characteristics of the SCW. It can be used as a proxy of the magnetic substorms onset. This index is widespread in the magnetospheric investigations. For example, a connection has been ascertained between the MPB index maxima and the appearance of plasmoids in the magnetospheric tail [17].

The MPB index was computed by data of 35 low- and midlatitude stations, globally distributed [14, 15]. A correlation between the MPB index and the SuperMag SML index was found out [16]. But data from the INTERMAGNET European stations and of the midlatitude stations of the Scandinavian longitudinal chain NAL-BRZ were not examined in these studies. The MPB index by data of the European midlatitude stations was not computed up to now. To fill up this gap, the European midlatitude data should be processed to obtain the horizontal magnetic power (main field and mean Solar quite magnetic variations removed) and to use them for the MPB index evaluation.

In this work the development of a catalog including the midlatitude positive bays registered at the midlatitude station Panagjurishte, Bulgaria, and the concomitant studies are presented. The catalog data could be useful in the studies of the relation between the appearence of midlatitude positive bays and the solar wind and Interplanetary Magnetic field (IMF) parameters and for the determination of the concrete drivers in the solar wind responsible for the positive bays at midlatitudes.

2. Structure of the Panagjurishte substorm catalog

The Panagjurishte substorm catalog was envisaged to be FTP positionned. The planned structure of the catalog is presented schematically in Fig.1. The main part of the catalog consists of data, derived from the Panagjurishte magnetic measurements. A complimentary part including some additional data about the substorm days is planned, as well. The main part comprises every day data (graphics and data files of the processing results) and data about every substorm (some parameters of the magnetic positive bay). The complimentary part (concomitant data) includes information as the X bay sign convertion latitude, the structures in the solar wind, some geomagnetic indices for the day and the presence or no of a geomagnetic storm.

3. Processing procedure: calculation of the horizontal power of the magnetic field

The midlattitude horizontal power of the magnetic field, main field and mean Solar quite magnetic variations (Sq) removed (MPB) is a usefull measure for substorm identification by midlattitude magnetic data. The midlattitude positive bay index can be evaluated by the average of the horizontal power computed for a number of midlattitude stations.

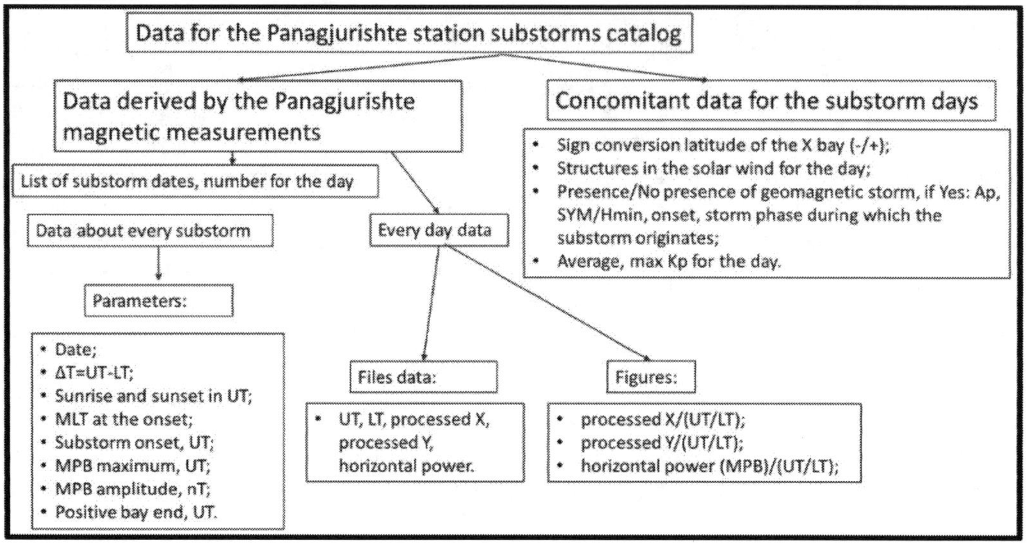

Fig. 1. Scheme of the structure of the Panagjurishte substorm catalog.

3.1. Data used

The processing was performed for January and February 2013. Data from the International Real-time Magnetic Observatory Network (Intermagnet) Data Download (intermagnet.org) for the Bulgarian station Panagjurishte (PAG) (42.5°N, 24.2°E, 556 m altitude; ~37° GMLat, ~97° GMLon) were used. The one minute mean data for the magnetic field components X, Y, Z and the total field intensity are given in nT for moments in UT.

The magnetic field data for 25 successive days centred at the day under consideration were processed. Following the McPherron's algorithm [15] this time interval is sufficient to estimate and remove the mean main magnetic field and the mean field under solar quiet conditions.

3.2. Preprocessing

The preprocessing of the observed magnetic data includes procedures for removal of data gaps, for peak detection and for detection and removal of days with very high magnetic disturbances (called for shortness "disturbed days"). Data gapes and peaks are usually randomly distributed and are not concentrated within a day. Therefore as a first step, time series of the whole interval of the considered consecutive 25 days was constructed from the measurements of every magnetic component. Because the data are given for UT, the time series are converted to LT by a corresponding to the geographical longitude data shifting.

3.2.1. Data gaps removal

Data gaps have mostly technical reasons as electrical power outage or required maintenance. If no data was registered, then the value 99999 was written in the original data file. In the gapes the values of an additional series are set to a constant many greater than the magnetic variances and so they are well graphically presentable. The interval borders of the data gapes are then easy to found by means of the first differences. The no data values were replaced by the result of the linear interpolation between the last and first measurements outside of the no data interval. In the special case when no data were measured at the beginning or end of the 25 days series, the no data were replaced by the nearest observations. For the illustration of the gap and peak detection and removal the data for the substorm on 20.02.2017 were used. In the time series from 08.02.2017 up to 04.03.2017 five gaps were identified. One example is shown in Fig.2.

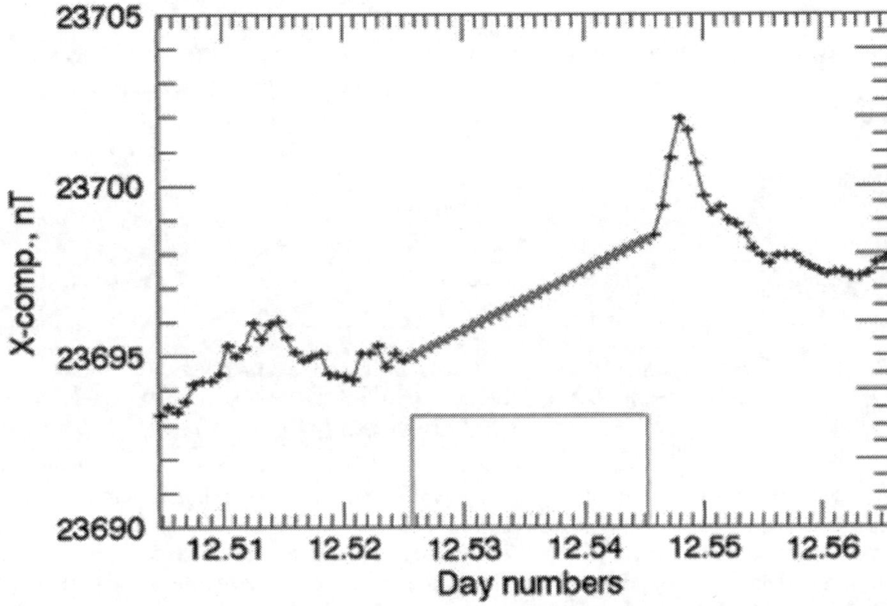

Fig.2. The figure demonstrates the procedure for the gape removal. One gape, observed at the day 13 (20.02.2017) of the time interval is marked by the bar plot. The original data are presented by plusses, connected by lines, and the interpolated data are shown by asterisks, also connected with a line. The sequence of days begins with number one for 08.02.2017.

3.2.2. *Peak detection and removal*

Under a peak here we understand a sudden increase or decrease of the magnetic component data. Such type of peaks are often referred spikes. The reason of spikes can be running engines of cars passing the station periphery. The sudden change can be characterised by the absolute value of the first derivation greater than a given threshold. Based on this assumption a simple and fast procedure of peak detection was developed. The peak borders at the peak base were determined by the first undercut of mean standard deviation of the first derivation at left and right of the peak. The values between the peak borders were replaced in the same manner as for the gape removal, by linear interpolation. The peak detection and removal is organized in a program loop, where in every loop the peak with the maximal first derivation is removed. The detection stops when the first derivations do not exceed the given threshold. An example of detection and removal of a single peak and of a sequence of peaks is presented in Fig. 3.

3.3. Data processing

The calculation of the MPB follows in general the algorithm developed by McPherron [15].

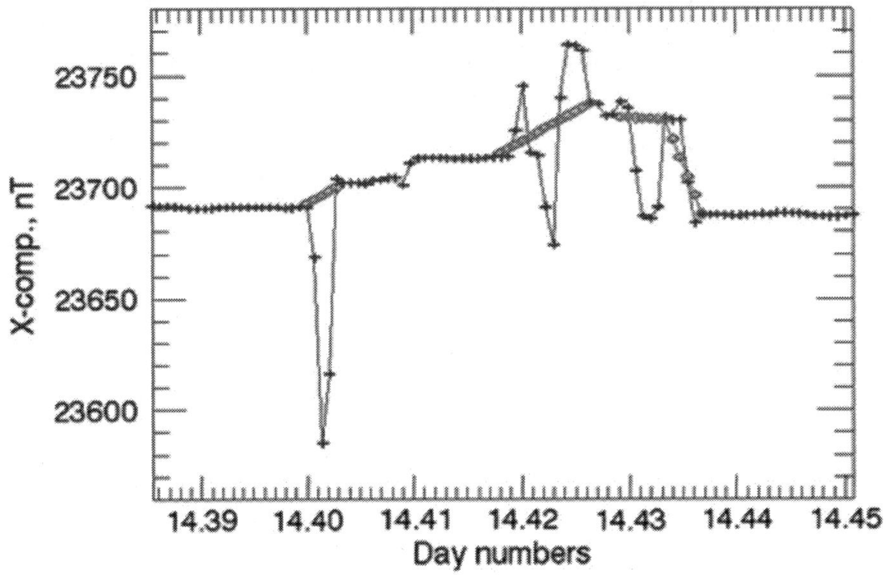

Fig.3. The figure demonstrates how the peak removal is working. The original data are shown by plusses connected by a line. The interpolated data are drawn by diamonds, connected also by a line. In the left side near the day number of 14.40 (22.02.2017) a single peak, detected and removed, is presented. In the right side a sequence of peaks was found and removed by linear interpolation. The sequence of days is the same as in Fig.2.

At first, to eliminate the main magnetic field and the slow changes due to the ring current and partial ring current [15] a base line was constructed by a smoothed spline interpolation through the midnight points (at 24h in LT) (see Fig.4). The resulting spline was subtracted from the time series. The remaining series were splitted again in daily sequences but with three additional hours at the beginning and at the end of every day. (For the PAG station data this allows later the easy conversion again to UT). The time segments with the length of 30 hours hereinafter are called extended days. After the time series shift and splitting from the original 25 days (in UT), 23 extended days (in LT) remain. To calculate the mean Solar quite magnetic variations (Sq), extended days with strong magnetic disturbances were removed. The strong variations of the field components during a disturbed day are signalized by their high standard deviation. The standard deviations are calculated for all 23 extended days. The results are considered as random quantities. By the Lilliefors test [18] it was verified, that they are approximately normally distributed. We have determined outliers, meaning disturbed days, applying the Grubbs test, using the one side significance level of 0.12 [19]. The procedure is asymptotic and in average up to two, three disturbed extended days were found. These days were excluded in further calculations. From the remaining days Sq is determined by averaging of the daily variations of the field components like in the procedure applied in the Chu's algorithm of the MBP calculation [14]. High frequency variations in Sq were smoothed using running mean. Tests to calculate Sq with the help of Principal component analysis (PCA) were also performed as in the McPherron computations [15]. This procedure is much complicated and slower. Much more, the result is very close to the one obtained by a simple averaging and smoothing.

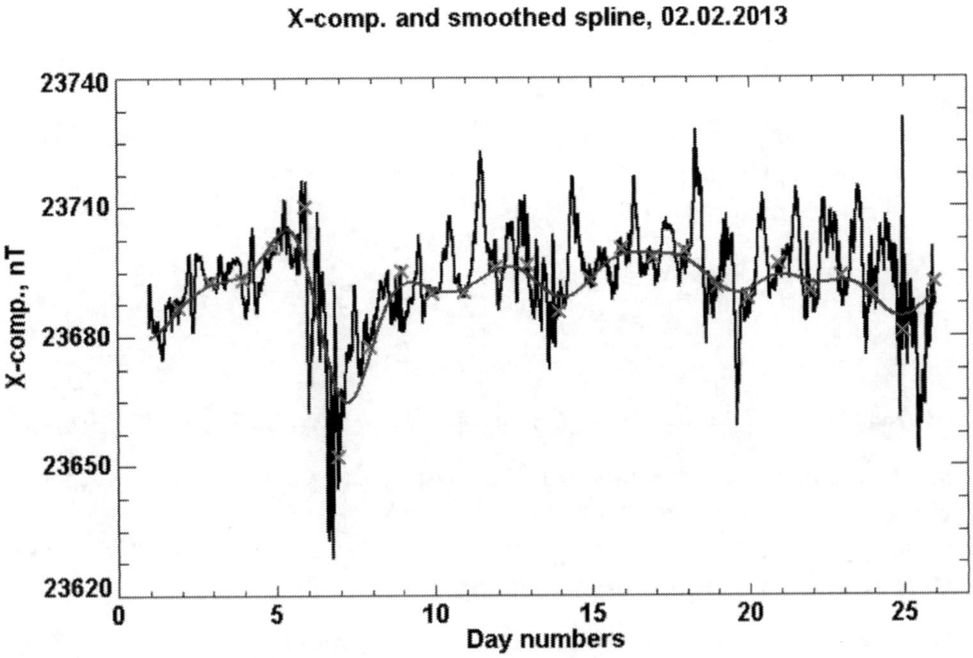

Fig.4. X-component series for 25 days, centered on 02 February 2013. The midnight points (24 LT) are marked by "x" sign. The computed smoothed spline through the midnight points is drawn by a continuous line.

Analogous to the McPherron's method the X and Y components with removed Sq were filtered by a high pass filter to suppress the high frequencies and to keep the low frequency changes. Signals with frequencies longer than about 3 hours were filtered out. Thus only substorm variations remain. In Fig.5, the resulting X (left panel) and Y (right panel) magnetic components for 02 February 2013 are presented. On the x-axis below the figure, the universal time (UT) is indicated, and on the abscissa above the figure – the local time (LT) is shown. The horizontal power of the magnetic perturbations is obtained by the sum of the squared processed X and Y magnetic components (fig6). On the abscissae UT and LT are presented, as in Fig.5.

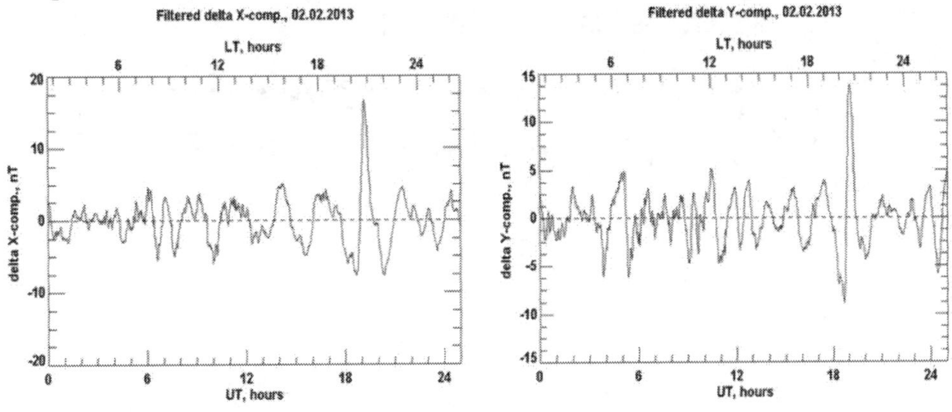

Fig.5. Processed X (left panel) and Y (right panel) components of the magnetic field, measured on 02 February 2013 at the Panagjurishte station.

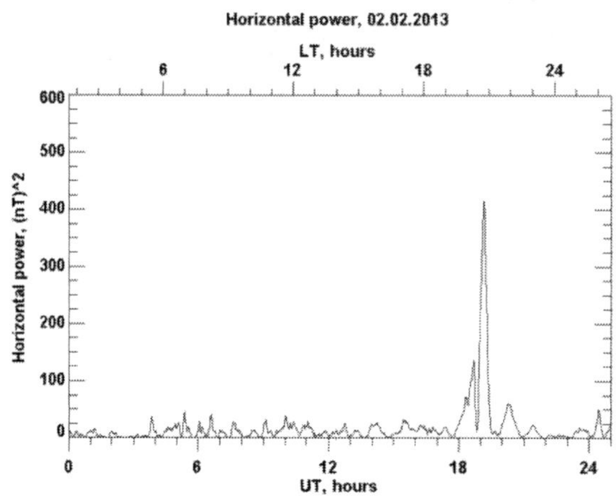

Fig.6. Horizontal power of the magnetic field perturbations on 02 February 2013 at Panagjurishte.

The obtained maximum at about 19:10 UT corresponds to the substorm, observed in the auroral zone, on the IMAGE longitudinal chain PPN-NAL.

3.4. Every day data included in the catalog

In the catalog, data files and graphics of some results are included.

The every day files comprise one minute data for the processed X and Y magnetic components, and the horizontal power of the magnetic field (main field and mean Solar quite magnetic variations (Sq) removed), called MPB by McPherron [15].

The graphics present the processed X and Y components, and the horizontal power of the magnetic field.

The consecutive windows in the Catalog to see the graphics for a given day, are presented in Fig. 7.

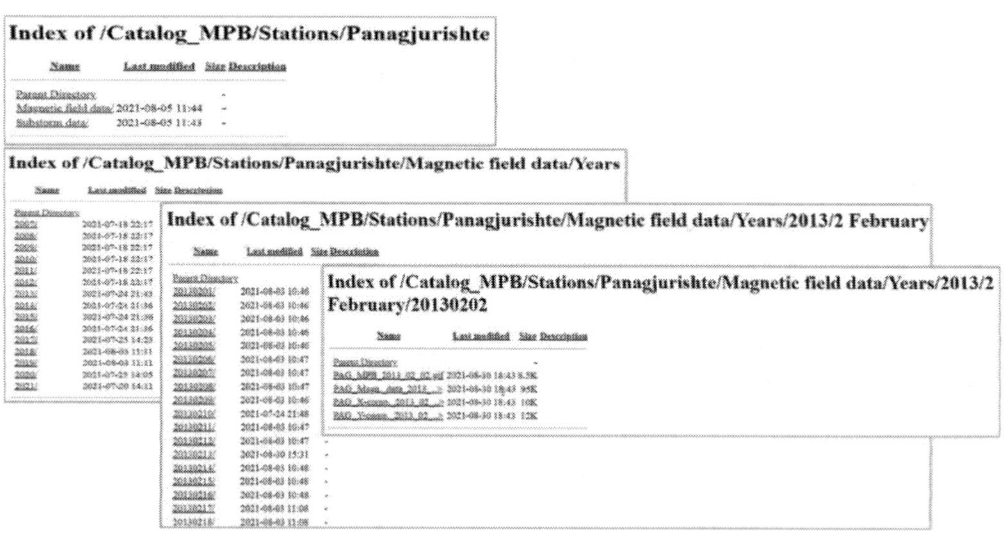

Fig. 7. Consecutive windows to look at the graphics for a given day.

4. Detection of the substorms

We considered all substorms observed during the time interval from January 1 to February 28, 2013 in the auroral zone, i.e. on the IMAGE longitudinal chain (PPN-NAL stations) For each substorm in the auroral zone their midlaltitude displays – positive bays – on the station PAG (Bulgaria) was calculated. We used also data from SuperMAG and INTERMAGNET networks at geomagnetic longitudes near the longitude of the PAG station (~97° GMLon), in the longitudinal band 90° - 104° GMLon, to follow the substorms development. For the chosen period 67 substorms were identified, this number includes all observed substorms.

5. Characteristics of the positive bays

For the identified substorms, the following parameters of the X component magnetic positive bay were determined from the PAG station data and included in the catalog (see Fig.8):

- Time of the onset of the positive bay (vertical line (1));
- Time of the positive bay maximum (vertical line (2));
- Amplitude of the X component (difference between the maximum and the minimum at the onset of the X component) (arrowed line (3));
- Time of the positive bay end (vertical line (4)).

Except these parameters, the following information is also included in the catalog:

- Sunrise and sunset at Panagjurishte for the substorm day;
- Magnetic local time (MLT) at the substorm onset.

In the beginning of every file the date and the difference between local and universal time (LT-UT) for Panagjurishte are written.

The consecutive windows to see the characteristics of the positive bays are similar to these for the day graphs.

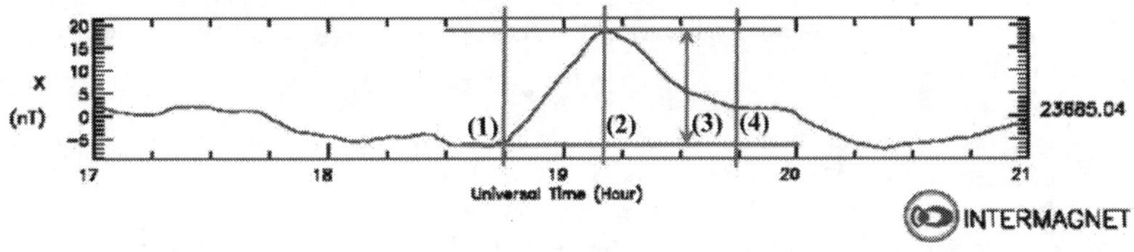

Fig.8. An example of the determination of the MPB parameters. We have chosen an isolated substorm, namely the one at 18:43 UT on 02 February 2013. The way of estimation the MPB parameters is presented on the plot from INTERMAGNET by straight lines, enumerated as follows: (1) time of the substorm onset; (2) time of the positive bay maximum; (3) amplitude of the X-component; (4) time of the positive bay end.

6. Summary

A substorms catalog including the horizontal power of the magnetic field, main field and the mean field under solar quiet conditions subtracted, and the positive bays characteristics by data of the midlattitude Bulgarian station Panagjurishte (PAG) is under construction. The catalog is FTP positioned with preliminary address https://magnetic.nuclearmodels.net/Catalog_MPB/Stations/Panagjurishte/. The tools indispensable for data preprocessing and processing have been developed. The substorms during January and February 2013 have been identified. The PAG magnetic data for these two months have been processed. The characteristics of the positive bays were determined. The obtained results have been uploaded in the catalog.

In the future work we intend to process in such a way all one minute magnetic data from Panagjurishte (since 2007) and to complete the substorm catalog.
Such catalog is being made for the first time. This work could be useful for further studies. The processing tools could be used for other European stations data to obtain a dataset. This will allow to implement a complex analysis of the space-temporal characteristics of the magnetic substorms at midlattitudes and auroral latitudes in order to reveal the peculiarities of the spreading of the geomagnetic disturbances to midlattitudes depending on the conditions of the near-Earth space weather.

Acknowledgements. The authors are grateful to the creators of the databases International Real-time Magnetic Observatory Network (Intermagnet) Data Download (intermagnet.org) and IMAGE (http://space.fmi.fi/image/), for the opportunity to use them in this work. We thank the experts from Panagjurishte observatory (NIGGG-BAS) for providing data and support for their processing.
We are grateful to X.N. Chu for the explanations of some processing procedures and the fruitful discussions.
This study was supported by the National Science Fund of Bulgaria (NSFB) (project number КП-06-Русия/15) and by the RFBR (project number 20-55-18003Болг_а); the work of I.V. Despirak and A.A. Lubchich carried out within the framework of the RFBR grant No. 20-55-18003_Bulg_a.

References

1. R.L. McPherron, C.T. Russell, M.G. Kivelson, P.J. Coleman, Jr., Radio Sci., **8**(11), 1059-1076 (1973)
2. J.L. Kisabeth, G. Rostoker, J. Geophys. Res., **79**(7), 972–984 (1974)
3. O.A. Troshichev, B.M. Kuznetsov, M.I. Pudovkin, Planet. Space Sci., **22**, 1403–1412 (1974)
4. M.I. Pudovkin, O.A. Troshichev, Planet. Space Sci. **20**, 1773-1779 (1972)
5. I.V. Despirak, A.A. Lubchich, H.K. Biernat, A.G. Yahnin, Geomagn. Aeron., **48**(3), 284-292 (2008)
6. C.E. McIlwain, Dordrecht-Holland:D. Reidel, 143-154 (1974)
7. S.-I. Akasofu, S. Chapman, C.-I. Meng, J. Atmos. Terr. Phys., **27**, 1275–1305 (1965)
8. S.-I. Akasofu, C.-I. Meng, J. Geophys. Res., **74**, 293–313 (1969)
9. C.-I. Meng, S.-I. Akasofu, J. Geophys. Res., **74**, 4035–4053 (1969)
10. V.A. Sergeev, R.J. Pellinen, T.I. Pulkkinen, Space Sci. Rev., **75**, 551 (1996)
11. X.N. Chu, T.S. Hsu, R.L. McPherron, V. Angelopoulos, Z.Y. Pu, J.J. Weygand, K. Khurana, M. Connors, J. Kissinger, H. Zhang, O. Amm, J. Geophys. Res. Space Phys. **119**, 1909–1924 (2014)
12. L. Kepko, R.L. McPherron, O. Amm, S. Apatenkov, W. Baumjohann, J. Birn, M. Lester, R. Nakamura, T.I. Pulkkinen, V. Sergeev, Space Sci. Rev., **190**, 1–46 (2014)
13. S. Ohtani, J. Gjerloev, J. Geophys. Res., Space Physics, **125**(9) (2020)
14. X. Chu, *Configuration and generation of substorm current wedge*, Los Angeles: University of California Los Angeles (2015)
15. R.L. McPherron, X. Chu, Space Sci. Rev., **206**(1–4), 91–122 (2017)
16. R.L. McPherron, X. Chu, J. Geophys. Res. Space Physics, **123**, 2831–2850 (2018)
17. T. Nagai, R. Nakamura, T. Mukai, T. Yamamoto, A. Nishida, S. Kokubun, Adv. Space Res., **20**, 961–971 (1997)
18. H.W. Lilliefors, Journal of the American Statistical Association, **62**, 399-402 (1967)
19. F.E. Grubbs, G. Beck, Technometrics, **14**, 847-857 (1972)

Lidar for atmospheric transparency monitoring

Boris Shevtsov[1,*], *Vasily* Bychkov[1,**], *Andrey* Perezhogin[1,***], and *Ilya* Seredkin[1,****]

[1]Institute of Cosmophysical Research and Radio Wave Propagation, FEB RAS, 7, Mirnaya st., Paratunka, Kamchatka, Russia, 684034

Abstract. The transparency of the atmosphere affects the quality of astronomical observations and optical communications, but above all, it directly controls the fluxes of radiation, which is of particular interest for the study of weather and climate changes. The transparency of the atmosphere also affects the success of observing thermospheric lidar reflections. Since they are a consequence of the high transparency of the atmosphere, the reflections can be used as characteristics of atmospheric transparency.

1 Introduction

The thermospheric lidars were created to determine the concentration of the components of the upper ionosphere of molecular nitrogen, atomic helium, nitrogen and oxygen and their ions. But the successful implementation of the thermospheric lidar is known only for the ions of atomic nitrogen and oxygen [1], and these are the components of the upper ionosphere with the lowest concentrations. Perhaps this result is due to low concentrations, at which giant resonant scattering cross sections are realized [2]. And at high atomic concentrations, scattering appears as in a continuous medium on fluctuations of the refractive index, but this is a less effective scattering mechanism.

In the ionosonde version, the thermospheric lidar can also be used as a detector of precipitations of charged particles into the ionosphere [3], as a result of which additional gas ionization occurs to the level determined by the ultraviolet fluxes. Here we will consider another application of the thermospheric lidar for atmospheric transparency measurements. This value affects the quality of astronomical observations and optical communications, but above all it directly controls the radiation flux, which is of particular interest for the study of weather and climate changes.

The transparency of the atmosphere also affects the success of observing thermospheric lidar reflections. Since thermospheric lidar reflections are a consequence of the high transparency of the atmosphere, they can be used as characteristics of this transparency.

The occurrence of thermospheric lidar reflections can serve as an indicator of the high transparency of the atmosphere. Below we will consider the features of lidar reflections at the heights of the thermosphere.

*e-mail: bshev@ikir.ru
**e-mail: vasily@ikir.ru
***e-mail: andreyp@ikir.ru
****e-mail: seredkin@ikir.ru

2 Thermospheric lidar reflections

In the exploration of great heights, the following stages can be distinguished. After the start of the lidar station in 2006, the first resonant lidar reflections in the thermosphere were obtained in 2008 [4], figure 1.

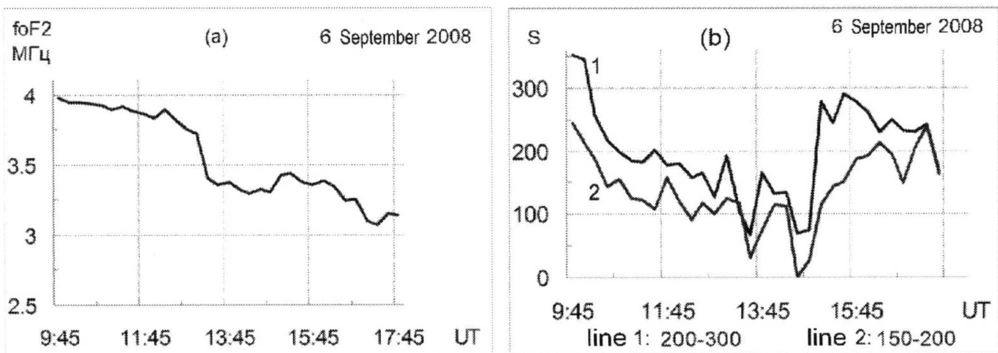

Figure 1. Critical frequency F2 of the ionospheric layer (a) and reflection signals at a wavelength of 532 nm from layers 200-300 and 150-200 km (b).

Note that with the departure of the thermosphere into the shadow, the concentration of electrons decreases, and, accordingly, the lidar signal also decreases. From this we can conclude that the reflected signal was mainly determined by the ionization of the thermosphere; however, after midnight local time, fluctuations in the signals begin to increase, which cannot be explained by changes in ionization alone. This is better seen in figure 2.

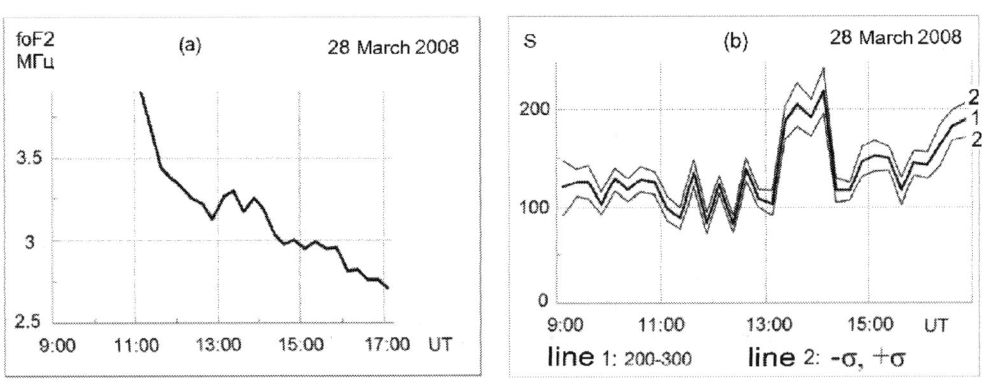

Figure 2. Critical frequency F2 of the ionospheric layer (a) and the signal of reflections at a wavelength of 532 nm from a layer of 200-300 km (b).

In 2014, it was concluded that reflections in the thermosphere are resonant scattering by excited ions of atomic nitrogen [4], and in 2017, simultaneous reflections in the thermosphere were obtained at wavelengths 561 and 532 nm, corresponding to resonance scattering by excited ions of atomic oxygen and nitrogen [1], figure 3.

In figures 3a and 3b, the dependence of lidar reflections on the ionization of the thermosphere is visible as in figures 1 a and b. In figure 3c, lidar signals oscillate with a two-hour pe-

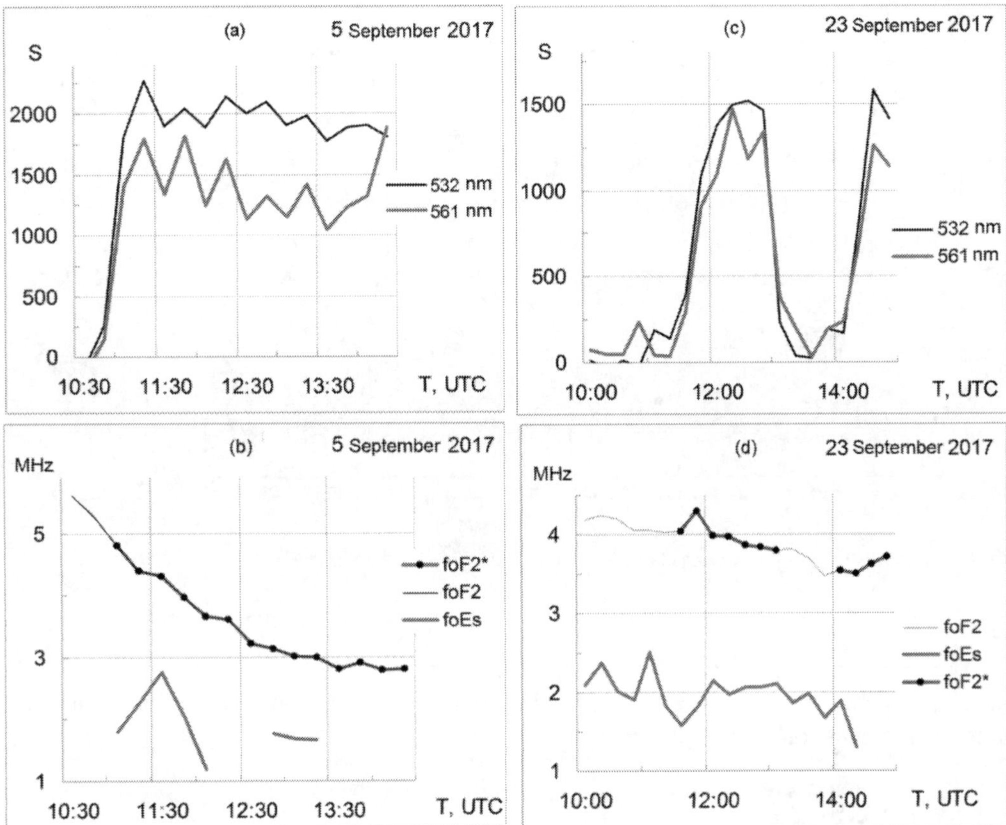

Figure 3. Reflection signals from a layer 200–400 km (top) and $foF2$ and $foEs$ (bottom).

riod corresponding to the frequency range of internal gravitational waves in the atmosphere. This can explain the reason of oscillations in figures 1 and 2.

The significant oscillations of lidar signals in figure 3c give an understanding of how high-altitude observations depend on the state of the lower atmosphere. Perhaps this explains the extreme rarity of thermospheric observations. To start work at high altitudes, it is necessary to determine the state of the lower layers of the atmosphere, figure 4.

Obviously, with such an ionization environment in the middle atmosphere, high-altitude lidar observations become impossible. These data can be used to assess the transparency of the atmosphere and predict the stability of the operation of communication optical systems.

It is noteworthy that the cases of successful sounding of the thermosphere took place during the decline of solar activity, when its radiation effect on the neutral atmosphere reaches a minimum, which creates favorable conditions for high-altitude observations.

Strong fluctuations in the amplitude of lidar reflections due to changes in the conditions of signal transmission along extended paths are of a multiplicative nature. This explains the wide range of changes. Before the start of thermospheric lidar observations, the influence of atmospheric transparency on the fading of lidar signals was not known, so it took a long time to understand why resonantly reflecting layers in the thermosphere are observed so rarely.

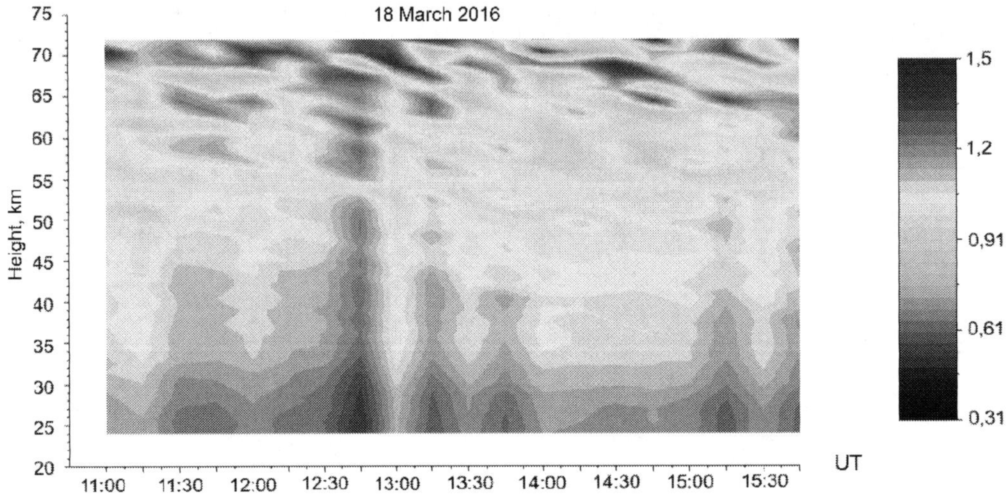

Figure 4. Hours of lidar observations in the lower and middle atmosphere.

The radiative effect of solar activity on various layers of the atmosphere is of interest in studies of weather and climatic changes. And lidar observations provide an opportunity for an integral estimation of this influence.

3 Lidar calibration

There is a problem with the calibration of the lidar. In figure 3, we see that the signal of atomic oxygen ions is slightly less than the signal of atomic nitrogen ions, while the concentration of atomic oxygen ions is two orders of magnitude higher than the concentration of atomic nitrogen ions. This is inconsistent with the theory of resonant scattering, in which the density of scatterers is decisive.

With an increase in the average density of scatterers, the relative density fluctuations decrease, the medium becomes weakly inhomogeneous. In homogeneous media, the scattering also becomes relatively weak. With an increase in the density of scatterers, the statistical characteristics of the scattering change [2]. Perhaps this explains the absence of scattering signals from neutral plasma components, which are five orders of magnitude more than ions.

But there is one more very important circumstance, ions participate in recombination processes, the frequency of which is many orders of magnitude higher than the collision frequency, and this significantly affects the width of resonance lines and reduces the requirements for the coherence of radiation generators.

According to Langevin's theory, the rate of recombination of an atomic oxygen ion depends on the electron concentration, which coincides with the concentration of atomic oxygen ions where atomic oxygen prevails over all other neutrals. This relationship takes place at the operating altitudes of the thermospheric lidar.

The nonlinear effect in recombination must be taken into account, first of all, when we explain the anomalously low lidar signal on atomic oxygen ions. Thanks to it, the dependence of the scattering coefficient on the density of the scatterers changes.

Atomic oxygen of the thermosphere plays an important role in the calibration of a high-altitude lidar due to the fact that the concentration of its ions can be compared with the density of electrons, which is measured by an ionosonde.

And the nonlinear effect in recombination makes it possible to compare the signals of atomic oxygen and nitrogen ions. The atomic oxygen ion signal can be used to calibrate the lidar and determine the recombination parameters. To obtaine the characteristics of the medium, one can use the solutions of the problem of non-stationary reflections [5] which take into account the dispersion properties of the medium and are presented in figure 5.

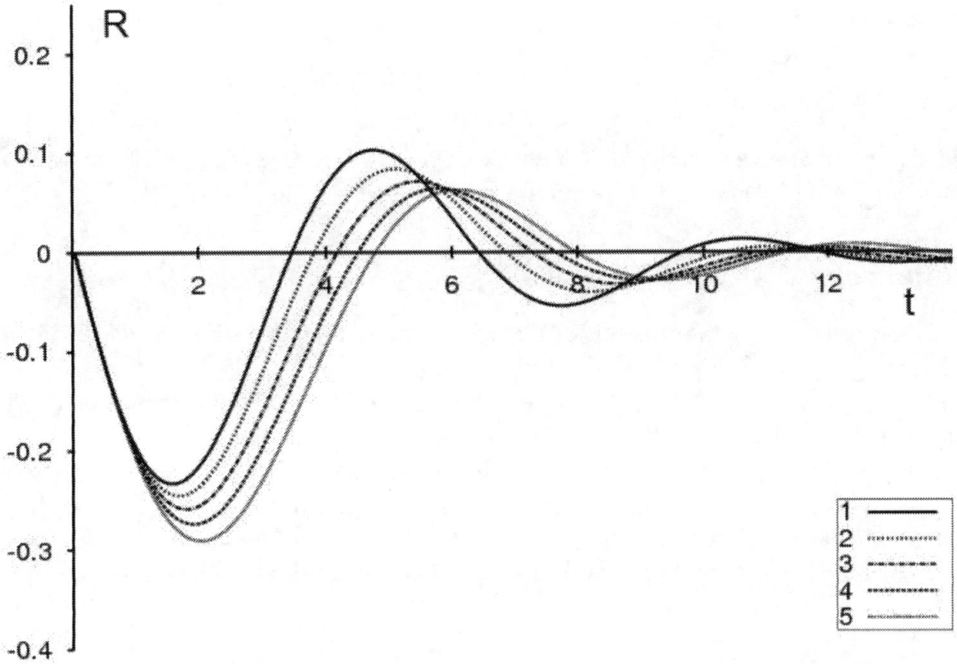

Figure 5. Reflections $R(t)$ in the case of fractional oscillations at different values of the resonance curve parameter b: 0.3 - 1; 0.4 - 2; 0.5 - 3; 0.6 - 4; 0.7 - 5.

The last three years have been decisive in understanding the physical principles of thermospheric lidar operation. This was largely facilitated by multifrequency lidar observations [1].

Conclusions:

1. Thermospheric lidar reflections can be used as an indicator of high transparency of the atmosphere. This will find application for assessing the astroclimate and the efficiency of communication systems.
2. Resonant reflections occur under the usual conditions for the thermosphere, the level of its ionization is quite sufficient. Charged particle precipitation assumptions are not required. They can be considered as a source of additional ionization.
3. Strong fluctuations and difficulties in the detecting of resonant reflections in the thermosphere are caused by unfavorable conditions for signal transmission in the underlying atmosphere.
4. The differences in resonance scattering on neutrals and ions are determined by a significant difference in their concentrations and ion recombination processes.

5. The anomalously weak signal of scattering by atomic oxygen ions can be explained by the nonlinear recombination effect.

6. The signal of atomic oxygen ions can be used to calibrate the lidar and determine the recombination parameters.

7. The obtained solutions for non-stationary reflections can be used for diagnostics of the thermosphere.

Acknowledgments

The paper was carried out on the subject Physical Processes in the System of Near Space and Geospheres under Solar and Lithospheric Impacts: AAAA-A 21-121011290003-0.

References

[1] V. V. Bychkov, A. S. Perezhogin, I. N. Seredkin, B. M. Shevtsov, Proc. SPIE **10833**, 264 (2018)

[2] Francesco Andreoli, Michael J. Gullans, Alexander A. High, Antoine Browaeys, Darrick E. Chang, Phys. Rev. X **11**, 011026 (2021)

[3] V. V. Bychkov, A. S. Perezhogin , I. N. Seredkin, E3S Web Conf. **62**, 01011 (2018)

[4] V. V. Bychkov, Yu. A. Nepomnyashchiy, A. S. Perezhogin, B. M. Shevtsov, Earth, Planets and Space **66**, 150 (2014)

[5] A. S. Perezhogin, B. M. Shevtsov, Radiotechnika i electronica **59**, 46 (2014)

Mid-latitude effects of “expanded” geomagnetic substorms: a case study

Veneta Guineva[1,*], *Irina* Despirak[2], *Rolf* Werner[1], *Rumiana* Bojilova[3], *Lyubomira* Raykova[1]

[1]Space Research and Technology Institute (SRTI), Bulgarian Academy of Sciences, Stara Zagora Department, Bulgaria
[2]Polar Geophysical Institute, Apatity, Russia
[3]National Institute of Geophysics, Geodesy and Geography (NIGGG), Bulgarian Academy of Sciences, Sofia, Bulgaria

Abstract. The goal of this work is to examine the effects of the “expanded” or “high-latitude” substorms at mid-latitudes. These substorms are generated at auroral latitudes and propagate up to geomagnetic latitudes above ~70° GMLat. They are usually observed during reccurent high-speed streams (HSS) from coronal holes. To identify the substorm activity, data from the networks IMAGE, SuperMAG and INTERMAGNET, and data from the all-sky cameras in Lovozero were used. To verify the interplanetary and geomagnetic conditions, data from the CDAWeb OMNI and from the WDC for geomagnetism at Kyoto were taken. We analyzed one substorm event on 20 February 2017 at ~18:40 UT, it developed during HSS, in non-storm conditions. Some features of mid-latitude positive bays (MPB) at the European and Asian stations, and in particular at the Scandinavian meridian have been studied: the bay sign conversion from negative to positive values, the longitudinal and latitudinal extent of the MPB. The central meridian of the substorm was determined.

1. Introduction

Substorms are a specific phenomenon, related to a number of processes in the magnetosphere and ionosphere, generalized by Akasofu [1]. Magnetospheric substorms are the reason for the main magnetic disturbances in the Earths magnetosphere. It is known that substorms are a typical phenomenon at auroral latitudes (~ 60°- ~ 71° MLAT) [2]. Although, according to the conditions in the solar wind and the magnetic activity, substorms can propagate to very high latitudes (above 70° MLAT) (>70° MLAT) (e.g. [3-5] as well as to spread to middle latitudes (~ 50° MLAT) [6]. It has to be noted, that substorms, appeared during different solar wind conditions, usually differ considerably from each other (eg., [7], [8], [9], [10]). Consequently, the substorms have been classified into different types, as: “limited” and “extended” [11], “localized” and “normal” [12], “substorms on the contracted oval” and “normal” [13], "polar" and "usually" [14], “high latitude” and “normal” [5], “extended” and “polar” [15].

*Corresponding author: v_guineva@yahoo.com

It should be noticed, that the midlatitude effects of substorms, by contrast to the ones at auroral latitudes, are expressed by positive bays in the X-component of the magnetic field at ground [16]. At first, this effect was associated with the low latitude return currents of the westward electrojet [17]. Later, the presence of positive bays was connected to the upward field-aligned currents [18]. The accepted theory at present is, that the midlatitude positive bays are the result of a current system, namely the Substorm Current Wedge (SCW) [19,16]. Therefore, taking into account the variety of substorms types, the positive bays development at middle latitudes during substorms can also have some peculiarities, depending on the different conditions. So, the midlatitude effects of different types of substorms should be investigated. In this work, we focused our efforts to estimate the specificity of midlatitude bays development during "expanded" or "high-latitude" substorms. The expanded substorms occur at auroral latitudes, and spread to very high latitudes, above ~70° GMLat. Two expanded substorms have been selected to be presented in detail: on 20 February 2017 and on 30 March 2013.

2. Data used

For the substorm identification and for the study of the substorms development, data from the magnetometer networks IMAGE, SuperMAG and INTERMAGNET, and data from the all-sky cameras in Apatity and in Barentsburg were used.

To verify the interplanetary and geomagnetic conditions, data from the CDAWeb OMNI (http://cdaweb.gsfc.nasa.gov/), from the catalog of large-scale solar wind phenomena (http://www.iki.rssi.ru/omni/) and from the WDC for geomagnetism at Kyoto (http://wdc.kugi.kyoto-u.ac.jp/index.html) were taken.

3. Interplanetary and geomagnetic conditions

One case of expanded substorm, the substorm at 18:40 UT on 17 February 2017 was considered. The interplanetary and geomagnetic conditions during the substorm are presented in Fig.1. From up to down the following quantities are drawn: the magnitude of the IMF vector, the IMF By component, the IMF Bz component, the velocity V, the dynamic pressure P, the PC index, the SYM/H index and the SML index. The boundaries of the structures in the solar wind are marked by rectangles. The dashed vertical line indicates the substorm onset.

From the left upper panel it is seen that during the time interval 16 – 21 February 2017 two consecutive structures in the solar wind were observed: CIR in front of a High-speed stream (HSS). The CIR begun at 7 UT on 16 February 2017 and lasted 24 hours. The HSS was observed from 7 UT on 17 February to 24 UT on 20 February 2017.

There was no magnetic storm developed, SYM/H was ~-18 nT. The considered substorm was registered during HSS in the solar wind. The SML index was ~ -800 nT. The behavior of the solar wind parameters before the substorm onset is presented in more detail in the right panel of Fig.1.

4. Substorm development

The substorm development at ~18:40 UT on 20 February 2017 is presented in Fig.2. In the upper panels the ground-based magnetic disturbances in the X (left upper panel) and Z (right upper panel) magnetic field components from 12 UT to 24 UT on 20.02.2017 by the IMAGE magnetometers chain (NAL-PPN) are shown. The substorm disturbances are marked by ellipses. The substorm onset was at auroral latitudes, 66°-68° GMLat (the region between

MAS-SOR stations) and the disturbances spread up to NAL station (76.6° GMLat), i.e., above 75° GMLat. The center of the westward electrojet was between the stations LYR and NAL (75.6° – 76.6° GMLat). So, this substorm was high-latitude, or expanded. From the X component variations it is seen, that the sign conversion latitude of the magnetic bays lays at about 60° GMLat (between the HAN and RAN stations). In the bottom panels, the substorm development by data of the Polar Geophysical Institute (PGI) is given. The left panels present the magnetic field components at Barentsburg (BAB) and Lovozero (LOZ) stations. It is seen that at these longitudes, the substorm is observed as at auroral latitudes (~64° GMLat), as well as at higher latitudes (above ~75° GMLat). The right bottom panels present the substorm development by aurora, measured in the 5577Å emission at LOZ. The substorm bulge expansion from South to North is clearly seen.

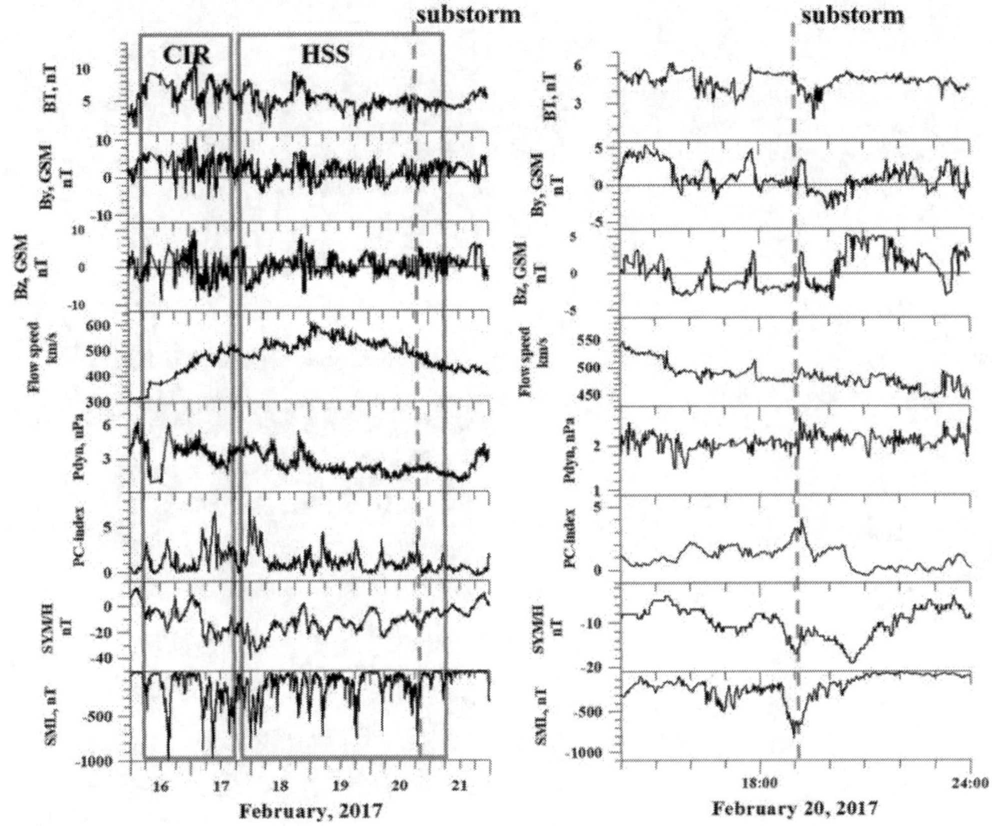

Fig.1. Solar wind and the interplanetary magnetic field components and some geomagnetic indices during the examined event on 20 February.

5. The substorm effect at midlatitudes

To study the substorm effects at midlatitudes, the spread region of the positive bays has been verified. The presence of positive bays at the IMAGE magnetometer network, in the longitudinal band 90°-105° GM longitude, at all European and a number of Asian and African stations has been ascertained. The latitudinal dependence of the positive bays amplitudes has been examined, the longitudinal and latitudinal extent of the positive bays and the central meridian of the substorm, have been estimated. Positive bays were observed at auroral, middle and even equatorial latitudes.

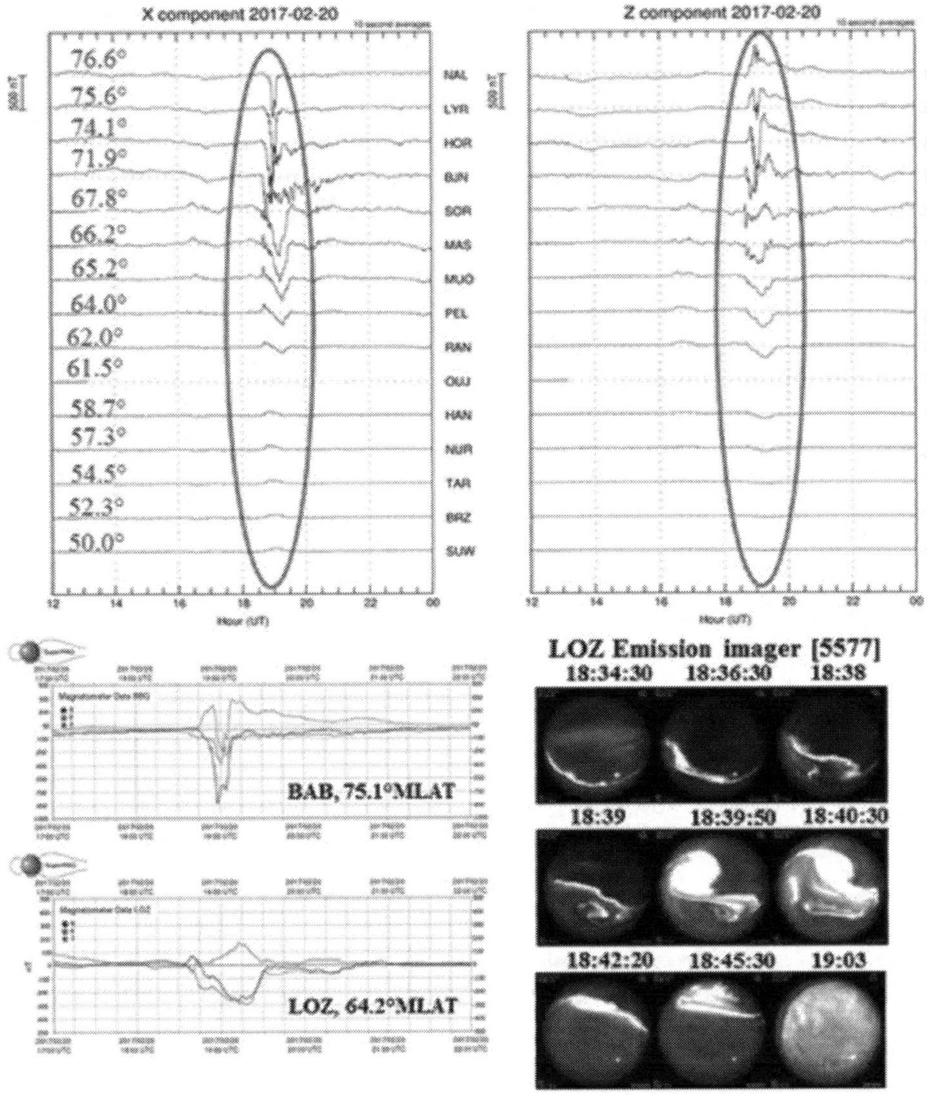

Fig.2. The development of the substorm at 18:40 UT on 20 February 2017 by IMAGE NAL-PPN chain X and Z magnetic components data (upper left and right panels, respectively), and by PGI data: magnetic field components at Barentsburg and Lovozero (left bottom panels) and 5577Å emission data at Lovozero (right bottom panel).

5.1. Positive bays by data of the IMAGE magnetometers network

The positive bays are observed at all IMAGE stations down from MEK (58.7°) and DOB (59.6°). In Fig.3 the X and Y components of all IMAGE stations with positive X bays (the two left panels) are presented, and the locations of these stations are shown (the right panel), indicated by stars.

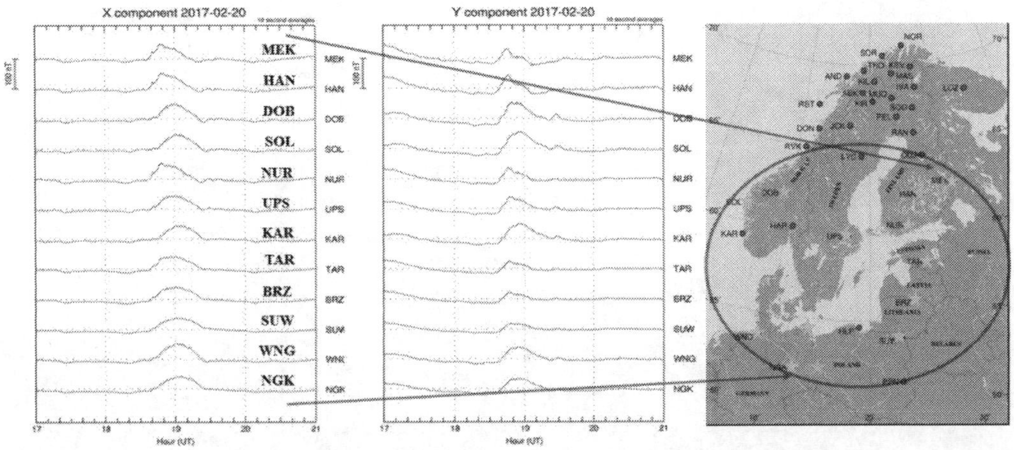

Fig.3. Two panels to the left: X and Y components of the magnetic field by all IMAGE stations where positive bays were observed. Right panel: locations of the IMAGE satations with positive X-bays, marked by stars.

5.2. MPB latitudinal dependance

To examine the MPB latitudinal dependence, data from the stations in the longitudinal band 90° - 104° GMLon, which is round the longitude of the Bulgarian station Panagjurishte (~97° GMLon) are used. In Fig.4, the X and Y components of chosen INTERMAGNET stations in this interval are shown. The substorm onset is marked by red vertical lines.

To build the MPB latitudinal dependence, stations data from the largest possible interval of geomagnetic latitudes with positive bays: 31.8° - 75.25° GMLat in the mensioned above longitudinal band were used. For the considered event on 20 February 2017, there were 20 stations with positive X bays in this area. The relationship between the MPB amplitude and the geomagnetic latitude is presented in Fig.5. The sign conversion latitude is marked by vertical line. Taking into account that there is a strong dependence on the geomagnetic longitude, as well, the longitudinal band was divided into three narrower strips (90°-95°, 95°-100°, 100°-105° GMLon). The data grouped in this way, are indicated by different symbols. It is seen, that the amplitude of the MPB decreases towards the lower latitudes.

5.3. Latitudinal extent of the positive bays

To determine with more precision the latitudinal extent of the MPB and the sign conversion latitude, the behaviour of the magnetic field components at more stations than the examined so far was verified, too. It was obtained that the highest latitude at which positive bays were registered, was 60.3° GMLat (HOV, Foroe Island station) (see Fig.6, left panel). Thus, 60° GMLat can be assumed as upper boundary of the latitudinal extent of MPB and the latitude of conversion of the bay sign. During the substorm positive bays were observed to equatorial stations as ABG (Alibag, 12° GMLat) and TAM (Tamanrasset, 8.9° GMLat) (shown in Fig.6, right panel). So, the latitudinal extent of the positive bays was ~51°.

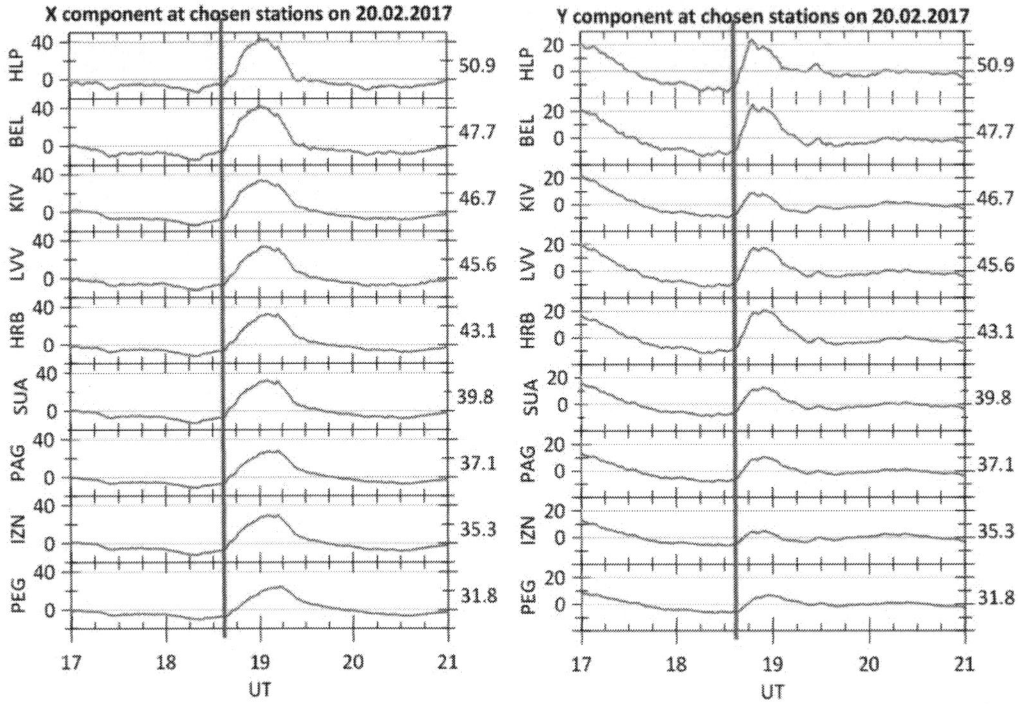

Fig.4. X (left panel) and Y (right panel) magnetic components at chosen INTERMAGNET stations from 17 UT to 21 UT on 20 February 2017.

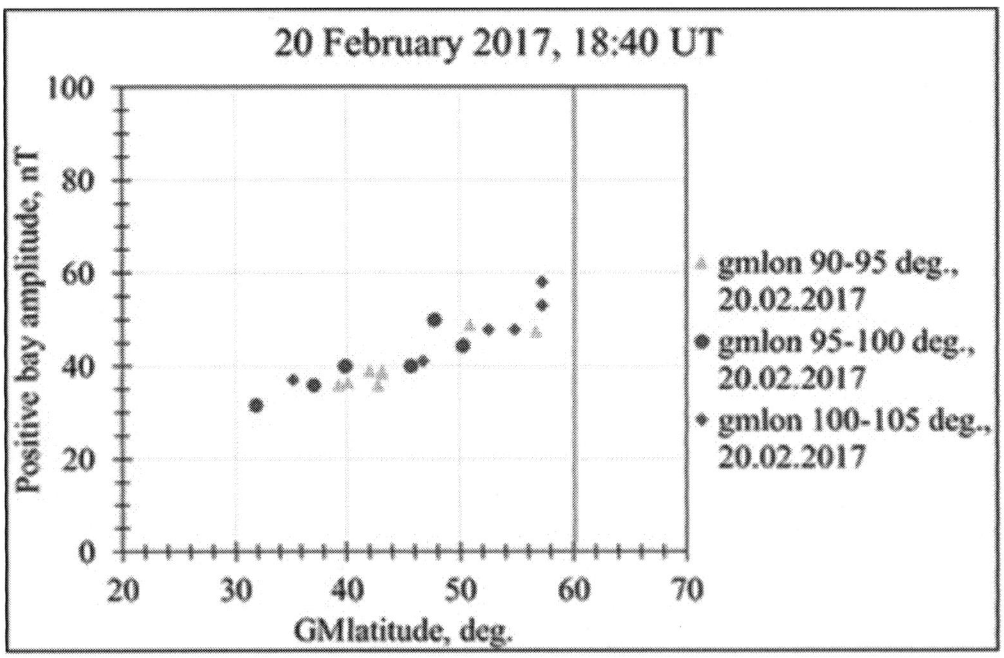

Fig.5. The dependence of magnetic positive bays (MPB) amplitude on the geomagnetic latitude for the substorms on 20 February 2017. The results for the examined longitudinal intervals are marked as follows: 90°-95° GMLon – by triangles, 95°-100° GMLon – by circles, and 100°-104° GMLon – by diamonds (shown in the right part of the panel).

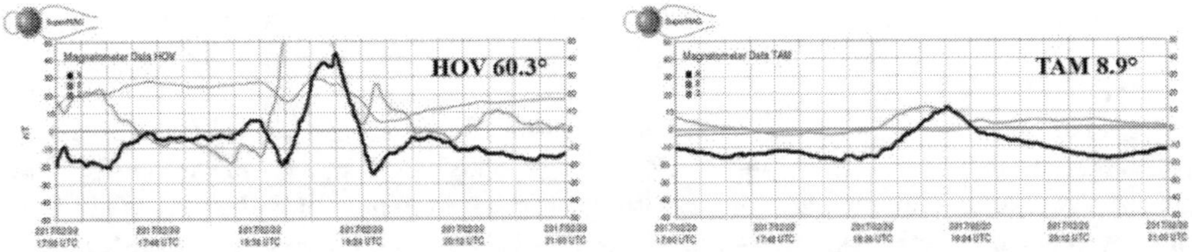

Fig.6. Midlatitude positive bays, observed at the highest geomagnetic latitude (HOV, Faroe Island station, 60.3° GMLat) (left panel) and at the lowest geomagnetic latitude (TAM, Tamanrasset, 8.9° GMLat) (right panel). The X component is drown by a thick line.

5.4. Estimation of the longitudinal extent of the positive bays

During the event on 20 February 2017, the positive bays were observed from ~70° GMLon to ~155° GMLon. In the left panel of Fig.7, the magnetic field components at the station with the most West location with registered positive bay, VAL (Valentia, Ireland, 70.3° GMLon) are shown. VAL is the most West European station. In the right panel, the magnetic components at the most East station with observed positive X-bay, NVS (Novosibirsk, Russia, 155.7° GMLon) is presented. So, the longitudinal extent of this substorm was ~85°.

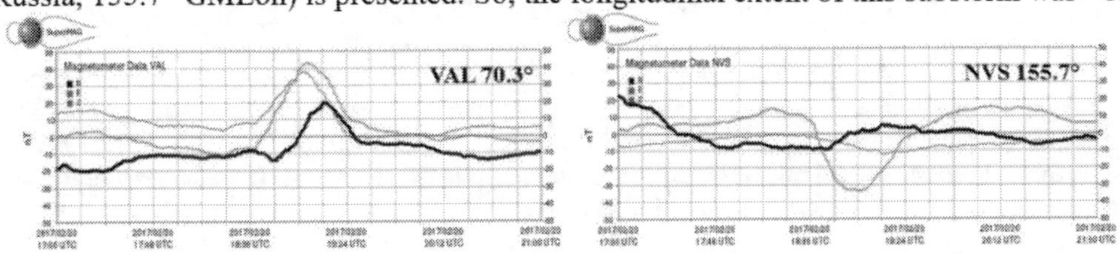

Fig.7. Records of the magnetic components at the most western (left panel) and most eastern (right panel) locations where MPB were observed. The X component is drown by thick line.

5.5. Determination of the central meridian of the substorm

It is known, that based on the peculiarities of the midlatitude positive bays a method to determine the geomagnetic longitude of the auroral substorm onset was worked out [21, 22]. It was established, that at middle latitudes, the azimuthal component of the magnetic field is positive to the West from the electrojet center, and negative – to the East from it. At the central meridian of the substorm the Y component passes through zero from positive to negative values.

The central meridian of the examined substorm was estimated by data of the IMAGE and SuperMag databases.

For the concidered event, positive Y values corresponded to the positive X bays of the IMAGE magnetometers network (see Fig.3, middle panel), hence, the IMAGE magnetometers were to the West from the center of the electrojet. In Fig.8 some examples of registered components to confirm the determination of the substorm central meridian are presented. The most eastern stations with positive sign of Y are MEK (Mekrijärvi, 59.1° GMLat, 108.5° GMLon), SPG (Saint Petersburg, 57.2° GMLat, 105.8° GMLon), MNK (Minsk, 50.8° GMLat, 103.3° GMLon), KIV (Kiev, 46.7° GMLat, 104.0° GMLon) and ODE (Odessa, 42.1° GMLat, 104.5° GMLon). So, the region of 103.3° - 108.5°, and of all latitudes below 59° GMLat was to the West from the center of the electrojet. As an example, the magnetic components at KIV are presented in the left upper panel of Fig.8.

The first station to the East of this region with negative Y component during the substorm, is KLI (Klimovskaya, 57.5° GMLat, 115.5° GMLon). Therfore this longitude was to the East of the central meridian of the substorm. The magnetic components at KLI are shown in the left bottom panel of Fig.8.

The change of the sign of the Y component was observed at the stations MOS (Moskow, 51.9° GMLat, 112.1° GMLon) and BOX (Borok, 54.6° GMLat, 114.2° GMLon). The registered magnetic components at these stations are presented in the middle panels of Fig.8. Thus, the central meridian of the substorm was located at about 112°-114° GMlon.It is indicated by a continuous line in the right panel of Fig.8, on the map of the magnetic vectors during the substorm by SuperMAG. The determined longitudinal boundaries of the substorm are indicated by dotted lines in the same panel.

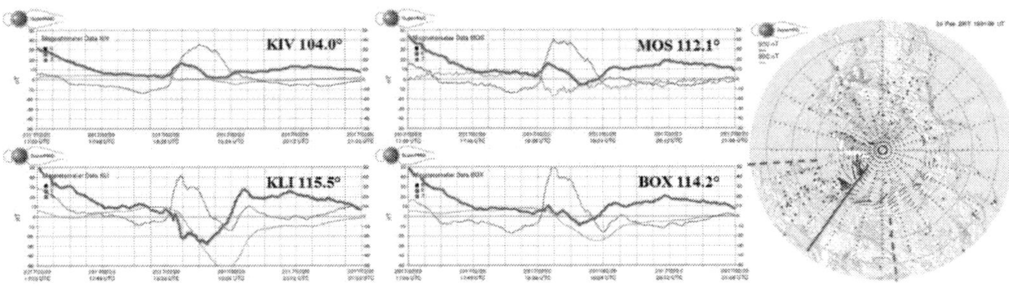

Fig.8. Examples of positive Y component (upper left panel), negative Y component (bottom left panel), and of cases, where Y passes through zero (middle panels). The Y component is marked by a thick line. The right panel presents the magnetic field vectors during the substorm (at 19:01 UT) by SuperMAG. The determined central meridian of the substorm is indicated by a continuous line, and the longitudinal boundaries of the positive bays extent – by dotted lines.

6. Discussion

We compaired some characteristics of the considered here substorm with our earlier results about the midlatitude effects of substorms.

The substrm, examined in [23], at 29:12 UT, on 27 September 2020, resemble to the studied here event. This substorm occurred also during HSS, in non-storm conditions, but just before the substorm, Sheath and Ejecta in the solar wind were observed. Thus, the interplanetary conditions were some more disturbed. The substorm on 27.09.2020 started at auroral latitudes, ~67° GMLat (SOR), and the center of the electrojet propagated to ~75° GMLat (HOR-LYR). The sign conversion latitude of the magnetic bays in this case was observed at lower latitudes, ~56° GMLat. Probably this is due to the more disturbed conditions in this case. Moreover, the conversion latitude during the other event, studied in [23] – the substorm at 21:25 UT on 6 February 2018, a usual substorm, generated during quiet conditions, when slow solar wind flow was observed, appeared at higher latitude, ~63° GMLat. In the study of usual and expanded subtorms during large geomagnetic storms the sign conversion latitude was observed in the range 50°-56° GMLat [24]. So, from these examples it follows, that the more disturbed are the interplanetary conditions, the more low is the conversion latitude of the substorms X magnetic bays.

The obtained dependence of the positive bays amplitudes has a similar shape, as the obtained in [23]: as a hole, it decreases with the latitude. A slight maximum at about 50° is observed here, as in the cases considered in [23]. The rise after the conversion latitude, obtained in [23], is not clearly expressed in the considered here substorm on 20 February 2017.

The mean amplitude of the positive bays during the substorm on 20 February 2017 was 42.4 nT, against the result of 55 nT for 27 September 2020 and 13.7 nT for 06 February 2018. The amplitudes of the positive bays during large geomagnetic storms were ~100-150 nT [24], and they were larger for the expanded substorm than for the usual ones. This result confirms, that the positive bays amplitude is greater in more disturbed conditions.

7. Summary

Expanded substorms are accompanied by midlatitude positive bays, the maximum amplitude of which is observed in the midnight sector. For the considered substorm the sign conversion latitude (~60°GMlat), the central meridian (~112°-114°GMlon), the latitudinal extent (~51°) and the longitudinal extent (~85°) of the positive bays were determined.

The observed conversion latitude for the examined substorm is typical for "expanded" substorms (~60° GMlat), it is higher than for the storm-time substorms or usual substorms.

For "expanded" substorms the amplitude of the bays is higher than for usual substorms. The amplitude of the positive bays as a whole, decreases with the latitude. A slight maximum at about 50° GM latitude is obtained.

A difference of about 50% between the minimal and maximal positive amplitude at different latitudes in the interval 30°-60° GMlat and 95°-105° GMlon was obtained.

More data about the midlatitude positive bays behaviour are needed to make statistical conclusions.

Acknowledgements. The authors are grateful to the creators of the databases OMNI (http://omniweb.gsfc.nasa.gov), SuperMAG (http://supermag.jhuapl.edu/), IMAGE (http://space.fmi.fi/image/), WDS at Kyoto (http://wdc.kugi.kyoto-u.ac.jp/index.html) and the solar wind large-scale phenomena catalog (http://www.iki.rssi.ru/omni/) for the opportunity to use them in this work.
This study was supported by the National Science Fund of Bulgaria (NSFB) (project number КП-06-Русия/15) and by the RFBR (project number 20-55-18003Болг_а).

References

1. S.-I. Akasofu, Space Sci. Rev. **113**, 1-40 (2004)
2. S.-I. Akasofu, Planet. Space Sci., **12(**4), 273–282 (1964)
3. M.I. Pudovkin, O.A. Troshichev, Planet. Space Sci., **20**, 1773-1779 (1972)
4. E. Nielsen, J. Bamber, Z.-S. Chen, A. Brekke, A. Egeland, J.S. Murphree, D. Venkatesan, W.I. Axford, Ann. Geophys. **6**(5), 559-572 (1988)
5. I.V. Despirak, , A.A. Lyubchich, H.K. Biernat, A.G. Yahnin, Geomagn. Aeron., **48**(3), 284–292 (2008)
6. Y.L. Feldstein, G.V. Starkov, Planet. Space Sci., **15**(2), 209–229 (1967)
7. E. Tanskanen, T.I. Pulkkinen, H.E.J. Koskinen, J.A. Slavin, J. Geophys. Res., **107(**A6), SMP 15-1 – SMP 15-11 (2002)
8. N.E. Turner, W.D. Cramer, S.K. Earles, B.A. Emery, J. Atmos. Sol. Terr. Phys., **71**, 1023–1031 (2009)
9. V. Guineva, I. Despirak, B. Kozelov, Sun and Geosphere, **11**(2), 125-130 (2016)
10. V. Guineva, I.V. Despirak, R. Werner, J. Atmos. Sol. Terr. Phys., **177**, 63-72 (2018)
11. A.T.Y. Lui, , S.-I. Akasofu, E.W. Hones, Jr., S.J. Bame, C.E. McIlwain, J. Geophys. Res., **81**(7), 1415–1419 (1976)

12. R.L. McPherron, C.T. Russell, M.G. Kivelson, P.J. Coleman, Jr., Radio Sci., **8**(11), 1059-1076 (1973)
13. Y. Kamide, S.-I. Akasofu, S.E. Deforest, J.L. Kisabeth, Planet. Space Sci., **23**(4), 579-584 (1975)
14. N.G. Kleimenova, E.E. Antonova, O.V. Kozyreva, L.M. Malysheva, T.A. Kornilova, I.A. Kornilov, Geomagn. Aeron., **52**(6), 746–754 (2012)
15. I.V. Despirak, A.A. Lubchich, N.G. Kleimenova, J. Atmos. Sol. Terr. Phys., **177**, 54-62 (2018)
16. R.L. McPherron, C.T. Russell, M. Aubry, J. Geophys. Res., **78**(16), 3131–3149 (1973)
17. S.-I. Akasofu, S. Chapman, C.-I. Meng, J. Atmos. Terr. Phys., **27**, 1275–1305 (1965)
18. C.-I. Meng, S.-I. Akasofu, J. Geophys. Res., **74**, 4035–4053 (1969)
19. R.L. McPherron, Planet. Space Sci., **20**(9), 1521-1539 (1972)
20. V. Guineva, I. Despirak, B. Kozelov, Aerospace Res. in Bulgaria, **26**, 145-154 (2014)
21. G. Rostoker, S.-I. Akasofu, J. Foster, R.A. Greenwald, Y. Camide, K. Kawasaki, A.T.Y. Lui, R.L. McPherron, C.T. Russell, J. Geophys. Res., **85**(A4), 1663-1668 (1980)
22. В.А. Сергеев, Н.А.Цыганенко, Магнитосфера Земли, М., 1980. 174 с.
23. V. Guineva, R. Werner, I. Despirak, R. Bojilova, L. Raykova, C. R. Acad. Bulg. Sci., **74**(8), 1185-1193 (2021)
24. V. Guineva, I. Despirak, N. Kleimenova, Aerospace Res. In Bulgaria, **31**, 27-39 (2019)

Observations of resonant scattering in the thermosphere and upper mesosphere in the winter season of 2021

Vasily Bychkov[1,*], Ilia Seredkin[1], Alexei Dmitriev[2,3]

[1]Institute of Cosmophysical Research and Radio Wave Propagation FEB RAS, p. Partunka Kamchatka Territory, st. Mirnaya 7

[2]National Central University, Chongli 320, Taiwan

[3]Moscow State University, Moscow, Russia

Abstract. An analysis of the Kamchatka lidar station data obtained in January-February 2021 is presented. Pronounced light scattering layers at altitudes of 300 km have been detected. Layers at these heights appear during precipitation of electrons with energies of about 300 eV. Analysis of the lidar data showed that increased light scattering was observed at the heights of the mesosphere, accompanied by an increase in the ionospheric parameter fmin. It was concluded that the observed aerosol formations in the mesopause region could be caused by precipitation of charged particles with energies of ~ 500 keV. Aerosol formations in the area of 60-75 km should be recognized as imaginary. This is a manifestation of resonant scattering by excited ions of atomic nitrogen

Introduction

Lidar station Kamchatka was put into operation in autumn 2007. In March 2008, resonance scattering by excited ions of atomic nitrogen at a wavelength of 532 nm was detected. Backscattering signals at a wavelength of 532 nm at altitudes of 100-300 km are presented in [1]. In [2-3], the results of two-frequency sounding in the area of 25-500 km are presented. In 2021, lidar sounding of the atmosphere was carried out at a wavelength of 532 nm on two receiving channels in the range of 10-600 km.

*Corresponding author: vasily.v.bychkov@gmail.com

Equipment and measurement methods

Recording of reflected signals was carried out on two receiving channels - for the main (25-600 km) and reflected (10-600 km) signals. Several percent of the main signal was separated by a glass plate installed in the fiber at an angle of 45 degrees to the light flux and directed to another photomultiplier [5]. A continuous signal profile in the range of 10 - 600 km was obtained by stitching the profiles of the main and reflected signals. Figure 1 shows the working scheme of the lidar.

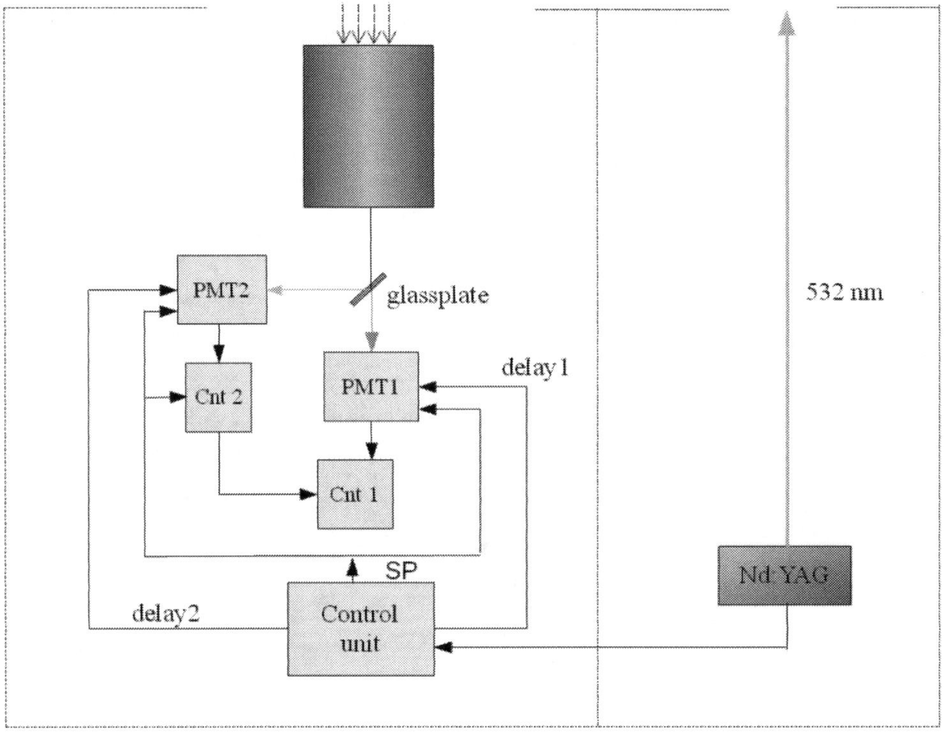

Fig. 1. Lidar scheme: SP - sync pulse generated by laser, Cnt - photon counter. PMT - photomultiplier tube, delay - blocking pulse duration μs, set in the control unit.

The aim of the research was to observe the appearance of precipitation of superthermal electrons into the thermosphere and the appearance of aerosol in the middle atmosphere.

The precipitation of superthermal electrons similar to the events of the autumn of 2017 [4, 5] was not registered in 2020-21. This is due to the very low geomagnetic activity at the minimum of the solar cycle. During quiet times, the precipitation of energetic electrons from the outer radiation belt practically disappears.

Equipment, as well as methods of measuring and processing signals are given in [4, 5]. A lidar with a Nd : YAG laser was used to generate radiation at a wavelength of $\lambda = 532.08$ nm. To exclude photomultiplier illumination from near-field signals in both receiving channels, the photomultiplier was electronically locked. A dipole transition was studied for an excited nitrogen ion with a wavelength of 532 nm. (Table 1).

Table 1. Dipole transition of an excited nitrogen ion in the laser emission band

	Component	Wavelength (nm)	Lower level	Term	J	Top level	Term	J
1	NII	532,0958	$2s2p^2(^4P)3p$	$^5P^o$	1	$2s2p^2(^4P)3d$	^5P	2

The state of the ionosphere was monitored according to the results of measurements of the Parus A ionosonde. The computers controlling the operation of the ionosonde and lidar are synchronized by GPS.

Experimental data

As already noted, precipitation of superthermal electrons into the thermosphere in 2020 was not registered. In January 7, 13 and 27 and in February 5, 12 and 15, mild precipitations appeared. Figure 2 shows the values of the total "signal-background" modified by multiplying by the coefficient $k = (H / 100)^2$, where H is the height in km [4].

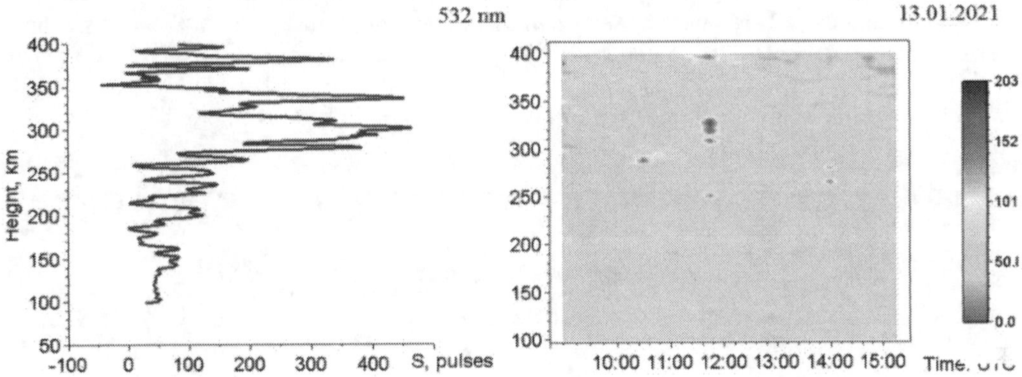

Fig. 2. Lidar "signal-background" plotted according to data from January 13, 2021.

Figure 3 shows the results of atmospheric sounding by the Parus-A ionosonde. (3a), the profile of the scattering ratio for the first 2.5 hours of observations (3b) and the total overnight profile (3c). Attention is drawn to the behavior of the ionospheric parameters fmin and foEs. The time intervals for the accumulation of lidar signals were chosen taking into account the fact that at the beginning of observations the fmin values were close totheir usual values of about 1.

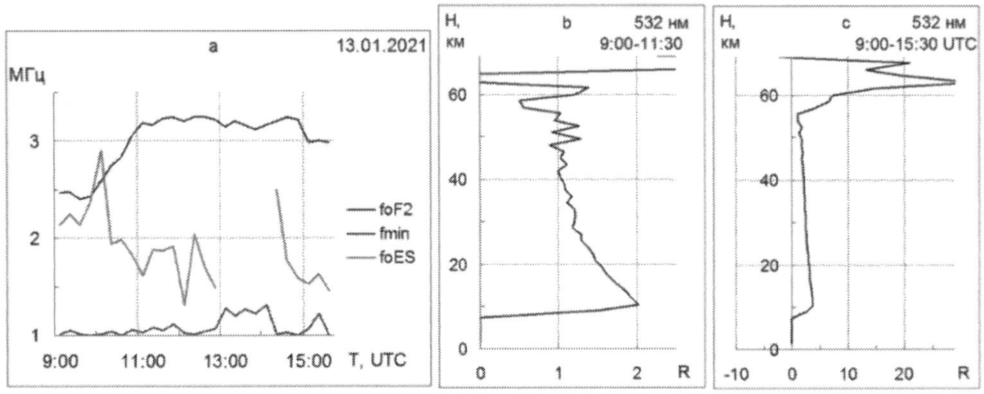

Fig. 3. Values of foF_2, fmin, foEs (a) and profiles of the scattering ratio at the beginning of the sounding session (b) and total over the night of 01/13/2021 (c).

The geomagnetic situation during measurements in January is represented by the Dst index in Figure 4. It can be seen that the observation intervals on January 7, 13, and 27, 2021, fall on the recovery phase of recurrent magnetic storms. It is known that during these periods the electrons of the outer radiation belt are actively accelerated to energies of hundreds of keV - MeV and , and can precipitate into the atmosphere at high and middle latitudes, causing additional ionization of the upper and middle atmosphere [6].

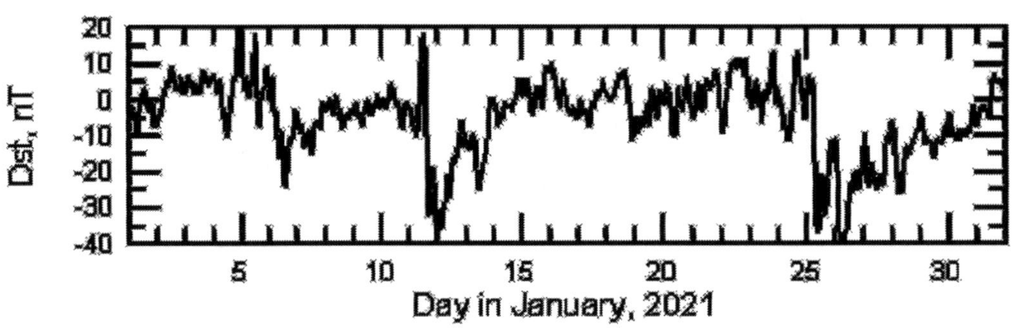

Fig. 4. Dynamics of the geomagnetic activity index Dst in January 2021 Small negative variations are associated with recurrent magnetic storms.

The discussion of the results

An increase in the nighttime values of fmin means an increase in the concentration of free electrons in the upper mesosphere. At night, this is caused by precipitation of electrons with energies of hundreds of keV. As noted in [5], the maximum of the scattering ratio R is located above the maximum of ionization produced by precipitating electrons.

The profiles of lidar signals in the range of 50–110 km for the same accumulation intervals as in Fig. 3 were compared with the ionization rate profile for a monoenergetic electron flux constructed according to the method described in [7]. It is shown that the height of the maximum of the ionization rate coincides with the height of the maximum of the signal for an electron flux with an energy of 500 keV.

Thus, we can conclude that on January 13, 2021,near 12:00 UTC, precipitations of electrons with energies of 300-330 eV were observed, according to the results of [4], and in the interval 13: 00-15: 30 UTC with energies of about 500 keV.

The Es layer was in the area of 100-110 km. Its disappearance at the moments of anomalous growth of fmin can give rise to the assumption that the layer is corpuscular, and its disappearance is associated with an increase in the energy of precipitating electrons. This is not true. The absence of a resonant scattering signal in the range of 100-110 km indicates that the layer is correctly defined by experts as type f - flat.

Analysis of ionograms showed that the absence of foEs values, which coincides in time with an increase in fmin, can be associated with a large absorption of radiowave. It means the absence of a sufficient return signal, but not real Es disappearance. Large fmin values, close to foEs values, lead to the fact that weakly pronounced fragments of the Es track are reflected in ionograms at an altitude of 100 km, which do not allow determining the exact value of foEs.

The estimates of the energies of the precipitated electrons obtained from the lidar data are confirmed by the results of satellite measurements. Figure 5 shows the data of measurements of electron fluxes and energies on the Metop 1 meteorological satellite. It flew both east and west of Kamchatka during lidar observations.

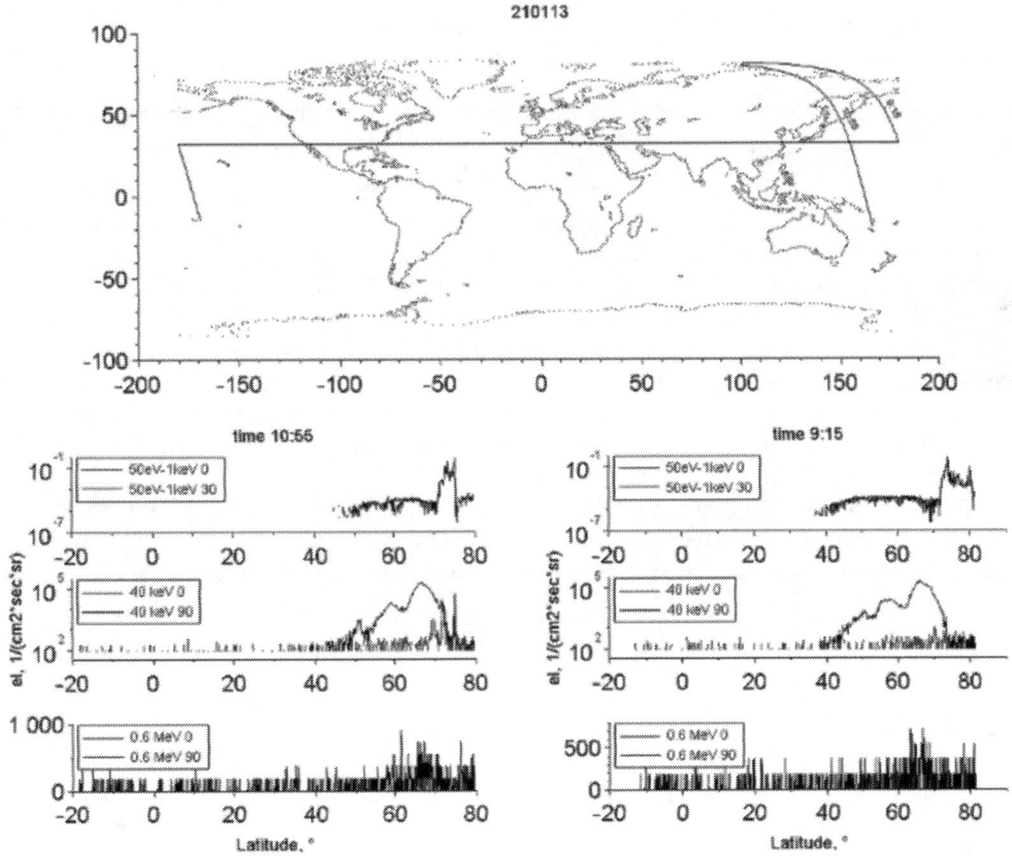

Fig.5. Electron fluxes recorded on the Metop 1 satellite during lidar observations on January 13, 2021.

The energy of 500 keV, determined from the lidar data, fits well into the energy range 40-600 keV, shown in the middle panel of Fig. 5. Very low values of superthermal electron fluxes also do not contradict the lidar data, since both satellite passages occurred before the appearance of the layer at an altitude of 300 km. Verification of satellite data for other days with Dst index values close to zero showed the absence of fluxes of both superthermal electrons and with energies in the range of 40-600 keV.

We also present the data of studies of the aerosol distribution in the middle atmosphere of Kamchatka, presented in [8], in which the presence of "non-aerosol" scattering in the lidar data is discussed. Figure 6 present data showing the correlation of the ionospheric parameter fmin with the scattering ratio R, total over 5-6 km layers in the mesopause region.

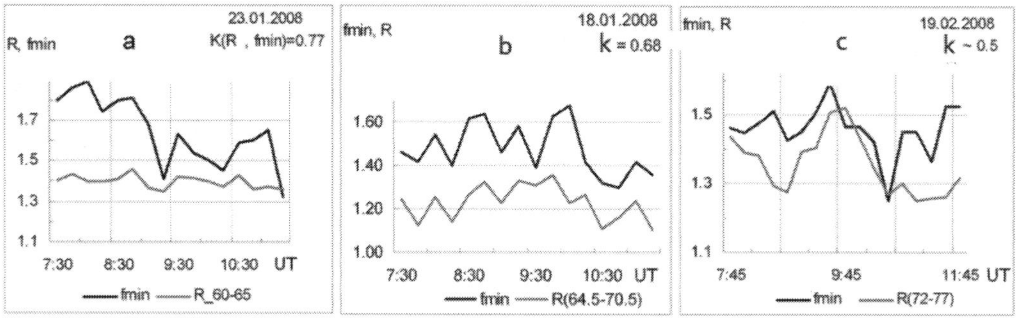

Fig. 6. Correlation of the ionospheric parameter fmin with the scattering ratio R in the mesosphere

It was shown in [8] that at the lowest temperatures observed from 2007 to 2011, the content of water vapor in the mesosphere was not enough by more than an order of magnitude for their condensation with aerosol formation. In this case, aerosol layers in the mesosphere were observed.

The abrupt termination of the correlation at the end of observations on February 19 (Fig.6c) means that the phenomenon is controlled by a fast-flowing process. The correlation of the signal with fmin suggests that this process is the precipitation of electrons. Comparisons with the results of measurements from the Demeter satellite showed that precipitations of relativistic electrons were observed and these electrons could ionize in the region of 65-80 km.

The data presented in 2008 show that the precipitations of electrons into the mesosphere observed in January-February 2021 are not unique events and can occur, for example, during the restoration of the magnetosphere after geomagnetic disturbances.

Conclusions

Imaginary aerosol formations in the mesosphere when probed at a wavelength of 532.08 nm is an expected and ordinary event. The reason is resonance scattering by excited ions of atomic nitrogen, which are formed during the ionization of the atmosphere.

Funding

The work was carried out within the state assignment on the theme (2021–2023) Physical Processes in the System of Near Space and Geospheres under Solar and Lithospheric Impacts. Registration no. AAAA-A21-121011290003-0 with the use of equipment of the Collective Use Center of the Institute of Cosmophysical Research and Radio Wave Propagation, Far Eastern Branch, Russian Academy of Sciences, North-Eastern Heliogeophysical Center (SKR_558279, UNU 351757). The work was partly financially supported by the Russian Foundation for Basic Research (grant nos. 19-05- 00543 A and 19-5-700008 r_a).

References

1. V.V. Bychkov, Y.A.Nepomnyashchiy, A.S. Perezhogin, B.M. Shevtsov, Earth, Planets and Space (2014)
2. V.V. Bychkov, Y.A.Nepomnyashchiy, A.S. Perezhogin, B.M. Shevtsov, Atmospheric and Oceanic Optics, **28** (2015)
3. V.V. Bychkov, A.S. Perezhogin, I. N. Seredkin, B.M. Shevtsov, SPIE. Proc., **10466** (2017)
4. V. V. Bychkov, I. N. Seredkin, Atmospheric and Oceanic Optics, **34** (2021)
5. V. V. Bychkov, I. N. Seredkin, V. N. Marichev, Atmospheric and Oceanic Optics, **34** (2021)
6. A.V. Dmitriev, J.-K. Chao, J. Geophys. Res., **108** (2003)
7. M.G. Deminov, Earth's Ionosphere, in Plasmennaya geliogeofizika (Plasma Heliogeophysics), Fizmatlit, **2** (2008)
8. V.V. Bychkov, A.S. Perezhogin, A.S. Perezhogin, B.M. Shevtsov, V.N. Marichev, G.G. Matvienko, A.S. Belov, A.A. Cheremisin, Atmospheric and Oceanic Optics, **25** (2012)

A new quantum model of the magnetic field of the Hot Earth, Moon and terrestrial planets

Vladimir Kuznetsov*

Institute of Cosmophysical Research and radio wave propagation FEB RAS,
684034 Kamchatka, Paratunka, Mirnaya str., 7, Russia

Abstract. Distribution of P- and S- waves velocities in the Earth's inner core suggests that its matter has been quantum entangled since the origin of the Earth and the Solar system. This assumption we made allows us to develop the quantum model of the geomagnetic field evolution from its start to its disappearance. Unlike the generally accepted dynamo our model provides an obvious source of energy which is a phase transition and the thermal, mechanical and electrical energy released during it. The latter generates a double electric layer which rotation gives rise to the initial dipole field. Changing its direction the phase transition causes a reversal of the magnetic field. Magnetic and paleomagnetic data on the Earth, Moon, Mercury and Mars analyzed in the frameworks the Hot Earth model and features of their gravitation recorded at NASA project offered as conditions for the planets formation and evolution and so predictions for the further evolution of the Earth and its magnetic field.

1 Introduction into the Hot Earth model

The temperature T at the centre of the star, planet, satellite at the time of their formation is the main parameter of our model. During evolution of the Solar system, the temperature estimated by: $T \approx 3/5GM/Rc_p$, where c_p is the heat capacity of the substance, G is gravity, and R is the radius of the cosmic body is kept constant (somewhat decreased) only in the central part of the body. Temperature dependence on mass of a star, planet, satellite $T \sim M^{2/3}$ includes main-sequence stars, brown dwarfs, the planets of the Solar system and their seven large satellites. The temperatures estimated by this formula are somewhat overstated since the actual value of c_p is increasing when temp's rising [1].

Here the planets and the satellites of the Solar system are qualified by their mass dependence of gravitational energy $W = 3/5GM/R$ (kJ/g), see, Fig.1. It should be noted that all bodies, ranging in size from the Sun to the Europa, a satellite of Jupiter, are spherical. The Europa's gravity potential GM/R of about 2 kJ/g along with the average value of its heat capacity c_p = 0.3 cal/g. deg provides the temperature of its matter of about 1.5×10^3 K. All other cosmic bodies including asteroids and satellites with their mass less than that of Europa are not spherical and apparently, they hadn't melted completely.

* Corresponding author: vvk@ikir.ru

The heat of the evaporation (condensation) for silicon dioxide, as the main "building" material of terrestrial planets, is $U_e = 15$ kJ/g, and its heat of melting (crystallization) $U_m = 1$ kJ/g. Divided into three classes the cosmic bodies at the Solar System are these with $GM/R < U_m$, $U_m < GM/R < U_e$ and those objects that have $GM/R > U_e$. The small satellites and asteroids belong to the first class (I), the large satellites and the small (smaller than Venus) planets fall into the second one (II), and the planets with their mass greater than that of Venus are among the third class (III), see, (Fig. 1). Confirming U_1 and U_2 we access that Europe is the smallest spherical body and Titania is the largest non-spherical one. They differ ten times in mass and three times in size. The specific energy appears to be equal to the melting heat of the intermediate size cosmic body with $M \approx 2\times10^{25}$g and $R \approx 10^8$cm, then $GM/R = 1$kJ/g. According to our model the Earth's matter heated to high temperatures through the planet's formation had evaporated. Criterion for the total evaporation [2] is the compliance with the condition $GM/R = 2\ U_e$ or $U_e = 15$ kJ/g.

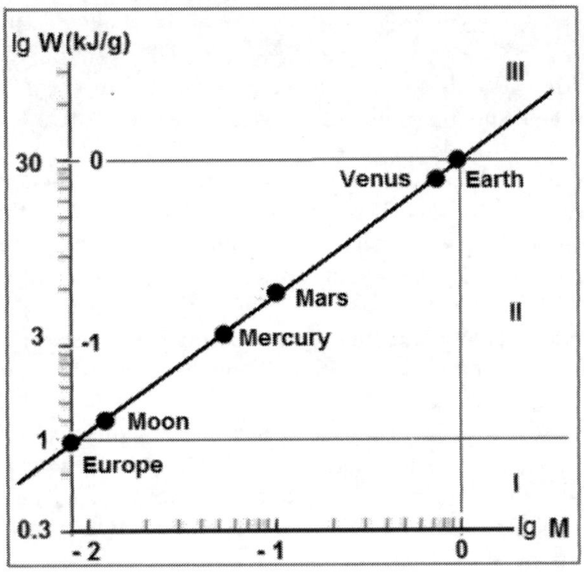

Fig.1. Specific gravitational energy *W* of the planets related to the Earth is shown as function of their mass *M*. The bodies are divided into three classes: I, II and III.

Table 1. Planets parameters.

		M/M_{Er}	R/R_{Er}	E/E_{Er}	p/p_{Er}	E/U_m	E/U_e
1	Mercury	$5{,}5\times10^{-2}$	0,375	0,14	0,13	4.2	
2	Mars	10^{-1}	0,53	0,19	0,14	6.0	
3	Venus	0,815	0,93	0,85	0,76		1.7
4	Earth	1 (6×10^{27} g)	1($6{,}4\times10^8$cm)	1 (30 kJ/g)	1 (360 GPa)		2.0
5	Moon	$1{,}2\times10^{-2}$	0,26	0,038	0,026	1.3	

The Earth's gravitational energy $E = 3/5GM^2/R = 2.25\times10^{39}$ erg (here: G is gravity, M – the Earth's mass and R - its radius) [1] necessary to smash the Earth's mass into dusty particles far from each other would come out in the reversal process of the Earth matter compression by selfgravitation yielding its heating. Dividing E by the heat capacity of the Earth's substance ($c_p = 0.3$ cal/g·grad) and its mass we see the temperature of the Earth's matter of about 30 000 K [3]. Estimating the initial radius of the Earth R_o one can see that the

continents can be laid one to another with virtually no gaps on the ball with the radius equal to that of the outer core of 3.5 thousand kilometers which was the Earth's initial radius.
Our model of the Hot Earth proceeds from the assumption which I couldn't not validate in my earlier research that at its beginning the Earth's matter was a tight ionized vapour (35 g/cm^3) which after its cooling became quasi-solid [4]. The fact that according to the model, the matter located in the inner core was really solid, did not serve to prove this indisputable fact. The problem why P-and S-waves don't accelerate as they approach the core center, since both the density of the core and its pressure increase became solved after I got acquainted with a relatively young concept of physics – the quantum entanglement (entanglements).

2 Quantum entanglement

By "Science in Focus "(according to the materials of BBC Science Focus, No. 2, 2014) the quantum entanglement as a new promising scientific direction was included into "20 GREAT IDEAS". Needing shocks and the research excitement the science is fueled by Great Ideas. At any time some of them are wrong, others lead to new breakthroughs. Science moves on formulating these ideas together with the attempts to refute them. By-products are the advances in technology and the new branches of science... Among these 20, quantum entanglement is N6.

The fact that it is a completely new trend in physics is indicated at least by the fact that it does not have an established Russian name. In English, everything is unambiguous: Quantum Entanglement. But in Russian... Quantum nonlocality, quantum entanglement, entangled quantum states... It seems that the term of quantum entanglement is gradually winning.

It is the quantum entanglement (QE) that forms the basis of quantum teleportation, quantum computer and quantum cryptography. By analogy I entitled the quantum geophysics (www.vvkuz.ru).

A question arises, why is the QE as a new area of physics, relevant to such long-established field of science as the geomagnetic field. The geomagnetic field is well known to arise in a liquid iron core under the magnetic dynamo effect. The dynamo mechanism definitely applies to all other space objects that have (or used to have) a magnetic field, for example, to the Moon or Mars. One wonders why was the magnetic field on the Moon and Mars and why had it disappeared? Was there a magnetic field on Venus?

About 40 years ago, I became interested in the problems of the Earth physics and, of course, the nature of geomagnetic field (EMF). The first thing I did was to uniquely determine the location of the source of EMF generation. Using various methods, I confirmed that it is generated inside the F-layer at the boundary of the Earth's inner core (the green label in Fig. 2) rather than that is prescribed in [5]. Secondly, I ascertained that ionization, the formation of electric charges, their separation in the F-layer and daily rotation provide a weak dipole magnetic field which may be strengthened to the required value by the Hall dynamo. In the 80-90 years I proceeded from this idea, but the scientific community did not accept it. Recently I made sure that my concept is correct after I discovered that a non-dynamo mechanism for generating a geomagnetic field is considered in the encyclopedia of geomagnetism [6].

Solving problems in physics of the Earth and, especially, of its magnetic field, I was interested in why do the velocities of P-and S-waves in the inner core behave so (red lines in Fig. 2) as if they are not affected by the fact that the external pressure and the matter density increase towards the center of the G-core, which should cause an increase in the values of these velocities. Only if a matter is quantum entangled it does not obey the laws of classical physics. I made sure after I had published a dozen of papers about various effects in geophysics that have the QE (books, papers and reports on website www.vvkuz.ru).

3 Quantum phase transitions

Quantum phase transformation is a matter transition from one quantum thermodynamic phase to another when external conditions change, and it occurs in the absence of thermal fluctuations, that is, at $Tc \to 0$. Thus, the system is rebuilt under the influence of any non-thermal parameters (for example, they are pressure or magnetic field).

Quantum fluctuations controlled by non-thermal parameters, such as pressure or particles concentration, are responsible for the phase transition in a quantum-mechanical system.

As estimated here the temperature of the Earth's matter at the time of its formation was: $T =$ 30 000K which is higher than the critical point (for example, the critical point of Au is about 6000K). At the critical point (CP) the quantum entangled substance [7] in a state of the quantum vapour is able to huge density fluctuations, which ultimately lead to an increase in the concentration of Au and other precious metals [8].

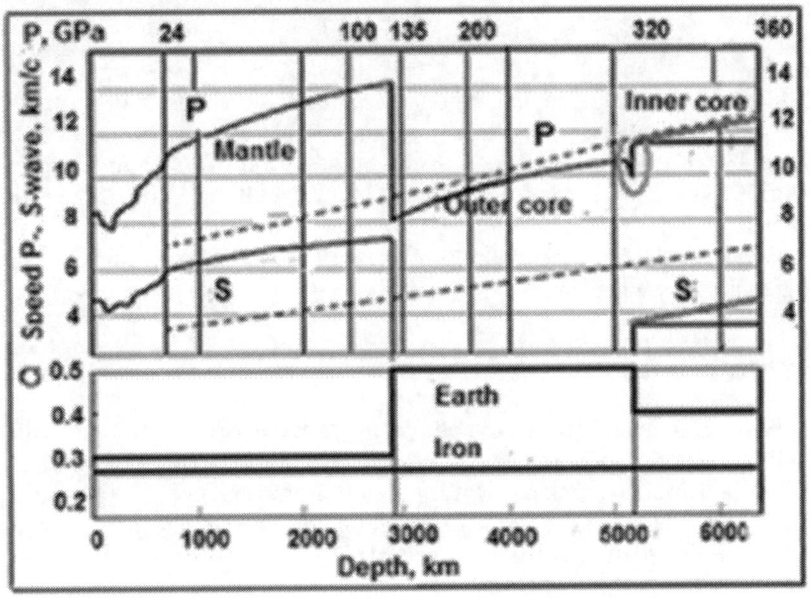

Fig. 2. The velocities of P- and S-waves in the Earth's shells and, in particular, in the G-core and in iron (dashed line) at the external pressure equal to that of the Earth's one. Poisson's coefficients σ of Earth and iron [1].

4 Generation of the geomagnetic field

Fig. 3 shows the mechanism of the geomagnetic field generation with two modes in the F-layer which cause the Earth's expansion or its compression.

Condensation (C) of the superheated substance alternates its evaporation (E). Starting from the QE our concept of the geomagnetic field generation suggests that the acquisition of the QE (coherence) is replaced by its destruction (decoherence). The first mode results in the solid state of the Earth's substance (in the inner core or over the whole Earth just after its origin). The modes change if the internal core becomes overheated relative to the external one ($T_G > T_E$) or backwards if $T_G < T_E$.

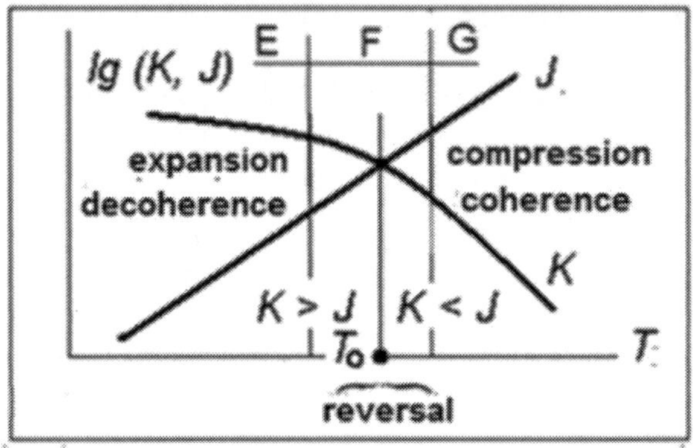

Fig. 3. Scheme of the phase transition in the F-layer after the classical model: K - the rate of condensation of hot vapour, J -the rate of evaporation. Quantum entanglement considers the decoherence and coherence. Essentially there is matter compression or its expansion. When modes alternate, the reversal of the EMF occurs. Top: E-layer, F-layer, G-core.

Fig. 3 shows the change in the model parameters corresponding to the Earth's expansion (left) and the compression mode is given at the right. During the last 800 thousand years of Brunhes chron the Earth has been in the quiet period of its compression. Before Brunhes, the Earth was mostly expanding for one and a half million years of the Matuyama chron. Sooner or later, a period similar to the Matuyama chron (New Matuyama) will ensue [7]. This assumption will be discussed later, now we will justify the possibility for the EMF to arise without dynamo mechanism.

5 Nondynamo theory

At the encyclopedia of geomagnetism the nondynamo theories [6] offer four mechanisms: 1) Thermoelectric effect, 2) Thermomagnetic effect, 3) Hall effect and 4) Rotating electric field, based on: 1) Dipole associated with a rotating ferromagnet, 2) Differences in temperature and composition, 3) Currents arising in presence of the magnetic field, 4) Currents arising from the rotating charged capacitor.

The paper «Rotating electric fields and charges» ([5], P. 705, David J. Stevenson) placed here, states that EMF may be generated by the rotation of charges, the «problem is that in order to produce modest magnetic fields one requires very large voltage differences. The electric field E ($V \cdot m^{-1}$) required to provide a field B (T) is $\approx c^2B/v$, where v is the rotational velocity. This suggests $E \approx 10^{11}$ $V \cdot m^{-1}$, which is physically impossible (it will ionize atoms) ». As one sees the generation of the EMF requires an electric field of a huge value. D. J. Stevenson is quite true if the emergence of a magnetic field in the Earth's core is carried out only by the rotation of electric charges. A weak dipole field generated by the daily rotation of charges needs to be amplified by a current of Hall type directed along that caused by the rotation of the double electric layer (DEL).

6 Model parameters

In the F-layer the electric field intensity E_R: $E_R = q/(C \times \Delta R)$, here $q = NeV$ - the charge of the layer, ΔR - its thickness, N – the concentration of charges, e – the electron charge, $1/Ne$ is the Hall constant D, V – the volume of the spherical " capacitor": $V = 4\pi R^2_G \Delta R$, where R_G is the radius of the inner core, then C is the electric capacity of the DEL: $C = 4\pi \varepsilon_o R^2_G/\Delta R$, ε_o – is the electric constant and $E_R = \Delta R/(\varepsilon_o D)$.

As it is in the atmosphere the concentration of electric charges *Ne* seems to be unevenly distributed over the F-layer. As in the atmosphere clouds the charges separate in the region of the F-layer where condensation processes mainly occur resulting in the variation of E_R over the layer (along ΔR). For example, the electric field is maximum (10^5 V/ m) in clouds, where the generation of electric charges is commonly believed to occur. The field is high enough near the Earth's surface (10^2 V/m) and it is much less in ionosphere. Our model suggests that E_R should be maximal in thin layers at the boundaries of the F-layer, where, condensation and boiling are probably to proceed.
As discussed earlier, the daily rotation of the DEL with a frequency of ω may cause "quasi-currents " and generate the magnetic field. A positive charge is assumed here to be accumulated at one surface of the F-layer, and a negative - at the other one (Fig. 3). The Hall constant *D* is positive in the first case $D^{(+)}$, and it is negative in the second one $D^{(-)}$.

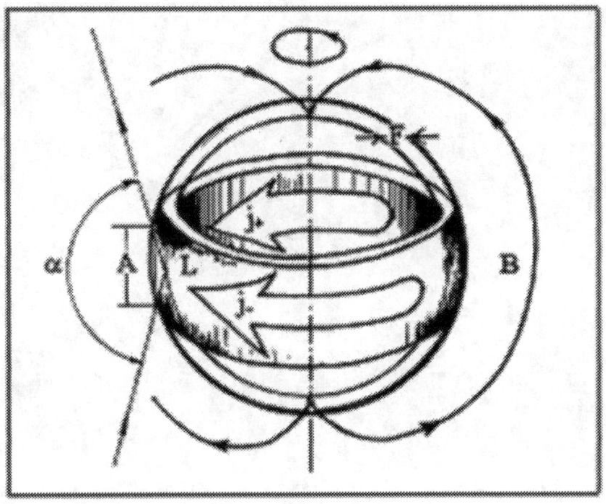

Fig. 4. Current rings at the F-layer [1].

The reversal of the DEL polarity, giving rise to the change of the "evaporation-condensation" mode, reverses the sign of E_R. The charges rotation at a velocity of ωR_G is equivalent to a "current" with a density of $j_\omega = \omega R_G/D$. The magnetic field (induction) of this "current" on the axis of the system rotation is:

$$B = \mu_o j_\omega S_X/2R_G,$$

here S_X is the effective area of charges assembling, $S_X \approx \delta \times A$, $\delta << \Delta R$ is the thickness of the layer where the current flows j_ω (μ_o - the magnetic constant, *A* - the width of the layer). Since the charges at the boundaries of the F-layer are opposite in sign, there are two opposite directed currents (related to the rotation direction and the sign of *D*). At the center (on the axis of the Earth's rotation) of axial circular "currents" the magnetic induction is: $B = B_G - B_F$, here $B_G = \mu_o j_\omega S_X/2R_G$, and $B_F = \mu_o j_\omega S_X/2R_F$. Substituting the corresponding values $j_\omega = \omega R_G/D$ (in the second case, we use R_F instead of R_G) we get $B = 0$. However, this is not the case for points external to the current rings. For example, the value of the induction (the horizontal component of the field) at the equator is:

$$B = \mu_o j_\omega S_X/h - \mu_o j_\omega S_X/(h + \Delta R) \approx \mu_o \omega S_X \Delta R R_E/h^2 D,$$

here $h = R_E - R_F$, R_E – the Earth's radius. Substituting ΔR:

$$B^* = (\mu_o \varepsilon_o \omega/R_E) \times S_X E_R.$$

The value of B^* is determined by E_R and the area occupied by the "current": $B^* \approx 10^{-28} E_R \times S_X$. Assuming S_X of about 10^9 m^2 and taking into account that E_R cannot be arbitrarily high, then the magnetic field generated by the rotation of the DEL is far less ($k_o >> 1$) than that required (at the equator the geomagnetic field $B \approx 3\times10^{-5}$ T, k_o is the required coefficient of the field increase). This assessment confirms the view of other authors that the geomagnetic field is impossible to be generated just due to the charges separation inside the DEL and its daily rotation.

The currents inside the F-layer are discussed here as those inside the spherical capacitor with two thin conductive layers-linings δ thick, here one of them is close to the inner core and the other one adjoins the outer core. Note that the current $j_\omega S_X$, induces a magnetic field B_o in the area of the opposite layer - δ where the radial field E_R acts and the current I_R is directed normally to the field B_o. Generated by the co-influence of B_o and the current j_R the Hall one is directed alike that of rotating charges j_ω which may be amplified due to the Hall effect

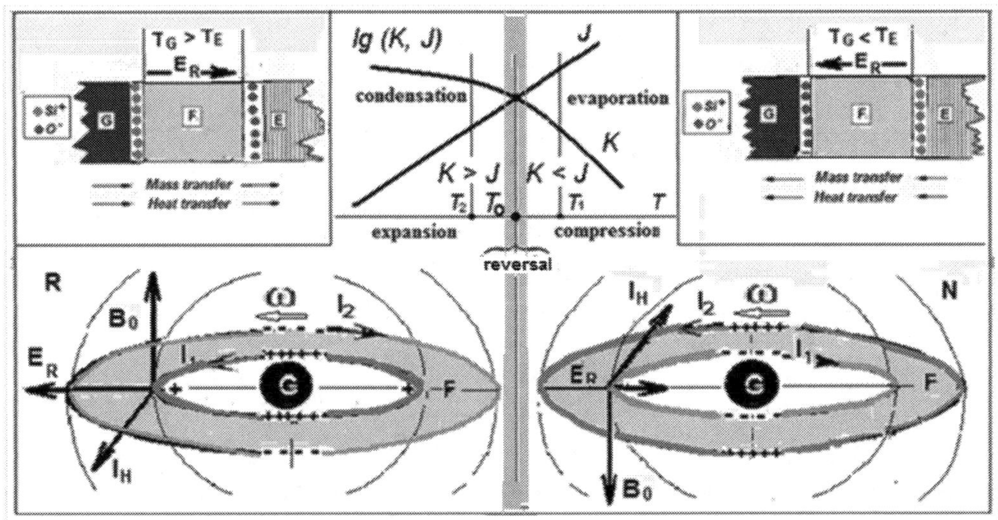

Fig. 5. Heat and mass transfer and charges separation in the F-layer. Left - the expansion mode (R), right - the compression mode (N). The formation of DEL and its daily rotation giving rise to the initial field B_0. The interaction of B_0 with the radial electric field of the DEL E_R generates the Hall current I_H directed along the current producing the field B_0.

Let us estimate B_o generated in the vicinity of a lining δ thick by the current $j_\omega S_X$ flowing along the other lining:

$$B_o = (1/2\pi)\mu_o j_\omega S_X/\Delta R = (1/2\pi)\mu_o\ \omega S_X R_G/D\,\Delta R.$$

Nearby the currents j_ω the directions and magnitudes of B_o are the same because the direction of the current is determined by the direction of the Earth's rotation and the polarity of the electric charges or the Hall constant D.

The electric charges D, the radial electric field E_R, the current I_R and the orthogonal magnetic field B_o cause the Hall voltage $U_X = DB_oI_R/A$ and electric field $E_X = U_X/2\pi R_G$, as well as the Hall current j_X:

$$E_X = D\,[B_o \times j_R],$$

$$j_X = \sigma_X E_X,$$

where σ_X is the Hall conductivity, $j_R = I_R/S_R$, $S_R = 2\pi R_G \times A$. As it follows from our model, the Hall currents are opposite directed in δ-layers where electric charges of opposite signs are concentrated. By our model a required field needs the current j_ω amplified by a factor of k_o. The possible value of k follows from $k = j_X/j_\omega$. Substituting above mentioned formulas we get:

$$j_X = \sigma_X D\,[B_o \times j_R].$$

Let's derive B_o from j_ω and j_R from E_R:

$$B_o = \mu_o S_X j_\omega / 2\pi\Delta R,$$

$$j_R = \sigma_R \Delta R / \varepsilon_o D.$$

Then k is:

$$k = (\mu_o / 2\pi\varepsilon_o)\, \sigma_X \sigma_R S_X.$$

The current amplification due to the Hall effect is limited by the values of the conductivities σ_X, σ_R. To obtain the required magnetic field (induction), the obvious condition $k \geq k_o$ should be met. The coefficient k_o should not be arbitrarily large, since it shows how much the velocity of the Hall current is greater than the speed of the charges daily rotation. The value of k_o is bound by the current (ions moving) speed which is not high.

Estimating σ_X and σ_R in the frameworks of our model we assume that if σ_X is large being close to the metal conductivity, then the small σ_R is corresponding to the conductivity of dielectric. (The matter polarization and the formation of DEL are actually possible only in dielectric). Let's determine the electrical characteristics of our system with geomagnetic field generation. The amplitude of the Hall current I_X and the voltage U_X can be estimated using the restriction of the geomagnetic field power (dissipation) $P_X = 10^{12}$ W. Assuming that in the F-layer there are two currents opposite directed, and the induction B in the region of the equator ($B \approx 30$ μT) is:

$$B = \mu_o I(1/h - 1/(h + \Delta R)) \approx \mu_o I \Delta R / h^2$$

then $I_X \approx 10^{10}$ A can be generated only if the inner core with its necessary ions charge is involved.

Similarly, the electrical characteristics of a system of radial currents and voltages can be evaluated. Apparently, the power of this system should be somewhat higher than that of the Hall one, since the Hall effect (if it is causing I_X) is secondary, but it should not be greater than the power of the heat flow $P_R = 10^{13}$ W (we assume the converting efficiency of thermal energy into magnetic field energy of about 10%).

The electric current "space – the Earth" of $(6.8\text{-}9.8)10^9$ A is quite enough to generate the geomagnetic field. Reaching the Earth's surface, the current is most likely "washed away" into the space by the currents of calm rains.

7 Evolution of the Earth's parameters in the context of our model of the EMF generation

To date, geologists have collected samples, the processing of which shows that the earlier Earth used to have another composition and density of atmosphere, surface temperature, oceans, the speed of the Earth rotation, gravity and magnetic field, etc. Almost all of them are in contradiction with the model of the cold Earth. The known facts of the Earth's evolution are discussed here in the context of our model.

Fig. 6 [9] shows stages of the Earth's evolution: (1) – formed in a quantum entangled state as a ball with a radius equal to that of the outer core the Earth consisted of gravity-constrained shells of silicon oxide (basalt) with a density of 35 g/cm^3 and a temperature of 30 000 K which is not observed now, (2) – loosing the QE through the first decoherence the matter (basalt) on the surface of the ball was heated (the red shell) to a high temperature, it evaporated then and during its further cooling turned into granite (Fig. 6) which formed the outer granite shell of the Earth. Escaped from the Earth the volatile impurities under the solar wind were split into hydrogen and oxygen, they combined then, rained and are preserved to our days in the form of fresh water, (3) – in the subsequent decoherences the liquid E was formed, where a layer of phase transition or the F-layer appeared, and the geomagnetic field was activated, (4) - after the repetitive decoherences, a liquid E-core and a solid D-mantle were formed, (5) - the inner G-core in the QE state is preserved inside the liquid core. After 5-10 million years, the G-core will melt due to the final decoherence and the geomagnetic field will disappear.

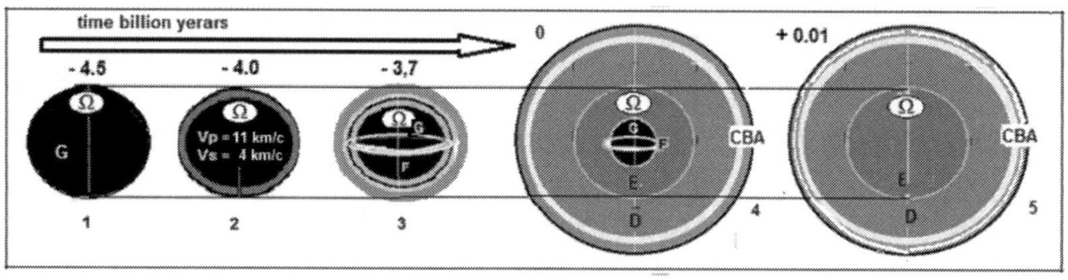

Fig. 6. Cooling of the hot, total quantum-entangled Earth from its formation 4.5 billion years ago up to nowadays approximated to the following 5-10 million years, after which the geomagnetic field will disappear.

8 The first decoherence

The velocities distribution of the seismic P - and S-waves inside the inner G-core and the Poisson's ratio (Fig. 2) related to the early period of the Earth's evolution, when all its matter was QE (Fig. 6-1, 6-2) provide the fact that in the inner core the wave velocities do not vary with the pressure increase along the radius of the Earth's core.

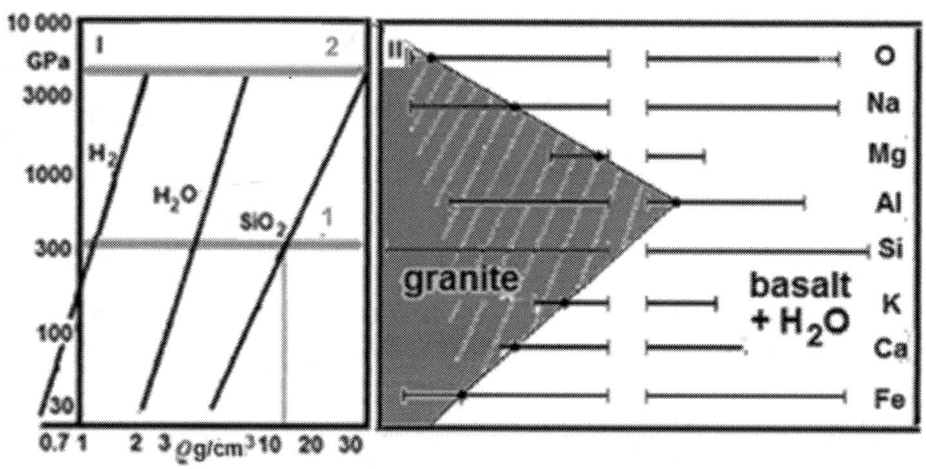

Fig. 7. I illustrates the main components of the Earth matter (pressure-depending density of H_2, H_2O and SiO_2) [1]. Blue lines (BL): 1 - pressure in the center of the modern Earth, 2 – at the time of its formation. Fig. 7-II shows how one part of granite (grey) can be obtained from two parts of basalt [1]. The chemical elements of basalt are indicated at the right. The length of the strip shows the relative concentration of basalt. Relative concentrations of Si in granite and basalt are assumed to be equal. The concentration of Al in granite is greater than in basalt. If we assume that when heated, a part of basalt will evaporate, then the darkened part which is granite will remain.

During the Earth cooling, as the pressure inside the sphere surrounded by the granite shell had been decreasing the concentrations of hydrogen, water, and silicon oxide (basalt) had been changing (Figure 7-I).The fundamental difference between the Earth and other planets is the fact that the total matter of the Earth was quantum entangled. At the Moon and

Mars, their quantum entangled matter was located in the center of the melt, which turned out at the Earth after a series of decoherences. It should be considered that the matter of the Earth had a high temperature and the traditional thermodynamics does not affect the quantum entangled matter. During the first decoherence the matter at the core boundary heated to a gaseous state was ejected into space, and the rest of it turned into the hot dense granite (Fig. 7-II), which formed a granite shell ten kilometers thick, where fresh water was raining. At the other planets where are no granites and fresh water this phenomenon did not occur.

9 Confirmation of our model from the Kola Superdeep Borehole (KSB)

The existence of granites and fresh water is known to be the main difference between the Earth and other planets of the Earth group. The Earth granites are wide described whereas I am not acquainted with any information about the origin of fresh water on Earth. There is no data why there are no granites (in large quantity) on the Moon, Mars. Here we discuss this fact in the frameworks of our model. The Russian physicist Ya. B. Zeldovich argued [3] his so-called "Principle of the total evaporation": the matter would evaporate entirely if the specific energy invested into it is twice the heat of the phase transition of evaporation. It is applicable to the Earth unlike other planets of the Earth's group. (The specific gravity energy of the Earth is $U = 30$ kJ/g, and the heat of evaporation of $SiO_2 = 15$ kJ/g). It follows that at the time of its formation the Earth's matter was entirely vaporized, otherwise, it was quantum entangled. The Moon's and Mars' matter was quantum entangled partly. Any breaking of the QE caused a huge heating of the Earth's matter: basalt, water, and other volatile components. The molten granite did not evaporate from the quantum entangled matter of the globe and cooling down it formed a granite shell as a base of modern continents.

The granite was produced from basaltic lava. The fact that in the absence of basalts the granites layer at the Kola Superdeep Borehole (KSB) was revealed caused a confusion among geologists. Nevertheless this fact proves our model.

A layer where the concentration of gold was elevated was found in granites at KSB. The Hot Earth model I am developing allowed me to tackle the problem how were mineral deposits generated. Basalt heated to about several tens of thousands degrees will turn into granite. If the temperature of granite is at the critical point, then, for example, gold at a temperature of about 6000 K acquires special quantum properties, called "giant density fluctuations", which lead to the fact that gold assembles into huge drops that will not disintegrate when cooling down [8]. This approach showed that temperatures and pressures at which diamond synthesis occurs were at a depth of hundreds kilometers at earlier stages of the Earth's evolution, whereas they correspond to ten kilometers at the depth of the modern Earth.

10 The magnetic field of the Moon, Mars and Mercury

Nature doesn't choose the mechanism of magnetic field formation for each planet aside. As Newton spoke such a waste is not inherent in nature. A person engaged into physics of a planet magnetic field does not worry that his model may not fit at another planet or satellite. Our model of the magnetic field of the Hot Earth can be legitimate only if it is applicable for every cosmic body. It is necessary to take into account that for a number of reasons the rates of planets and satellites evolution can vary greatly despite the fact that they were formed simultaneously in a single process. As follows from the Hot Earth model, our planet will soon spend the energy reserved in the inner core, and the geomagnetic field will extinct after further 2 – 3 million years of its evolution. Venus, Mars, the Moon and other large satellites as opposed to Mercury are assumed to implement this script.

11 Features of the Moon's evolution

Estimating the magnitude of the Moon's expansion we need the specific gravitational energy of the Moon: $e^*=1/2GM/R$ = 1.4 kJ/g. The heats of the phase transition (PT): U_2 is about 1 kJ/g for melting-crystallization and U_1 = 15 kJ/g for evaporation-condensation. About 0.4 kJ/g is left for the evaporation of the Moon's matter which is 0.4/15 = 3 % of the Moon's mass. With Zeldovich criterion (2 U_1) we get the half of the value that is (1.5 %) of the Moon's mass. Here we take 1.0 %. Since the Moon's mass is $7.3 \cdot 10^{25}$ g the mass evaporated is $\Delta M \approx 7 \cdot 10^{23}$ g. The density of the Moon's matter is 3.5 g/cm^3 and its volume $\Delta V \approx 2 \cdot 10^{23}$ cm^3. The Moon's surface $S = 38 \cdot 10^6$ km $\approx 4 \cdot 10^{17}$ cm^2 and the increase of the Moon's radius $\Delta R = \Delta V/S = 2 \cdot 10^{23}/ 4 \cdot 10^{17} \approx 5$ km.

Our estimate was confirmed by the GRAIL mission [9] aimed at studying the gravity and the internal structure of the Moon resulted in comprehensive pattern of the lunar crust saturated with igneous intrusions which formed during the ancient lithospheric extension. The Moon was revealed [9] to have been expanding during the initial 700 Myr of its evolution with the increase of lunar radius by 0.6–4.8 km.

Given the volume of the Moon's evaporated mass and the diameter of its inner solid iron core equal to 480 km, the initial density of the core matter was near to 10 g/cm^3 (Fig.8).

The initial density of the Earth was estimated earlier (35 g/cm3) and the current one is 12 g/cm^3. The initial density of the hot cores of Mars and Mercury should be estimated from the relationship between the mass of the planet (satellite) and the initial density of its matter (Fig. 8).

Through the increase of lunar volume on $2 \cdot 10^{23}$ the density of its core had dropped by half resulting in 5 g/cm^3. It should be noted that as the Moon had been increasing due to conductive heat transfer its radius evolution follows $\Delta R \sim t^{1/2}$.

12 The expansion of Mars and Mercury

The average densities are: the Moon - 3.3 g/cm^3, Mars - 3.9 g/cm^3 and Mercury - 5.4 g/cm^3. Let's assume that after some expansion the Moon, Mars and Mercury lost their hot cores formed at the earliest stage of their evolution. It is assumed here that unlike Mercury the hot matter in the cores of the Moon and Mars has relaxed by our time. Then the average density of Mercury should be about 3.6 g/cm^3. The mass of the planets is taken not to change during their relaxation. Therefore, after the end of Mercury expansion its final volume should be near to $9 \bullet 10^{25}$ cm^3, and its radius of 2.8 thousand km is more than the modern radius of Mercury by 360 km, and the volume increased on $3 \bullet 10^{25}$ cm^3 that is one and a half times of modern one.

20% of Mercury mass ($6 \bullet 10^{25}$ g) has turned into gas. If the density of the gaseous substance was 15 g/cm^3 then the volume occupied by the core should be approximately $4 \bullet 10^{24}$ cm^3, and the core radius should be $\approx$ 1000 km. Assuming that 80% of Mercury mass remaining was concentrated in a space between the initial surface of Mercury and its core, the planet's density was about 3.6 g/cm^3 and the initial radius was near to 1800 km. At its formation the density of the hot core of Mercury was about 15 g /cm^3 and the average density of the planet (Mercury) was slightly lower than the density of its core. During the evolution (4.5 billion years) the radius of Mercury has grown by about 600 km and its density has decreased to 5.4 g/cm^3. If the final density of Mercury's matter is 3.6 g/cm^3 then its radius will grow to 2.8 thousand km. As one can see Mercury has to increase its radius on about 400 km. It is difficult to estimate the time this process will take however it is clear that it should be close to that which has passed since the formation of the planet. It is obvious that the expansion of Mercury similar to that of the Moon follows: $\Delta R \sim t^{1/2}$.

During its formation about 50% of Mars' mass turned into gas which initial density is assumed to be ~16 g/cm^3 (Fig. 8). Mars' mass is 6.4•10^{26} g, its radius is 3.4 thousand km. We assume that the mass is equal to half of Mars' mass and the radius of its hot core is 1.7 thousand km that is close to that generally accepted (1694 km). (The radius of the core is believed to be about 1/3 of the planet radius). Just like at Mercury another half of the mass occupies the space between the initial surface and the core.

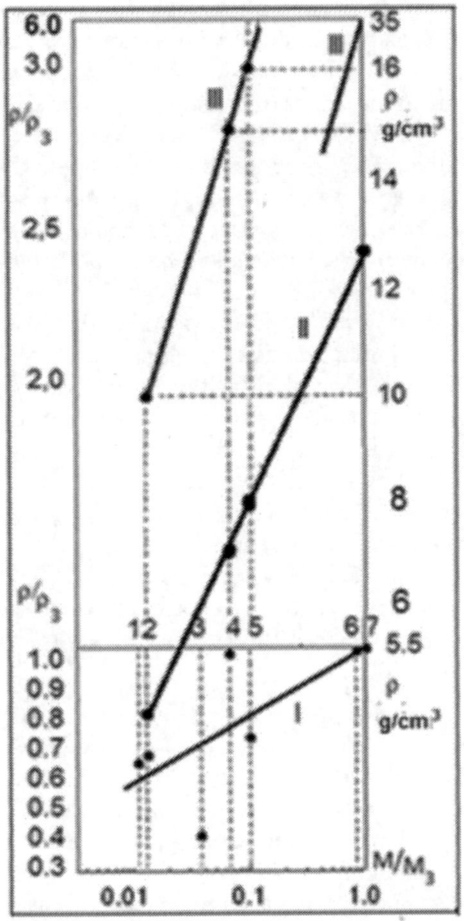

Fig. 8. Lines: I- the density of planets and satellites (right) and these relative to the Earth's mass (left). III - the density of gaseous matter at the time of planets and satellites formation. The density of the Earth and Venus is 30-35 g/cm^3, Mars-16 g/cm^3, Mercury-15 g/cm^3, the Moon-10 g/cm^3. II – the density of cores matter after the expansion stage: 1 - Europa, 2- Moon, 3 - Ganymede, 4 – Mercury, 5- Mars, 6 -Venus, 7 – Earth.

If the matter density was 3.9 g/cm^3 then the initial radius of Mars was about 2.9 thousand km. It follows that during its evolution Mars' radius has increased ($\Delta R \sim t^{1/2}$) on about 500 km (3.4 – 2.9), and the upper mantle composed of olivine 1113 km thick also corresponds to our result: 1200 = 2900 -1700 km. The thickness of the lower mantle is considered to be 561 km, which is also close to our estimate (3400 – 2900 km). It should be noted that the expansion of Mars due to the decompression of its matter inside the inner core occurred at the boundary of its hot and dense core, i.e. in the lower mantle. The convection did not arise here as well.

Nowadays Mars has finished its evolution with its radius of 3400 km. Mercury is at the middle stage of its evolution and its current radius is 2400 km. Mercury will expand further until its radius reaches 2800 km. At the end of Mercury's evolution the density of its matter intermediate between those of the Moon and Mars will be ≈ 3.6 g/cm^3.

Silicon oxide as the operating substance of the planets was useful because the behavior of the SiO_2 density as a function of pressure is well known, and we used this simplification when we were discussing the Earth. The condensation of SiO_2, its relaxation and transition into a condensed (liquid) state and, ultimately, the decompression of the compressed substance and planets expansion is a source of heat flow, tectonic phenomena and the energy necessary to generate the planet's magnetic field. In our model all these features are interrelated. The pressure at the center of Mercury and Mars is close to 0.13 and 0.14 of that at the center of the Earth (3.6 Mbar) or 0.4 and 0.5 Mbar. The corresponding density at the center of the planets, if they consist of SiO_2 (Fig. 7), is approximately 7-8 g/cm^3.

The density of the Earth's substance at the inner core is known to be about 12 g/cm^3. The initial density of the recompressed gaseous substance of the Earth was 35 g/cm^3. In the process of evolution the gaseous substance had been cooling and condensing, its pressure p and density had been decreasing (at the initial stage $p \sim T$). For example, at the time of the Earth formation its pressure was about 15 times higher than the modern one.

As shown above the initial density of the lunar inner gaseous core was about 10 g/cm^3. Let's assume that the difference between the planets (at least, of the Earth's group) including the density of their gaseous substance is determined by the magnitude of their mass. If we draw a straight line on the density – mass scale from 10 to 35 g/cm^3 (Fig. 8) it turns out that the density of the gaseous substance in the inner core of Mercury is about 15 g/cm^3, and that of Mars is 16 g/cm^3.

Should be repeated that the Mercury magnetic field indicates that its inner core had the initial density of about 15 g/cm^3 and it is noticeably lower now. The Mars' core functioning at a certain stage of its evolution which lasted according to various estimates from 1 to 2 billion years had already eliminated its energy reserve.

13 Tidal effects and slowing of Mercury's rotation

Powerful solar tides had reduced the speed of Mercury rotation from 8 hours up to 58.65 earth days. Due to the tidal braking as it was moving away from the Sun, Mercury was changing its angular velocity of rotation. Complete rotation around its axis Mercury makes in 2/3 of its year. The duration of Mercury sunny day is 176 earth days. It takes 87.969 earth days to orbit round the Sun. Mercury makes three rotations around its axis in two Mercury's years. Similar to the Moon the speed of Mercury's rotation around its axis should coincide with that of its rotation around the Sun.

The rotation of the Earth is too decelerated by the tidal friction, and the duration of the day gradually increases. Over the last 2.5 thousand years, it has increased by about 0.0024 seconds per century. Let's estimate the tidal effect for Mercury. If the initial period of Mercury's rotation around its axis was about 8 hours and after 4.5 billion years the modern one is 58 days then the average deceleration rate of Mercury is approximately 0.1 seconds per 100 years, that is about 100 times higher than that of the Earth.

Due to the tidal friction the rotational speed of Mercury decreased by 175 times. The rotational energy of Mercury at the initial stage of its evolution was $E\omega = M(\omega R)^2/2 = 7 \cdot 10^{26}$ J and the specific rotational energy was ≈ 2 kJ/g. The specific rotational energy of modern Mercury is $(175)^2 \approx 3 \cdot 10^4$ times less implying that almost all the rotational energy has been spent during the evolution of Mercury.

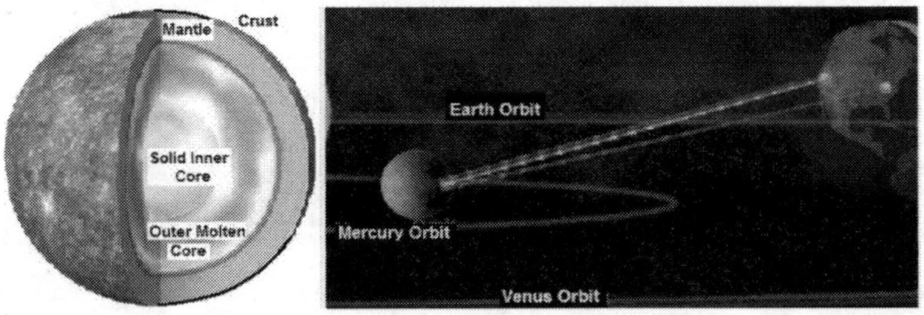

Fig. 9. The internal structure of Mercury (left) obtained by [10].

As estimated above, the evaporation and overheating of Mercury's matter should be spent to ≈ 3 kJ/g (see Fig. 1) which is close to E_1. It is clear that during its evolution Mercury had dissipated its rotational energy close to that necessary to expand its radius, generate its magnetic field and heat flow. The tidal dissipation increased the temperature of the upper layers at Mercury's mantle and since the temperature gradient dropped the thermal outflow from the hot core decreased. The heat flow is provided by the condensation of the gaseous matter and its further crystallization into mantle. The Moon and Mars are cooling without additional heating, their gaseous matter has already relaxed, and the generation of their magnetic field has stopped. Mercury, Mars and the Moon unlike the Earth and Venus had been cooling due to the thermal conductivity which is characterized by the conductive heat transfer with $\Delta R \sim t^{1/2}$. (Here t is the time of the planet's evolution). After the thickness of the liquid core at the Earth and Venus became more than 300 km the convection mode appeared, and further cooling was faster than $\Delta R \sim t^{1/2}$. This phenomenon did not occur at the Moon, Mercury and Mars.

14 Conclusions

A new model of the magnetic field of planets and large satellites is proposed. Unlike the generally accepted dynamo, our model proposes an obvious source of energy, it is a phase transition and the thermal, mechanical and electrical energy released during it. The last gives rise to a double electric layer, the rotation of which ensures an initial dipole field and the change of the phase transition mode causes a reversal in the polarity of the magnetic field.

In the frames of our model, such world problems of geomagnetism as reversals, the correlation between reversals and the intensity of tectonic phenomena, for example, the well-known paradox "superchron – superplume, the source of the geomagnetic field generation found a simple unambiguous solution. The lack of clear answers to these questions result in misunderstanding of such phenomena as: why was there a field at the Moon, and why did it disappear? Why does the same situation about the magnetic fields of other planets and satellites exist? Why has Mercury a dipole field? Is there any data on the paleomagnetic field of Venus?

References

1. V.V. Kuznetsov, Vvedenie v fiziku goryachei Zemli (KamGU, Petropavlovsk-Kamchatsky, 2008) (in Russian)
2. Ya.B. Zeldovich, Yu.P. Raizer, Physics of Shock Waves and High-Temperature Hydrodynamic Phenomena (Nauka, Moskow, 1966) (in Russian)
3. V.A. Magnitsky, Vnutrennee stroenie i fizika Zemli (Nedra, Moskow, 1965) (in Russian)

4. V.V. Kuznetsov, E3S Web of Conf., **196** (2020)
5. V. V. Kuznetsov, Vestnik KRAUNC. Phys.-Mat. Science, **2**, 9 (2014) (in Russian)
6. Encyclopedia of Geomagnetism and Paleomagnetism. Ed. D. Gibbins and E. Herrero-Bervera (Springer, Dordrecht, 2007)
7. V.V. Kuznetsov, Vestnik KRAUNC. Phys.-Mat. Science **35**, 2 (2021) (in Russian)
8. V.V. Kuznetsov, Geologia i MSR Sibiri **3**, 23 (2015) (in Russian)
9. J.C. Andrews-Hanna et al., Science, **339** (2013)
10. J. L. Margot et al., Science, **316** (2007)

Analysis of manifestation of joint acoustic and electric responses of near-surface sedimentary rocks on the deformation by earthquake seismic waves in the South of Kamchatka

Mikhail Mishchenko[1,*], *Oleg* Rulenko[2], and *Yuriy* Marapulets[1]

[1]Institute of Cosmophysical Research and Radio Wave Propagation FEB RAS, Mirnaya, 7, Paratunka, Kamchatkskiy kray, 684034, Russia
[2]Institute of Volcanology and Seismology FEB RAS, bul'var Piypa, 9, Petropavlovsk-Kamchatskiy, 683006, Russia

Abstract. We continue the investigation of earlier discovered joint acoustic and electric responses of near-surface sedimentary rocks on the deformation by earthquake seismic waves. The appearance and absence of this response in earthquakes with the energy class of more than 11.0, which occurred from June 2017 to May 2021 by the eastern coast of Kamchatka in the latitude band of 51.7-54.0° N, are under consideration. From those earthquakes we selected 52 events, the seismic waves of which had rock acoustic response at Karymshina site, IKIR FEB RAS. Occurrences of joint acoustic and electric responses of rocks and manifestations of their low-frequency and high-frequency acoustic responses were analyzed. Applying the nonparametric correlation Spearman analysis, statistically significant relation between the energy class of the earthquakes under consideration and the distance from hypocenter to the observation site was discovered.

1 Introduction

Transmission of seismic waves, generated by earthquake sources, causes rock deformation and is accompanied by different co-seismic effects [1]. These effects can manifest strongly enough as long as rock deformation occurs with the rate higher than that during the background regime. Among them, seismoelectric effect of the second kind [2] and geoacoustic effect [3], observed in sedimentary rocks, are known.

Disturbances of geoelectric and geoacoustic fields occurring during seismic wave transmission though the near-surface sedimentary rocks have their own features of genesis but common deformation nature. Joint seismoelectric and seismoacoustic responses of these rocks were firstly discovered in 2018 at Karymshina site of IKIR FEB RAS [4]. Investigation of these responses was continued in 2019–2020 [5, 6].

Based on drilling results, the near-surface rocks at the observation site region are sedimentary with the layer thickness of about 50 m [7]. The distance between the electrodes for geoelectric signal recording is 10 m and the depth of electrodes in the ground is 1 m.

*e-mail: micle@ikir.ru

According to the estimates in the paper [3], the sources of acoustic signals, generated in sedimentary rocks at the frequencies from the first hundreds of hertz to the first tens of kilohertz, are located at the distances up to 37 m from the hydrophone. Taking all that into account, we can say that the observed seismoelectric and seismoacoustic signals occur in the near-surface sedimentary rocks having the volume up to 10^5 m^3. This volume of rocks can be considered as two operating simultaneously natural seismometer, that transform the seismic wave energy into the energy of electric and acoustic signal.

In the previous investigations we considered the features of manifestation of joint seismoelectric and seismoacoustic responses for five [5] and six [6] earthquakes of different energy, which occurred in Kamchatka region. In the paper, 52 earthquakes of this region are considered during the investigation of such responses.

2 Analysis and results

Initial data. Seismoacoustic and seismoelectric observations were carried out at Karymshina site (52.83° N, 158.13° E) of IKIR FEB RAS. The observation method and the applied measuring-recording complex are described in the papers [4, 5]. To analyze the manifestation of rock joint response, we considered the earthquakes with the energy class $K_s > 11.0$. They occurred from June 2017 to May 2021 by the eastern coast of South Kamchatka in the latitude band of 51.7-54.0° N. Only those earthquakes were chosen, in which there was necessarily low-frequency, approximately 100 Hz, seismoacoustic response, since at the same distance from the hypocenter it occurs with weaker earthquakes than the seismoelectric response. The parameters of these earthquakes are taken from the catalogue of Kamchatka Branch of Geophysical Survey RAS (http://sdis.emsd.ru/info/earthquakes/catalogue.php) and are illustrated in Table 1.

Table 1. Earthquake parameters based on which the manifestation of seismoacoustic and seismoelectric responses of near-surface sedimentary rock on the deformation by seismic waves were analyzed; K_s is the earthquake energy class; D is the hypocentral distance to Karymshina site; A is the presence of joint seismoacoustic and seismoelectric responses of rocks (♦) or only their seismoacoustic response (○); B is the presence of low-, mid- and high-frequency clusters (♦) or only of low-frequency cluster (○) in seismoacoustic response

No.	Earthquake		Epicenter coordinates		H, km	K_s	D, km	A	B
	Date, UTC	Time, UTC	Lat.° N	Long.° E					
1	2017.06.27	02:23:18.1	52.58	160.80	40	12.1	186	○	○
2	2017.07.06	23:02:46.3	53.26	160.77	49	12.4	189	○	○
3	2017.07.25	16:24:51.5	53.85	159.74	124	11.1	199	○	○
4	2017.08.02	21:05:39.1	53.08	160.09	61	11.8	147	○	○
5	2017.08.19	00:59:56.4	52.90	160.25	87	11.1	167	○	○
6	2017.09.15	17:26:11.2	53.71	160.86	59	11.3	214	○	○
7	2017.09.16	21:00:54.7	52.82	160.00	53	12.8	136	♦	♦
8	2017.09.18	01:23:30.8	53.15	159.70	76	11.3	134	○	○
9	2017.09.29	19:24:59.1	53.10	160.33	51	13.4	159	♦	♦
10	2017.10.10	22:59:22.2	51.96	159.62	45	11.8	146	○	○
11	2017.12.02	15:11:38.1	52.54	159.88	44	11.3	130	○	○
12	2017.12.05	09:06:03.0	53.70	161.13	44	11.3	226	○	○
13	2017.12.22	14:44:16.2	53.68	160.88	69	14.2	217	♦	♦
14	2017.12.22	15:36:33.1	53.69	160.87	53	11.2	212	○	○

Continuation of **Table 1.**

No.	Earthquake		Epicenter coordinates		H, km	K_s	D, km	A	B
	Date, UTC	Time, UTC	Lat.° N	Long.° E					
15	2017.12.22	22:28:22.6	53.72	160.78	61	12.4	211	○	○
16	2018.01.04	02:44:54.4	53.12	160.20	56	12.9	153	○	♦
17	2018.01.14	14:04:50.3	52.72	159.88	50	11.3	128	○	○
18	2018.01.17	18:58:04.6	53.13	160.99	52	11.2	201	○	○
19	2018.03.05	15:42:41.0	52.38	160.80	48	13.1	193	○	○
20	2018.04.06	16:13:13.0	52.45	159.17	86	11.5	119	○	○
21	2018.07.21	04:19:16.7	51.73	159.40	36	12.3	153	○	○
22	2018.10.15	01:32:11.1	53.95	159.86	125	12.8	210	○	○
23	2018.11.02	09:39:27.7	51.98	158.72	72	13.1	125	♦	♦
24	2019.01.03	17:57:43.5	53.14	160.05	58	12.2	145	♦	○
25	2019.01.12	11:18:58.4	52.79	159.80	60	12.4	127	♦	○
26	2019.02.08	17:08:35.2	52.42	159.05	59	12.0	97	♦	○
27	2019.02.14	18:24:27.6	52.39	159.13	58	11.2	101	♦	○
28	2019.04.16	15:22:50.8	53.69	160.94	55	12.1	217	○	○
29	2019.07.18	12:45:38.6	52.43	159.99	51	11.8	142	♦	○
30	2019.07.28	09:53:04.0	53.70	160.88	52	11.5	213	○	○
31	2019.07.31	12:56:36.9	52.36	159.81	47	11.1	133	○	○
32	2019.09.23	21:54:55.9	52.01	159.93	21	11.8	153	○	○
33	2019.10.05	16:18:36.6	51.69	158.89	63	11.4	150	○	○
34	2019.11.09	11:16:57.2	52.33	161.05	54	11.6	212	○	○
35	2020.01.01	03:53:26.2	52.50	159.40	46	12.4	104	♦	♦
36	2020.01.16	16:31:09.5	52.01	159.84	29	12.8	150	♦	♦
37	2020.01.16	21:50:56.4	51.98	159.93	30	11.1	157	○	○
38	2020.01.28	04:19:45.8	52.40	159.56	47	11.5	117	♦	○
39	2020.02.20	18:57:34.4	53.44	160.92	52	14.3	205	♦	♦
40	2020.03.12	05:20:10.0	52.40	159.11	56	12.2	99	♦	○
41	2020.05.17	19:26:31.5	52.02	159.98	42	11.4	160	○	○
42	2020.05.30	11:01:32.1	52.83	159.40	115	11.9	143	○	○
43	2020.08.07	03:41:09.3	52.20	159.27	63	12.0	121	♦	○
44	2020.08.25	21:53:43.7	52.86	160.24	50	11.4	150	○	○
45	2020.09.04	21:10:12.9	52.08	160.56	49	12.7	190	○	○
46	2020.09.04	21:14:22.3	52.08	160.43	18	11.8	177	○	○
47	2020.09.05	00:21:21.0	52.09	160.59	46	11.8	191	○	○
48	2020.10.05	09:04:29.8	52.22	158.98	100	11.5	133	○	○
49	2020.11.27	22:57:13.1	52.04	159.98	40	11.8	158	○	○
50	2021.01.11	12:56:39.3	52.46	159.22	61	11.4	104	♦	○
51	2021.04.11	13:48:49.0	52.08	158.88	60	11.9	114	♦	○
52	2021.05.05	14:08:42.4	51.98	159.19	66	11.2	135	♦	○

Manifestation of joint acoustic and electric response of rocks. Fig. 1a shows the map of the considered earthquake (Table 1) epicenters and Karymshina site location. It shows the earthquakes in different ways, because during seismic wave transmission either joint acoustic and electric responses of rocks or only their acoustic response were observed. As the result, two groups of earthquakes were obtained. Using the same symbols, Fig. 1b illustrates comparison of their energy classes K_s and the hypocentral distances D to Karymshina site. To consider the relation between K_s and D, nonparametric correlation analysis was used in two selected groups. Rank correlation Spearman coefficient r_s was considered. It is less sensitive to any jumps and errors in observation results. Moreover, it allows us to estimate monotonous nonlinear relations between the variables and can be used for small sampling volumes [8] that is topical in our case.

Table 2 (Fig. 1b) shows the estimates of correlation coefficient r_s and its significance level p between K_s and D for the two groups of earthquakes illustrated in Fig. 1a. It is clear from Table 2, there is a statistically significant relation (r_s = 0.597, p = 0.009) between the energy class of earthquakes, during which joint seismoacoustic and seismoelectric responses of rocks were observed, and hypocentral distance D. Despite of a large sampling volume, there is no such a relation for the earthquakes during which only seismoacoustic response of rocks (r_s = 0.129, p = 0.468) was observed.

Manifestation of high-frequency acoustic response of rocks. It was discovered in the papers [4–6] that during the transmission of seismic waves from strong and relatively close earthquakes, rock acoustic response can appear at the frequencies of ~ 0.1–1 kHz and even ~ 1–11 kHz in addition to the low-frequency response up to about 100 Hz. Occurrences of these two higher-frequency, compared to seismic wave frequency, clusters were recorded during P-wave passage from a strong and close Zhupanovskoe earthquake on January 30, 2016 (M_w =7.2, D = 220 km) [4]. Occurrence of the cluster was observed only at the frequencies of ~ 0.1–1 kHz during P-wave passage from a strong but more remote earthquake on March 35, 2020 (M_w =7.5, D = 434 km) [6].

Fig. 2a, just like Fig. 1a, shows a map of the considered earthquake epicenters, and Fig. 2b illustrates the comparison of their energy classes K_s and hypocentral distances D to Karymshina site. Taking into the consideration the possibility of occurrence of higher-frequency clusters in the seismoacoustic response of rocks, all earthquakes were divided into two groups. First group included those, during which only low-frequency cluster up to about 100 Hz was observed in rock seismoacoustic response. Other group contained the earthquakes, during which besides that cluster, a mid-frequency cluster at the frequencies of ~ 0.1–1 kHz and even high-frequency cluster at the frequencies of ~ 1–11 kHz were observed in the seismoacoustic response. The results for the analysis of the relation between K_s and D for the cases of high-frequency acoustic cluster occurrences in rock seismoacoustic response and the cases of their absence are illustrated in Table 2 (Fig. 2b).

It is clear from Table 2 (Fig. 2b), there is statistically significant relation between K_s and D (r_s = 0.826, p = 0.017) for the earthquakes, during which besides the low-frequency cluster in the seismoacoustic response, mid-frequency and even high-frequency clusters are presented. The authors [5] think that occurrences of these two clusters indicate the transformation of rock low-frequency seismic wave energy into higher-frequency energy of acoustic signals. Such a transformation is likely to occur during intensive deformations of sedimentary rocks, formation of relative microdisplacements and interactions of small fragments. Based on the results from Table 2 (Fig. 2b), there is no correlation relation between K_s and D (r_s = 0.031, p = 0.841) for the earthquakes, during which only low-frequency cluster was presented in the seismoacoustic response.

We call attention to the statistically significant relation between K_s and D. The relation was discovered for a small number of earthquakes during which mid-frequency and even high-frequency clusters appeared in the seismoacoustic response (Table 2, Fig. 2b, r_s = 0.826, p = 0.017). However, if the dependence between the variables in a population is strong, it can be detected with high level of significance event on a very small sampling [8], that can, probably, be applied to our case.

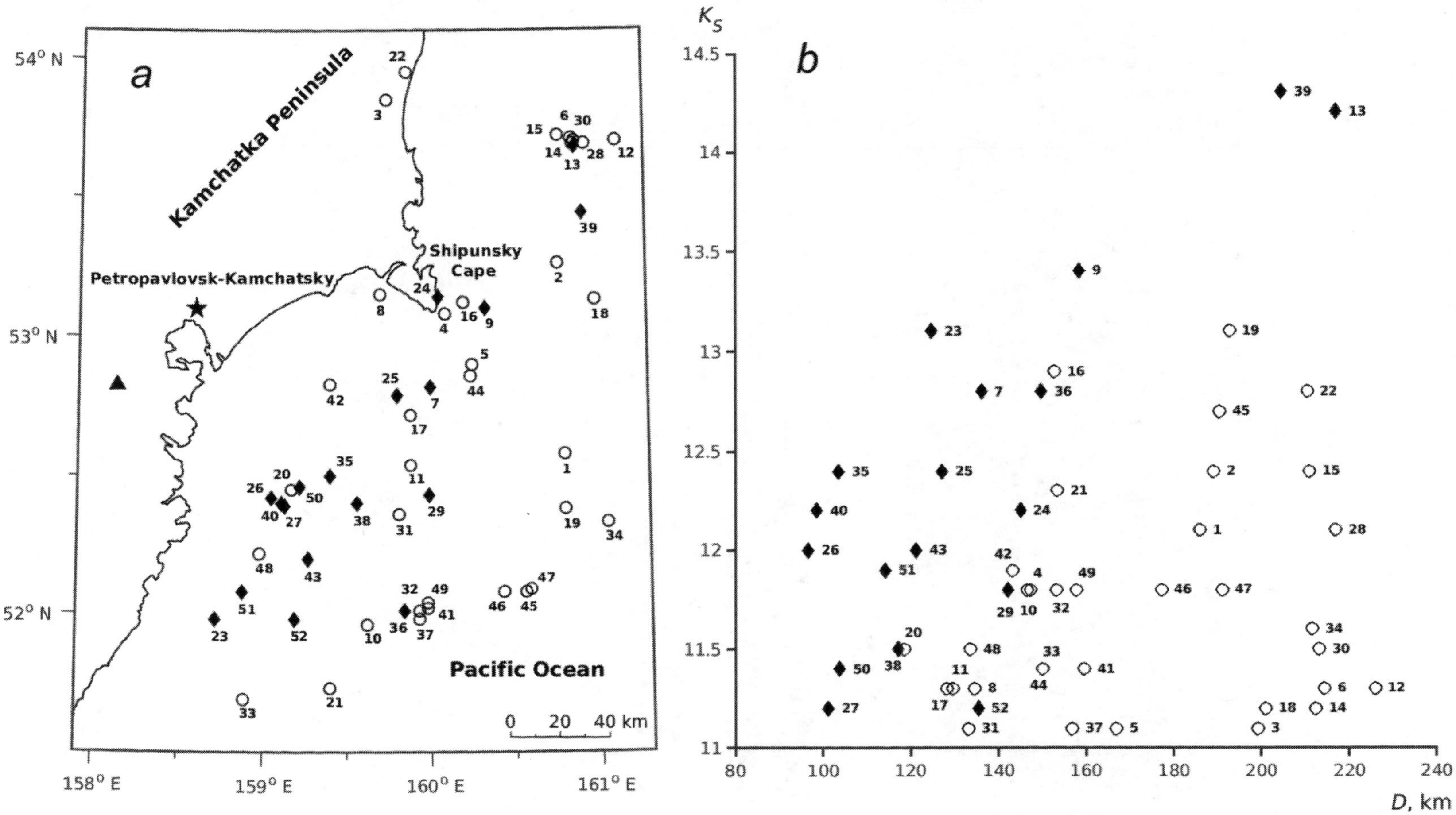

Figure 1. Map of the considered earthquakes epicenters (a) and comparison of their energy class K_s with the distance from the hypocenter D to Karymshina site (b). Designations: ▲ is the location of Karymshina site; ♦ are the earthquakes in which a joint seismoacoustic and seismoelectric response of rocks was observed; ○ are the earthquakes in which there was only a seismoacoustic response

https://doi.org/10.1051/epjconf/202125402002

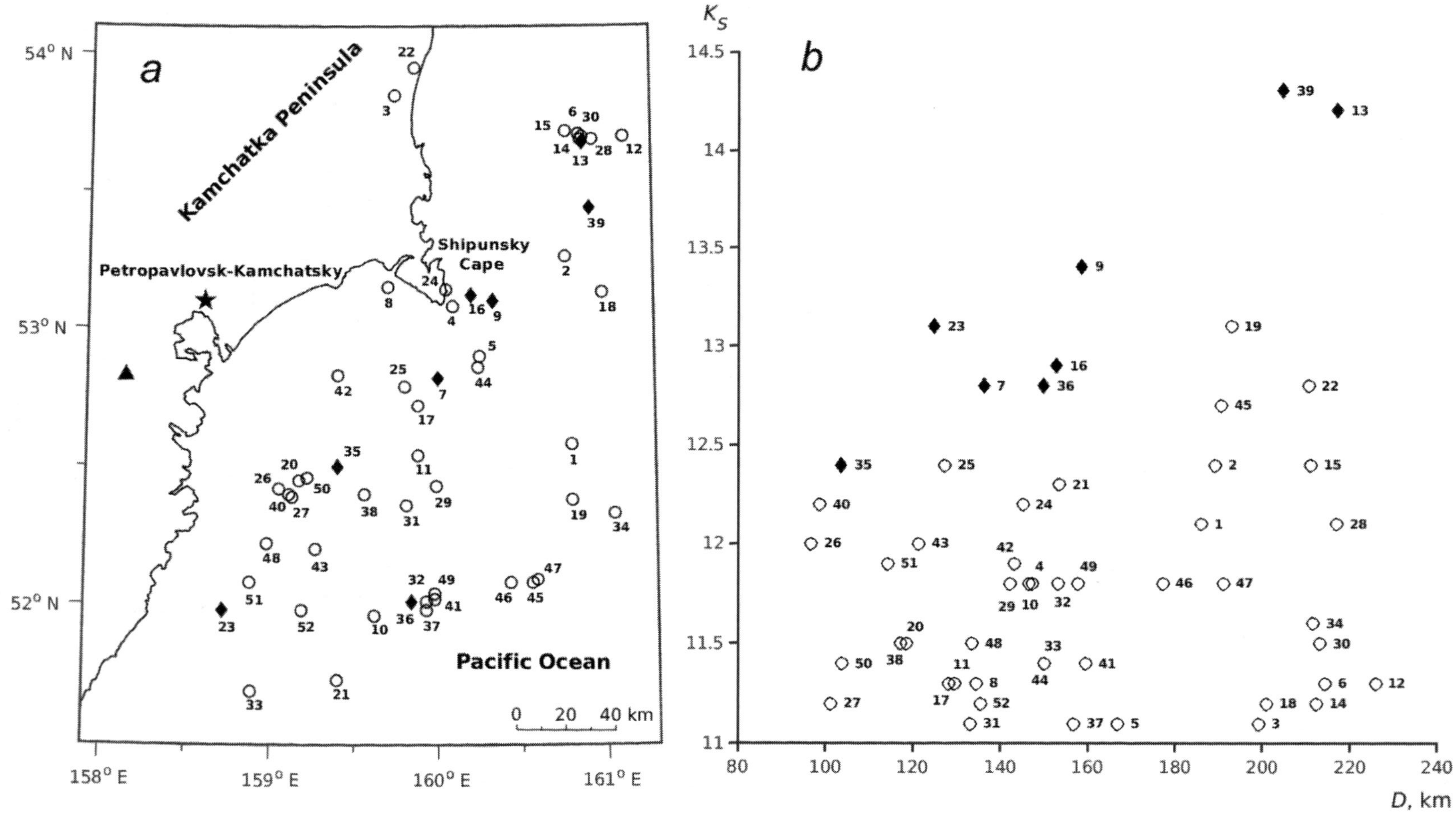

Figure 2. Map of the considered earthquakes epicenters (a) and comparison of their energy class K_s with the distance from the hypocenter D to Karymshina site (b). Designations: ▲ is the location of Karymshina site; ♦ are the earthquakes in which, in addition to a low–frequency cluster up to about 100 Hz, there was a medium-frequency cluster at frequencies of ~ 0.1–1 kHz and even a high-frequency cluster of ~ 1–11 kHz in the seismic-acoustic response of rocks; ○ are the earthquakes in which there was only a low-frequency cluster

Table 2. Estimates of correlation Spearman coefficient r_s and its significance level p between earthquake energy class K_s and distance from a hypocenter D to Karymshina site for each group of the detected earthquakes, n is the number of earthquakes in a group. Symbols (♦) and (○) are the same as for Table 1.

Parameter	Earthquake designation in figures			
	Fig. 1b		Fig. 2b	
	♦	○	♦	○
r_s	0.597	0.129	0.826	0.031
p	0.009	0.468	0.017	0.841
n	18	34	8	44

3 Conclusions

We have continued the investigation of earlier discovered joint acoustic and electric responses of the near-surface sedimentary rocks on the deformation by earthquake seismic waves. Occurrences and absence of such responses for 52 earthquakes with the energy class of more than 11.0 were considered. These events which took place from June 2017 to May 2021 by the eastern coast of South Kamchatka. We discovered the statistically significant relation between the energy class of the earthquakes, during which joint seismoacoustic and seismoelectric responses of rocks were observed, and between the distance from hypocenter to the observation site. Such a relation was discovered for rock acoustic response on seismic waves from the earthquakes, during which besides the low-frequency cluster of about up to 100 Hz, the cluster at the frequencies of ~ 0.1–1 kHz and even higher-frequency cluster at the frequencies of ~ 1–11 kHz were also observed. These clusters are likely to be generated by rocks during intensive deformations, formation of relative microdisplacements and interaction of small fragments. When there was no seismoelectric response of rocks and only low-frequency cluster was presented in the seismoacoustic response, we did not detect any statistically significant relation between earthquake energy class and distance from hypocenter.

The work was carried out within the framework of the State Task AAAA-A21-121011290003-0. We applied the data, obtained at a unique scientific unit "Seismo-infrasonic complex to monitor arctic cryolithic zone and a complex for continuous seismic monitoring of the Russian Federation, cross-border regions and the world" (https://ckp-rf.ru/usu/507436/).

References

[1] I.G. Kissin, *Fluids in the earth's crust: Geophysical and tectonic aspects* (Moscow, Nauka, 2015) 328

[2] A.G. Ivanov, Proceedings of the Academy of Sciences of the USSR, a series of geographical and geophysical, **5**, 18–21 (1940)

[3] Yu.V. Marapulets, B.M. Shevtsov, *Mesoscale acoustic emission* (Vladivostok, Dal'nauka, 2012) 126

[4] P.V. Muratov, O.P. Rulenko, Yu.V. Marapulets, A.A. Solodchuk, Bulletin KRASEC. Physical & Mathematical Sciences, **5**(25), 62–73 (2018)

[5] P.V. Muratov, O.P. Rulenko, Yu.V. Marapulets, E3S Web of Conferences, **127** (02015), (2019)

[6] M.A. Mishchenko, O.P. Rulenko, Yu.V. Marapulets, E3S Web of Conferences, **196** (02028), (2020)

[7] A.V. Kuptsov, I.A. Larionov, B.M. Shevtsov, Volcanology and Seismology, **5**, 45–59 (2005)
[8] V. Borovikov, *STATISTICA. The Art of Computerized Data Analysis: For Professionals* (St. Petersburg, Piter, 2003) 688

Analysis of the features of the spatio-temporal distribution of geoelectric inhomogeneities in the Earth's crust and seismic events

Elena Bataleva*

Research Station of the RAS in Bishkek, 720049, Kyrgyzstan. Bishkek-49

Abstract. The paper presents the results of experiments carried out at the regime points of magnetotelluric monitoring both on the territory of the Bishkek geodynamic test site (Northern Tien Shan) and on a series of monitoring profiles in various geological conditions. Previous studies indicate the relationship of variations in the electromagnetic and seismic fields, lunisolar tidal effects, seismic regime with the processes of fracturing. The purpose of this work is to establish the features of the relationship between the spatio-temporal distribution of seismicity and the distribution of geoelectric inhomogeneities in the Earth's crust (fault-block tectonics of the region). Based on the analysis of the results of the interpretation of magnetotelluric data (2D inversion) and new detailed seismotomographic constructions, the verification of geoelectric models was carried out, the analysis of the distribution of hypocenters of seismic events was carried out. Special attention was paid to the confinement of earthquakes to listric fault structures. The relationship between the distribution of the hypocenters of seismic events and the spatial position of the electrical conductivity anomalies is confirmed by the authors explanation of the physical nature of the identified conducting structures, based on hypotheses of fluidization and partial melt of the Earth's crust.

To date, significant progress has been achieved both in the study of the Central Tien Shan deep structure details (Fig. 1), and in the study of the seismicity distribution therein (Fig. 2) [1-5]. However, the confinement of earthquake sources to geophysical inhomogeneities of the Earth's crust, including fault structures, is still controversial. Based on the existing ideas about the confinement of earthquakes to tectonic faults, it is often impossible to find an explanation for the position of individual seismically active zones in relation to the geological structures observed on the day surface. Based on a comparison of the deep sections of the seismicity distribution field and the geoelectric structure of the zones of dynamic influence of fault structures, the author concluded that there is no compelling reason to connect the distribution of earthquake hypocenters to the near-surface fault zones for the study area. Moreover, on the geoelectric sections given in [1-3], subhorizontal structures are clearly manifested, to which most of the analyzed hypocenters are confined. Apparently, the structure of the stress-strain field is such that it is sensitive to

* e-mail: elena.bataleva@gmail.com

sub-horizontal inhomogeneities of the Earth's crust. Taking into account the presence of fluid heterogeneity of the Earth's crust [6] and the fact that the permeability of zones of dynamic influence of faults increases sharply before strong earthquakes [6,7], it is obvious that it is the geoelectric parameters of the geological environment that can serve as indicators of the ongoing geodynamic processes. This paper considers a comparison of the results of the main deep geophysical methods for studying the Earth's crust and the upper mantle - magnetotelluric sounding and seismic tomography, as well as the distribution of seismic events to identify blocks and layers of the Earth's crust with anomalous physical parameters (i.e., determining its stratification and block fault structure), identifying zones of fracturing or decompaction of rocks, which may be associated with fluid-saturated areas.

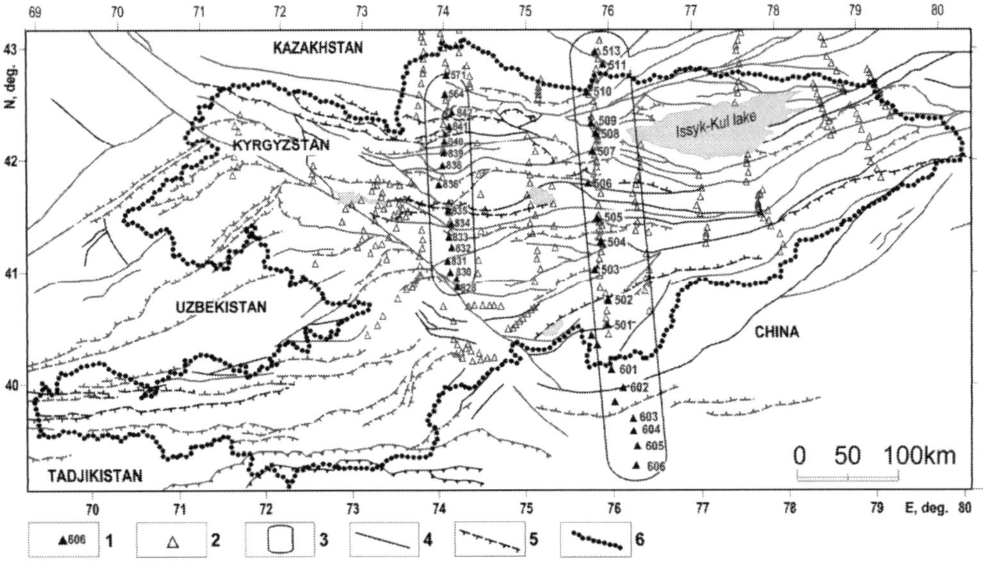

Fig. 1. Scheme of geoelectric exploration of the Central Tien Shan: 1 - points of deep magnetotelluric sounding (DMTS); 2 - points of magnetotelluric sounding (MTS); 3 - MTS profiles considered in the work; 4 - fault structures; 5 - thrusts; 6 - border of the Kyrgyz Republic

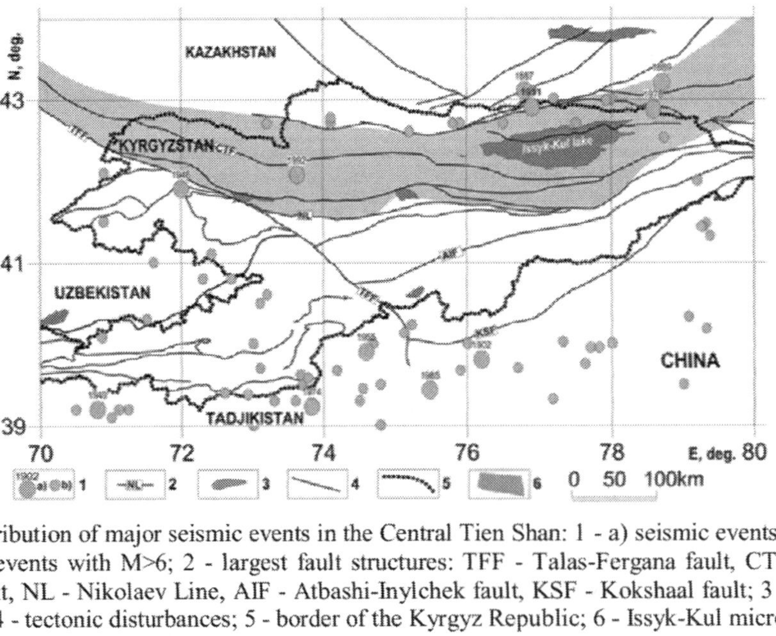

Fig. 2. Distribution of major seismic events in the Central Tien Shan: 1 - a) seismic events with M>7, b) seismic events with M>6; 2 - largest fault structures: TFF - Talas-Fergana fault, CTF - Central Terskei fault, NL - Nikolaev Line, AIF - Atbashi-Inylchek fault, KSF - Kokshaal fault; 3 - lakes and reservoirs; 4 - tectonic disturbances; 5 - border of the Kyrgyz Republic; 6 - Issyk-Kul microcontinent

Geoelectric models (Figs. 3,4) are built on the basis of magnetotelluric (MT) sounding data, carried out using a series of detailed and regional profiles by the Phoenix Geophysics stations (Toronto, Canada) [8]. Modern Phoenix MTU-5 equipment is distinguished by high performance, sensitivity and wide dynamic range, reliability, environmental friendliness and ease of use [9]. The recording units of the MTU-5 stations use 24-bit analog-to-digital converters and GPS receivers that provide time synchronization with an accuracy of 1 μs. The range of the periods of variations of the magnetotelluric field (MT-field) recorded with the MTU-5 is from 0.001 to 1600 s. Measurements of the magnetic field components are performed using standard MTC-50 induction sensors. The electric components of the MT-field are measured in the entire frequency range of registration using L-shaped measuring devices with electric dipoles 50 m long. The grounding of the dipoles when working with the MTU-5 was carried out using non-polarizable electrodes using manganese-carbon chips, developed and manufactured at the Research Station of the Russian Academy of Sciences. At each sounding point on the territory of the Central Tien Shan, 5 components of the MT-field were recorded in the frequency range 100 Hz - 1600 s. The magnetotelluric data were processed using robust analysis procedures, in the remote reference (RR) mode. Pseudo-sections ρ_k, phases and tippers were inverted using the software package [9].

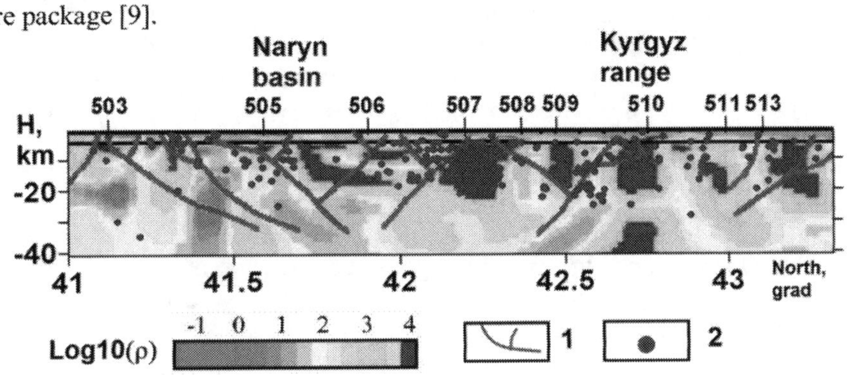

Fig. 3. Distribution of earthquake hypocenters (M>9) in the +1° band along the NARYN geophysical profile through the zone of the Issyk-Kul microcontinent. The position of the hypocenters is shown with asterisks, the numbers above the section show the numbers of the points of the DMTS. Fault structures are shown in [16].

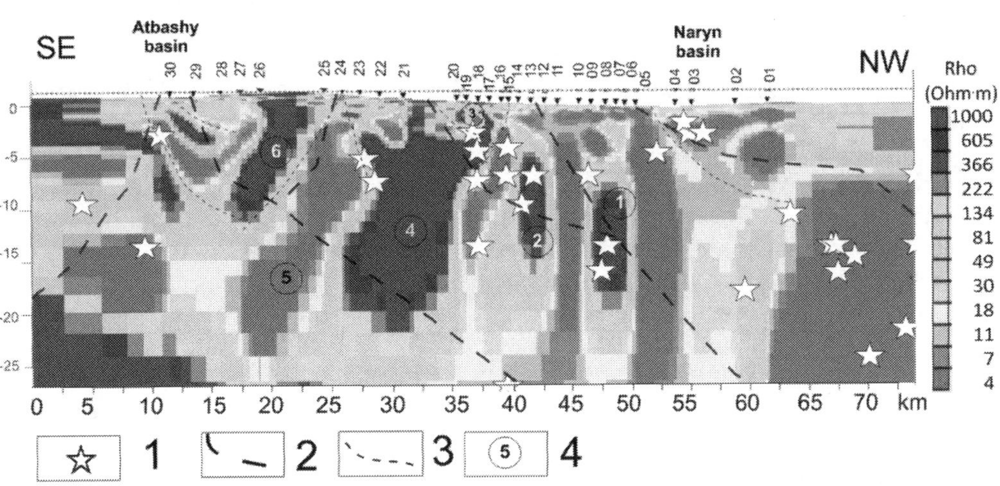

Fig. 4. 2D-inversion model for the "Karabuk" profile (75°), fault structures are shown in the edition [16]

When calculating the 2D model for the Naryn profile (76°), the bimodal inversion of the GEOTOOLS system was used, the inverted data ensemble included, in addition to the TE and TM modes, the components of the Wiese-Parkinson matrix. The final model obtained using parallel inversion is published in [10]. The advantage of this model is its good agreement with modern geological concepts, as well as consistency with seismic data, distribution of depth temperatures, behavior of mantle helium emanations and petrophysical studies [11]. Seismotomographic studies of the Central Tien Shan and adjacent territories were carried out by many authors using different databases and different details [12-15]. In this work, for comparison with the geoelectric model, the most modern and detailed model of the region under study, presented in [15], was selected. An example of the analysis of the comparison of the field distribution of the hypocenters of seismic events, the geoelectric model and the tectonic structure of the Central Tien Shan is shown in Figure 3 [11].

In [2, 3, 17] studies, the spatial position of earthquake hypocenters in a 100-km strip along the line of the "Naryn" profile was considered. It has been established that all earthquake sources lie in the upper and middle parts of the Earth's crust (at depths less than 30 km), while earthquakes in the Tarim area occur at depths of up to 100 km. This can be explained by the existence of various geodynamic settings in the Tien Shan lithosphere, which is reflected in the geodynamic models. The position of earthquakes of energy class M>9, which occurred on the territory of the Central Tien Shan, can be explained by the geoelectric features of the Earth's crust and, first of all, by the presence of high resistance gradients caused by high-resistance bodies and low-resistance zones. The nature of the distribution of geoelectric and seismic (velocity) inhomogeneities is a reflection of geodynamic processes occurring in the Tien Shan lithosphere, for example, seismicity (Figs. 3, 4). The process of collisional interaction continues at the present stage of tectogenesis of the intracontinental orogen. The main indicator of modern activity is seismicity, which is unevenly distributed in different tectonic segments. An analysis of the distribution of earthquake hypocenters may indicate the reactivation of ancient and weakened zones, for example, the Issyk-Kul microcontinent (Fig. 2), in the Earth's crust, which are manifested as pronounced inhomogeneities in geoelectric models [10, 11]. The location of foci of weak earthquakes within high-resistivity objects, which we can observe on the geoelectric model [5], built along the Naryn profile, obviously associated with deep manifestations of magmatism, indicates that stress relief occurs at the contact of blocks of basic rocks with rocks, with other petrophysical and rheological properties.

On the basis of a comprehensive analysis of the distribution of geoelectric and velocity inhomogeneities of the lithosphere of the Central Tien Shan along the Karabuk profile (75 ° E), it has been established that: a) areas with increased electrical conductivity correspond to areas of lower seismic velocities; b) zones with low electrical conductivity are areas of increased seismic velocities; c) rock massifs identified as a result of complex interpretation of geological and geophysical data as rocks of the Paleozoic basement are characterized by increased seismic velocities. The highest density of hypocenters of seismic events with M>9 is observed at the boundaries of high-resistivity blocks and within them. Noteworthy is the confinement of seismic events to zones of listric faults, which can be traced not only in geoelectric sections, but also in seismotographic constructions (Fig. 5).

Comparison of velocity anomalies in seismotomographic sections and objects of anomalously low electrical resistance in geoelectric sections of the Earth's crust indicates: a) their confinement to fault structures; b) about the listric form of deep faults; c) flattening of faults at depths of 10 - 20 km and depths of the top of the crustal conductive layer; d) the existence of zones of increased fracturing or plasticity in the indicated intervals of the Earth's crust; e) on the coincidence of waveguides and zones of increased conductivity.

In the geoelectric structure of the Earth's crust of the Central Tien Shan, a system of multidirectional conducting bodies has been identified, which are identified with fragments of deep faults. The structures of low resistivity are spatially confined to the zones of major faults, such as the Nikolaev Line, Atbashi-Inylchek and other fault structures of a lower rank. The conductive zones identified in geoelectric models seem to be mechanically weakened areas of increased fracturing, filled with fluid. For this reason, differences in the values and distribution of the resistances of these conductive zones in the sections under consideration may reflect different levels of fluid concentration or different degrees of connectivity of the pore-fracture space [18, 19].

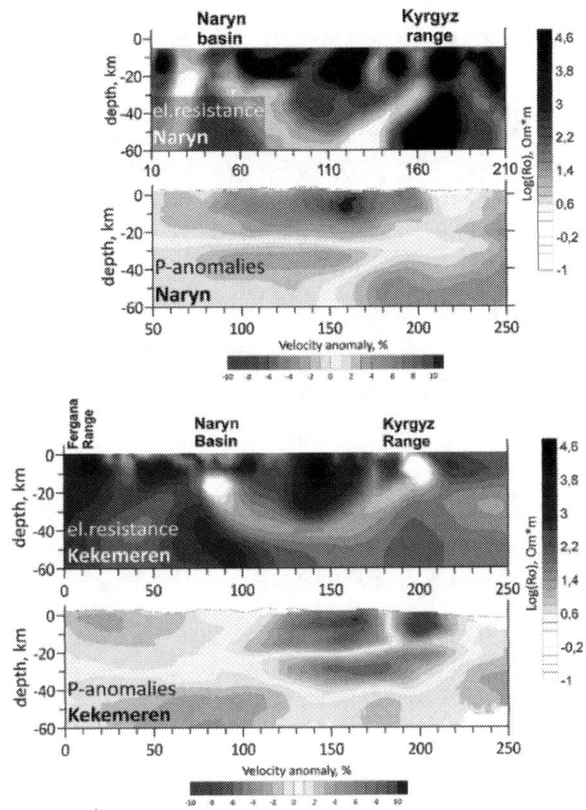

Fig. 5. Comparison of geoelectric models and seismic tomography along the Naryn (76° E) and Kekemeren (74° E) profiles (Fig. 1) [15]

The strongest earthquakes of the Tien Shan, taken out on the geoelectric models (Fig. 3, 4), are confined to the boundaries of blocks and weakened zones, which, as a rule, are also zones of maximum absorption contrast. The structure of the absorption field correlates with the tectonic structure of the area, in particular, the weakened zones correspond to the fault zones [7]. Analysis of the position of the hypocenters of seismic events and the distribution of velocity inhomogeneities in the earth's crust allows us to conclude that earthquakes are confined to high-gradient zones of geoelectric models [2-5]. Very often, these structures flatten out and have the form of listric faults (lateral compression structures) characteristic of areas of intracontinental compression. Flattening, as a rule, occurs at the boundaries of zones with anomalous physical properties - electrical conductivity or velocity characteristics of the medium.

In terms of the depth of occurrence, these seismic events are crustal and, as noted above, the hypocenters of earthquakes in the basement of the Tarim block lie at depths of up to 40 km, while the overwhelming majority of hypocenters of the Tien Shan earthquakes are localized at depths of up to 20 km. The crustal layer with low seismic velocities, corresponding to the zone of increased crustal electrical conductivity, practically does not contain earthquake hypocenters, which indicates its fluid saturation and plasticity, i.e. the inability to accumulate elastic energy sufficient for the occurrence of earthquakes.

When comparing the results of the interpretation of magnetotelluric data (2D inversion) and detailed seismotomographic constructions, the verification of geoelectric models was carried out, the nature of crustal electrical conductivity was assessed at various spatial-scale levels. This relationship between the distribution of seismicity and the geoelectric characteristics of the section is confirmed by the authors' explanation of the physical nature of the identified conducting structures, based on hypotheses of fluidization and partial melt of the earth's crust.

The work was carried out within the framework of the state assignment of the Research Station of the Russian Academy of Sciences on the AAAA-A19-119020190063-2 topic.

References

1. V.D. Bragin, Active Electromagnetic Monitoring of the Bishkek Prognostic Test Area. PhD Thesis (Candidate of Physics and Mathematics). 135 pp, Moscow (2001) (in Russian).
2. Yu.A. Trapeznikov, E.V. Andreeva, V.Yu. Batalev, M.N. Berdichevsky, L.L. Vanyan, A.M. Volykhin, N.S. Golubtsova, A.K. Rybin, Izv. Phys. of Solid Earth. **33**, 1-17 (1997).
3. A.V. Zubovich, Y.A. Trapeznikov, V.D. Bragin, O.I. Mosienko, G.G. Schelochkov, A.K. Rybin, V. Y. Batalev, Russ. Geol. Geophys. **42**, 1550-1557 (2001).
4. E.A. Bataleva, V.A. Mukhamadeeva, Geodynam. Tectonophys. **9 (2)**, 461-487 (2018) (In Russian).
5. A.K. Rybin, E.A. Bataleva, K.S. Nepeina, V.E. Matyukov, Geodynamics & Tectonophysics **12 (3)**, (2021) (In Russian).
6. I.G. Kissin Dokl. Earth Sci. **448(2),** 235-238 (2013).
7. Y.F. Kopnichev, I.N. Sokolova, Seism. Instr. **53**, 234–243 (2017).
8. L. Fox, Satellite-Synchronized 3-D Magnetotelluric System. U.S. Patent № 6 191 587 B1 (2001).
9. W. Rodi, R. Mackie, Geophys. **66**, 174–187 (2001).
10. R. A. Bielinski, S. K. Park, A. Rybin, V. Batalev, S. Jun, C. Sears, Geophys. Res. Let., **30,** 1806 (2003).
11. V.Yu. Batalev, E.A. Bataleva, V.E. Matyukov, A.K. Rybin, V.V.Egorova, Russ. Geol. Geophys. **52(12)**, 1592-1599 (2011).
12. A.A. Adamova, T.M. Sabitova, Izv. Phys. of Solid Earth. **40(5)**, 405-413 (2004).
13. Z.W. Li, S. Roecker, Z.H. Li, B. Wei, H. Wang, G. Schelochkov , V. Bragin, Tectonophys. **477**, 49–57 (2009).
14. I.Y. Koulakov, J. Geophys. Res. **116 (B04301)**, 1–22 (2011).
15. I. Medved, E. Bataleva, M. Buslov, Geosciences, **11**, 122 (2021).
16. S.C. Thompson, R.J. Weldon, C.M. Rubin, K. Abdrakhmatov, P. Molnar, G.W. Berger, J. Geophys. Res. **107 (B2203)**, 1–32 (2002).
17. E.A. Bataleva, V.Y. Batalev, A.K. Rybin Lithosphere. 5, 81-89 (2015).
18. I.G. Kissin, Izv. Phys. of Solid Earth. **32(4)**, 289-298 (1996).
19. I.G. Kissin, A.I. Ruzaikin, Izv. Phys. of Solid Earth. **33(1)**, 18-25 (1997).

Application of adaptive wavelet thresholding to recovery geoacoustic signal pulse waveforms

Olga Lukovenkova[1,*]

[1]Institute of Cosmophysical Research and Radio Wave Propagation FEB RAS, Laboratory of Acoustic Research, Mirnaya str., 7, Paratunka, Kamchatskiy kray, 684034, Russia

Abstract. Recorded geoacoustic signals often contain noise and interference. Their appearance is caused by various reasons, e.g. of propagation environment heterogeneity, weather condition influence, human activity, etc. So, geoacoustic emission signals contain a persistent background noise that changes in intensity over time. This noise significantly distorts the geoacoustic pulse waveforms and thus complicates analysis of the signal characteristics. The article presents results of estimating the geoacoustic signal background noise. On the basis of these estimates, a method of adaptive wavelet thresholding is proposed to remove noise from the signal and recovery the single pulse waveforms. In conclusion, the results of a computational experiment are presented. They confirm effectiveness of using the chosen method for the geoacoustic signal preprocessing. The work was carried out as part of the implementation of the state task AAAA-A21-121011290003-0.

1 Introduction

Some of the most common problems faced by researchers in applied science are distortion and noisiness of recorded signals. These problems are typical for different signals, e.g. for acoustic, electrical, electromagnetic ones. In practice, a signal waveform is influenced by natural noise, artificial interference, nonlinearity of receive path, dynamic range limitations, quantization errors, primary hardware processing, etc. At the same time, many promising analysis methods (time-frequency transformations, sparse approximation, classification, clustering, etc.) are sensitive to signal waveform distortions. It negatively affects quality and accuracy of analysis results.

The presented article is devoted to denoising of geoacoustic emission signals (Fig. 1*a*).

It is assumed that the methods of geoacoustic signal secondary processing and analysis are applied to undistorted or slightly distorted pulses [1]. Therefore, it is important to denoise the signal before its processing.

2 Estimation of background noise parameters

The author studied the geoacoustic background noise before choosing a denoising method. For this purpose, fragments without pulses were selected from the geoacoustic signal recorded in good weather conditions (no precipitation and strong wind) at "Karymshina"

*e-mail: o.o.lukovenkova@yandex.ru

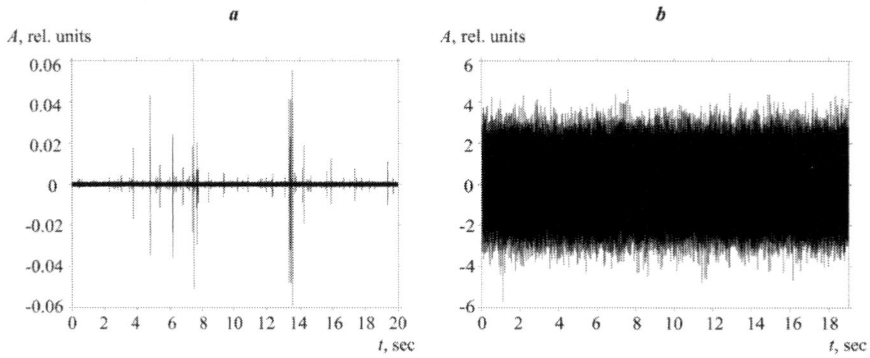

Figure 1. Geoacoustic emission signals: *a* — pulsed signal; *b* — background noise.

site (52.83° N, 158.13° E). Thus, a signal containing only background noise was generated (Fig. 1*b*). This signal was scaled and shifted so that the mean was $\mu = 0$ and the standard deviation was $\sigma = 1$.

There are several ways to estimate the distribution law of a random variable. The simplest ones used at the first stage of the study are a histogram and a quantile-quantile plot (Q-Q plot). Using them, we can visually assess similarity of the distribution with one of the known laws. Figure 2a shows histogram of the noise signal amplitude distribution. The number of intervals k is chosen according to the Sturgess rule

$$k = 1 + \log_2 N,$$

where N is a number of random variable measurements. Figure 2*b* shows Q-Q plot which compared the noise signal amplitudes on the vertical axis to a standard normal distribution values on the horizontal axis.

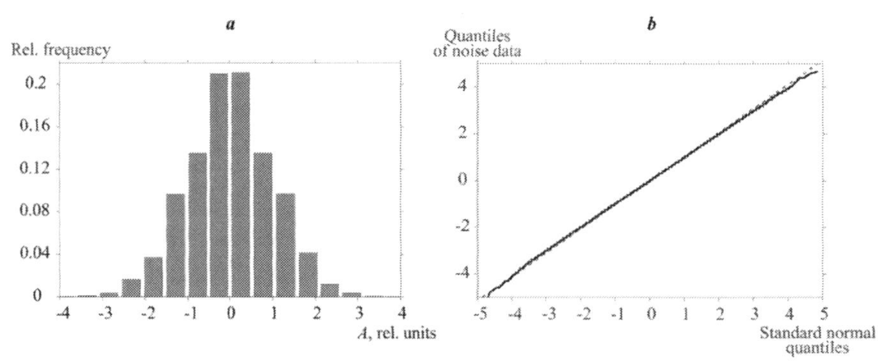

Figure 2. Estimation of noise signal amplitude distribution: *a* — histogram; *b* — Q-Q plot.

Visually, the distribution of noise signal amplitudes is similar to the normal one. However, the normality hypothesis H_0 requires confirmation. The results of testing the statistical hypothesis H_0 (the noise signal amplitudes have the normal distribution $N(\mu, \sigma)$, where μ and σ are estimated from the tested data) using Pearson's chi-squared test [2], Anderson–Darling test [3] and Lilliefors test [4] are given in Table 1. H_1 is an alternative hypothesis.

Table 1. Results of applying various statistical tests.

Test	Significance level, α	Sample size, N	Accepted hypothesis, H_0 or H_1
Pearson	0.05	500	H_0
	0.05	1000	H_0
Anderson–Darling	0.05	500	H_0
	0.05	1000	H_1
Lilliefors	0.05	500	H_0
	0.05	1000	H_1

When the sample size increases, most of the statistical tests reject the hypothesis H_0. This effect can be explained by quantization errors of the geoacoustic signal. The analog-to-digital converter used in the signal registration system has a resolution of 16 bits (65536 levels, values from -1 to 1). According to calculations, dynamic range of the background noise signal contains 112 levels (values from -0.00189 to 0.001526). It is supported by the Q-Q plot (Fig. 2*b*), its points form a stepped line.

To test this assumption, a Gaussian noise signal was generated. The signal was digitized so that its dynamic range contains 112 levels. The distributions of amplitudes of the real and model signals were compared using the Wilcoxon–Mann–Whitney test [5].

At significance level $\alpha = 0.05$ for sample sizes $N = 500$ and $N = 1000$, the Wilcoxon–Mann–Whitney test confirmed the hypothesis H_0 (the distributions of both samples are equal). This result is invariant to changes of the Gaussian noise generator seed. The conducted research shows that the background noise of geoacoustic signal is similar in structure to the roughly digitized Gaussian noise.

3 Wavelet thresholding as denoising method

A wavelet thresholding is one of the popular methods for removing noise, including Gaussian one, from a signal and separating an informative signal component. There is a large number of studies showing effectiveness of the wavelet thresholding method to remove high-frequency noise from signals of different nature [6–9]. The wavelet thresholding process consists of three stages: forward wavelet transform, thresholding, and inverse wavelet transform.

Denoising consists in thresholding of detail wavelet coefficients. The most common are soft and hard thresholds [10]. There are various strategies for choosing a threshold value: the Universal threshold [10], the Stein Unbiased Risk Estimator (SURE) [11], the Block James-Stein estimator (BlockJS) [12], the False Discovery Rate threshold (FDR) [13], the Empirical Bayes method [14], the Minimax principle [10], etc.

To select the most appropriate algorithm for calculating a threshold value, a computational experiment was carried out on a model geoacoustic signal. The signal containing 100 pulses with amplitudes from 0.05 to 1 relative units was generated. Gaussian noise of various amplitudes was artificially superimposed on the pulses so that the signal-to-noise ratio (SNR) was varied from -10 to 40 dB. The average relative error of wavelet thresholding was calculated for each SNR value by the formula

$$\varepsilon = \frac{1}{N} \sum_{i=1}^{N} \frac{||s_i(t) - \hat{s}_i(t)||}{||s_i(t)||}, \tag{1}$$

where $s_i(t)$ is the undistorted pulse; $\hat{s}_i(t)$ is the pulse processed by wavelet thresholding; N is the total number of processed pulses, $N = 100$.

Dependences of the error ε on SNR for the considered threshold calculation methods are shown in Fig. 3.

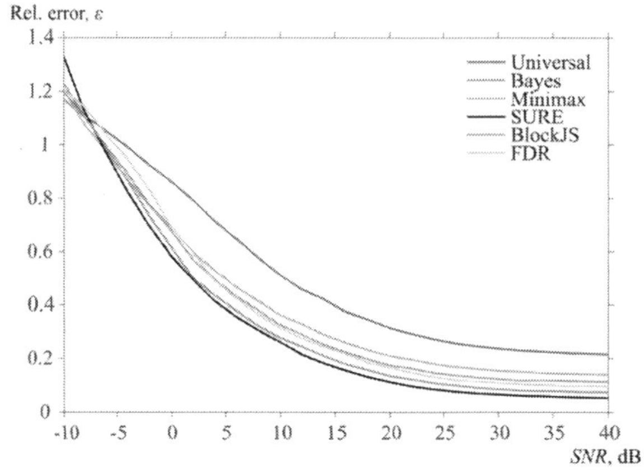

Figure 3. $\varepsilon(SNR)$ plots for different threshold calculation methods.

The best results were obtained using the Empirical Bayes and SURE methods. The Bayes method works more accurately on highly noisy signals (SNR less than -5 dB).

4 Wavelet family selection

Orthogonal and biorthogonal wavelets are most often used in noise reduction tasks. In the presented article, the following wavelet families are considered [15]: the Daubechies wavelets (Fig. 4*a*), the symlets (Fig. 4*b*), and the coiflets (Fig. 4*c*).

Experiment on model data show that the most accurate results are obtained using the fourth-order coiflets, and the high-order Daubechies wavelets and symlets (from the eighth order and higher). Figure 5 shows the $\varepsilon(SNR)$ plots obtained by the Empirical Bayes method using various wavelet families.

5 Approbation of adaptive wavelet thresholding on real data

According to the results obtained on model data, the Empirical Bayes method (the detail wavelet coefficients are estimated by the posterior mean value) and the fourth-order coiflets were chosen for geoacoustic signal recovering and denoising. Figure 6*a–e* shows the processing results obtained for the geoacoustic pulses of various waveforms (all ones were recorded in January, 2017). The processed signals are smoothed and denoised. The degree of denoising depends on the noise level. For example, the processing of highly noisy pulses is shown in Fig. 6*a,c*. The processing of undistorted one is shown in Fig. 6*b*. And Fig. 6*d–f* shows denoising of pulses with average noise level.

6 Conclusion

Registered geoacoustic signals are often highly distorted and noisy, while the noise level is constantly changing. It significantly complicates the processes of identification, analysis and classification of these signals. Study of the background noise shows that it is similar

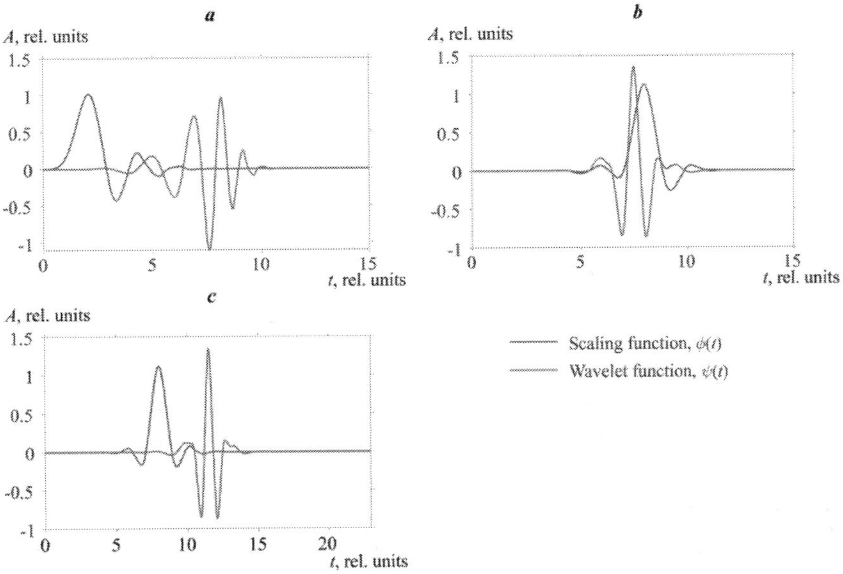

Figure 4. Mother wavelets $\psi(t)$ and scaling functions $\phi(t)$: *a* — eighth-order Daubechies wavelet, db_8; *b* — eighth-order symlet, sym_8; *c* — forth-order coiflet, $coif_4$.

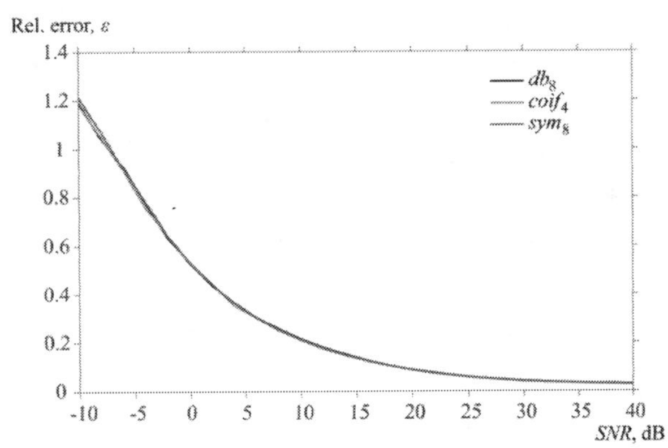

Figure 5. $\varepsilon(SNR)$ plots obtained by the Empirical Bayes method using various wavelet families.

in structure to the roughly digitized Gaussian noise. Therefore, the author proposes to use an adaptive wavelet thresholding as a signal preprocessing method. Experiments on model data show that the best noise reduction quality is achieved using the Empirical Bayes method and the fourth-order coiflets. The results of real data processing confirm effectiveness of the selected method for the geoacoustic signal preprocessing.

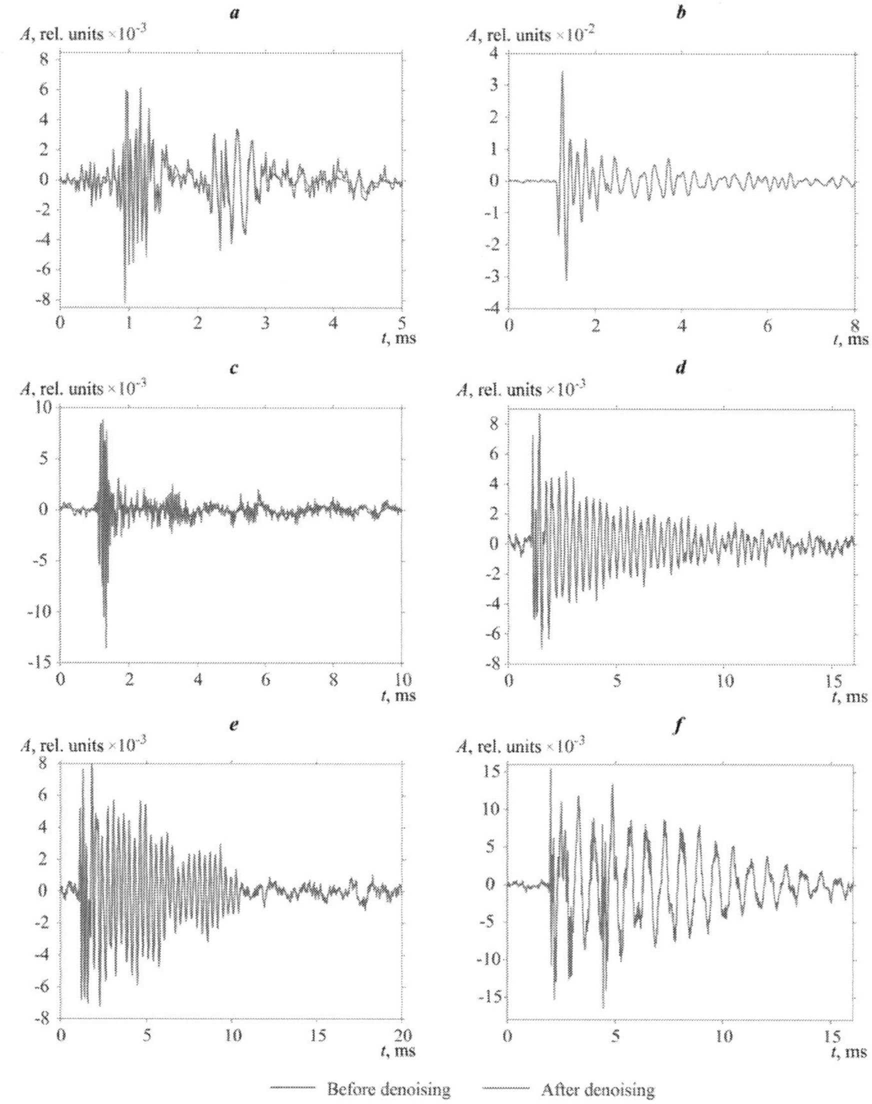

Figure 6. Applying Empirical Bayes method to geoacoustic pulses.

References

[1] Yu. V. Marapulets, Yu. I. Senkevich, O. O. Lukovenkova, A. A. Solodchuk, I. A. Larionov, M. A. Mishchenko, E. I. Malkin, A. O. Shcherbina, M. I. Gapeev, *Complex analysis of acoustic and electromagnetic signals to assess seismic hazard level* (Dalnauka, Vladivostok, 2020) 120 p.

[2] P. E. Greenwood, M. S. Nikulin, *A guide to chi-squared testing* (Wiley, New York, 1996) 280 p.

[3] T. W. Anderson, D. A. Darling, Ann. Math. Stat., **23**, 193–212 (1952)

[4] H. W. Lilliefors, J. Am. Stat. Assoc., **62**, 399–402 (1967)

[5] M. Hollander, D. A. Wolfe, *Nonparametric Statistical Methods* (John Wiley & Sons, Inc., Hoboken, NJ, 1999) 779 p.

[6] T. W. Anderson, D. A. Darling, IEEE Trans. Image Process., **9**, 1532–1546 (2000)

[7] M. Alfaouri, K. Daqrouq, Am. J. Appl. Sci., **5**, 276–281 (2008)

[8] M. Srivastava, C. L. Anderson, J. H. Freed, IEEE Access, **4**, 3862–3877 (2016)

[9] G. Srikar, Ch. R. Prasad, Adv. Comput. Sci. Tech., **10**, 3111–3121 (2017)

[10] G. Luo, D. Zhang, *Wavelet Denoising, Advances in Wavelet Theory and Their Applications in Engineering, Physics and Technology* (IntechOpen, 2012) 24 p.

[11] D. L. Donoho, IEEE Trans. Inf. Theory, **41**, 613–627 (1995)

[12] T. T. Cai, Stat. Sin., **12**, 1241–1273 (2002)

[13] F. Abramovich, Y. Benjamini, D. L. Donoho, I. M. Johnstone, Ann. Stat., **34**, 584–653 (2006)

[14] I. M. Johnstone, B. W. Silverman, Ann. Stat., **32**, 1594–1649 (2004)

[15] I. Daubechies, *Ten lectures on wavelets* (SIAM, Philadelphia, Pa, 1992) 377 p.

Characteristics of the catalogue according to the KNET network (Northern Tian Shan)

Naylya Sycheva*, *Vladimir* Sychev

Science Research Station of the RAS in Bishkek city, Kyrgyzstan

Abstract. Various characteristics of the catalogue obtained from the data of the KNET network, containing the parameters of more than 10,000 earthquakes that occurred from 1994 to 2020, are considered. A statistical analysis of the number of *P*- and *S*-waves registered by each station of the network is carried out. The activity of the stations during the localization of earthquakes has been determined. The most active station is AAK. The minimum number of P- and S-waves is recorded at the ULHL station. Statistical analysis of localization errors has been performed. The boundaries of the territory for which the minimum values of errors are observed are determined. The statistical characteristics of earthquakes in terms of magnitude, time, depth are presented. The distribution of the number of earthquakes by different energy classes in time and the spatial distribution of earthquakes by depths: 0-5 km, 5-10 km, 10-15 km and more than 15 km are constructed. The Gutenberg - Richter law was used to describe the energy distribution function of earthquakes. A representative sample of the catalogue was determined, which includes events with M ≥ 1.8. The b-value for the directory is 0.88.

Introduction

In 1991, a seismological network KNET was installed on the territory of the Bishkek geodynamic proving ground (BGPG, Northern Tien Shan), which includes 10 digital broadband stations, each of which is equipped with an STS-2 seismological sensor (Fig. 1, triangles). Various aspects of the operation of this network are described in more detail in [1]. The BGPG refers to the territory (see Fig. 1, indicated by the dashed line), which lies within the central segment of the North Tien Shan seismogenic zone and represents the junction zone of the Tien Shan orogen and the Turan plate. The main tectonic structures of the study area at the neotectonic stage are the Kyrgyz meganticline (Kyrgyz ridge) and the Chuya depression located to the north. To the south of the Kyrgyz ridge are the Suusamyr, Kochkor and Jumgal depressions, separated by small ridges. To the west are the Talas, Bolshoi and Maly Karatau ridges, and to the east - the Kungei-Alatau and Terskey-Alatau ridges with the Issyk-Kul intermountain depression separating them.
The stations of the KNET network make it possible to register earthquakes that occurred in the territory of the Northern and Central Tien Shan. To date, the catalogue of earthquakes according to the KNET network data (hereinafter referred to as the catalogue of

* Corresponding author: ivtran@mail.ru

earthquakes) contains the parameters of more than 10,000 events that occurred from 1994 to 2020 in the indicated territory. The stations of the network being on the territory of the BGPG maximally register weak earthquakes occurring here. Epicenters of earthquakes from the catalogue are shown in Fig. 1.

The processing of KNET data at the Research Station RAS (RS RAS) began in 1995, and it took time to gain experience in the localization of earthquakes, which affected the quality of the catalogue data for 1994-1995, but they can be taken into account when constructing maps of the epicentral location of earthquakes. The highest quality data in the earthquake catalogue have been presented since 1996, which can be used to solve various seismological problems. Since the start of the network in real time (July 1998), the network efficiency has reached 98-99% (data loss is 1-2%). Waveforms with a frequency of 100 Hz have been stored on the RS RAS since 1999, before this period digital records (with a frequency of 40 Hz) can be obtained from the IRIS website (Incorporated Research Institutions for Seismology).

The purpose of this work is to evaluate various characteristics of the earthquake catalogue.

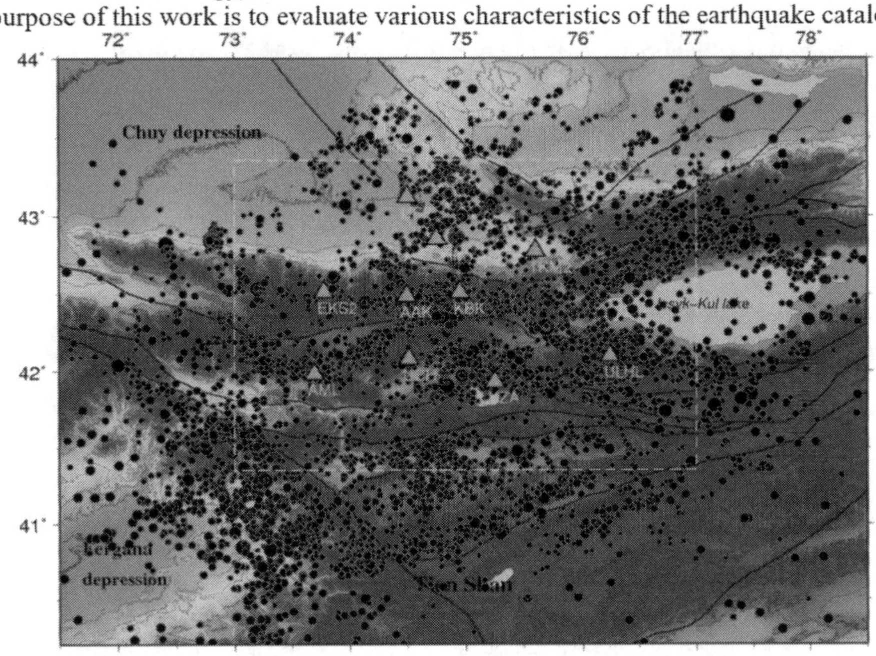

Fig. 1. Epicenters of earthquakes from the catalogue according to the data of the KNET network for 1994-2020 (10604 events). Δ - Stations position of the KNET network. A rectangle indicates the territory of the BGPG.

Methodology

Construction of a repeatability graph and the Gutenberg-Richter law. The Gutenberg-Richter Law [2] expresses the relationship between the magnitude and the total number of earthquakes in any given area and in a certain time period. This law is described by a linear function of the form:

$$\lg N(M) = a - bM, \quad (1)$$

where $N(M)$ is the number of earthquakes with magnitudes (or classes) not less than M, a and b are constants of the equation. Parameter a (a-value) formally describes seismic activity at $M = 0$, and b (b-value) is the angular coefficient of the linear part of the frequency distribution graph of earthquakes, which determines the rate of reduction of the relative number of events with an increase in their magnitude.

To construct the Gutenberg-Richter distribution, the function $N(M)$ is calculated. The graph of this function is plotted on a logarithmic scale on the axis of the number of earthquakes $\lg N$. Then it is determined M_c – *Magnitude of completeness*, M_{max} is the maximum magnitude for which enough events have occurred for statistics over the studied period of time. Then the portion ($M_c < x < M_{max}$) of the $y = \lg N(x)$ function is approximated by the $y = a - b{\cdot}x$ function.
In terms of the number of events with magnitude M exceeding a certain threshold value m (cumulative distribution), the law of distribution of earthquakes by energies takes the form:

$$N(M > m) \sim 10^{-bm}, \quad (2)$$

where N - denotes the number of events for a specified fixed time period and in a given geographical region, b - a constant (b-value), in most cases takes a value of about 0.9 [3].
Representative part of the earthquake catalogue. To correctly determine the representative part of the earthquake catalogue, the approach described in work [4] was used, which involves removing aftershocks from the catalogue before finding the representative part. The authors of this method are Molchan G.M., Dmitrieva O.E. and Smirnov V.B. A set of programs of the same authors was used to remove afteshocks [5].

Results

Program and Velocity Model. To calculate the parameters of the earthquake hypocenter recorded by the KNET seismological network, the Hypocenter program [6] is used, which uses a six-layer velocity model of the earth's crust calculated by Steve Recker [7]. The study of the velocity models of the Tien Shan region and the choice of the optimal model for calculating the parameters of the earthquake hypocenter were carried out in [8].
The program uses the arrival-file as an input file, where the times of arrival of P- and S-waves at the network station are indicated. An example of a seismogram (displacement velocity) of an earthquake (02/10/2016, K = 10.1) with a time stamp of the arrival of P- and S-waves is shown in Fig. 2.

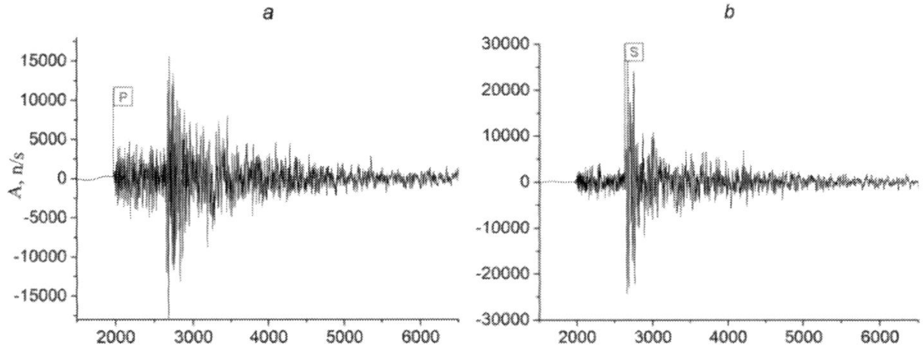

Fig. 2. An example of a recording of the earthquake on 02.10.2016 (K = 10.1) at the AAK station: *a* - Z component; *b* - E component. Flags - time stamps of volume waves arrival at the station.

It was noted above that the earthquake catalogue includes 10604 earthquakes. Each earthquake has a corresponding file of wave arrival times at the station. At present, there are 9717 *arrival*-files (files for 1995, as well as for individual earthquakes for 2011-2012 are lost).

The number of registered P- and S-waves. The statistical analysis of the number of P- and S-waves for each station of the network is presented in table. 1. The obtained quantitative data are presented in a sorted form, which makes it possible to assess the contribution of each station of the network in the process of localizing earthquakes.

Table 1. The number of *P*- and *S*-waves registered by the stations of the KNET network during the study period in sorted form (1994-2020).

***P*-wave**									
KBK	**AAK**	**USP**	**TKM2**	**UCH**	**CHM**	**EKS2**	**AML**	**KZA**	**ULHL**
8451	8236	7971	7676	7342	7146	6391	5872	5800	5538
***S*-wave**									
AAK	**KZA**	**KBK**	**CHM**	**UCH**	**USP**	**TKM2**	**AML**	**EKS2**	**ULHL**
7537	7342	7314	6296	6249	6131	5800	4781	4338	4054
(*P*+*S*) waves									
AAK	**KBK**	**USP**	**UCH**	**TKM2**	**CHM**	**KZA**	**EKS2**	**AML**	**ULHL**
15773	15765	14102	13591	13476	13442	13142	10729	10653	9592

The maximum number of *P*-waves was recorded at KBK, AAK and USP stations, the minimum - at ULHL, KZA and AML stations. The maximum number of *S*-waves is at stations AAK, KZA and KBK, the minimum is at stations ULHL, EKS2 and KZA. In terms of the sum (*P*+*S*) waves, the maximum number was recorded at the AAK, KBK and USP stations, the minimum - at the ULHL, AML, EKS2 stations. Stations AAK and KBK are located in the central part of the network and allow maximum registration of events that occurred in the territory controlled by the KNET network. The USP station is the extreme northern station, but the data for this station, as the results show (see Table 1.), are actively used for localization. The AML and EKS stations are located in the extreme western part, and the ULHL station is in the extreme eastern part of the network, which limits the registration of events occurring in the extreme opposite areas. Note that the AAK station is located in an adit (quiet station), and the ULHL station is located on the southwestern coast of Lake Issyk-Kul (the impact of waves on the coastlines on the noise level is possible). It is likely that the noise level at these stations also affected the result presented in Table. 1.

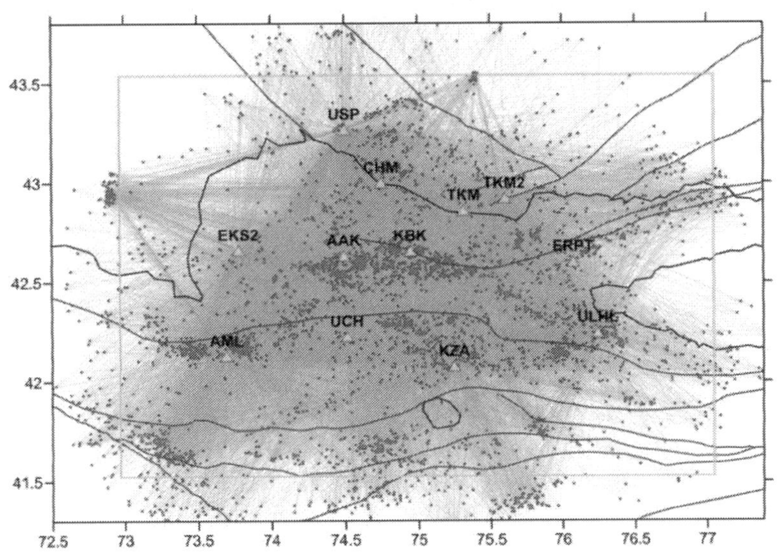

Fig. 3. Seismic wave traces (according to [9]). Δ is the position of seismic stations in the KNET network. Dots are earthquakes, brown lines are regional faults, black line is the northern border of Kyrgyzstan. The turquoise rectangle marks the territory of the BGPG.

In fig. 3, as an example, the epicentral position of earthquakes and seismic stations of the KNET network, as well as the *P*-wave traces, which were used in [9] when constructing the attenuation field of seismic waves and seismic tomography, is presented. The territory of the BGPG is covered with a good density of *P*-wave paths.
In total, during the study period, 130265 times of arrival of *P* and *S*-waves were received on the network stations: 70423 on the *P*-wave and 59842 on the *S*-wave. The arrival times of seismic waves can be used in solving various seismological problems: calculation of station spectra of earthquakes, seismic noise; during seismic tomography, when calculating the attenuation field, quality factor, etc. However, these data are not enough to study deeper layers of the earth's crust, since local earthquakes occur to depths of 30 km. In this case, it is necessary to attract additional data from distant earthquakes.

Errors in calculating earthquake hypocenters. When calculating the hypocenter parameters (latitude, longitude, depth), the *Hypocenter* program generates three types of calculation errors - the residuals of the seismic wave arrival times at the station (RMS), the horizontal error (ERH, epicentral position) and the vertical error (ERZ, depth). The estimation of the number of earthquakes is carried out, the parameters of which are determined with a given level of error. Table 2 shows the ranges of the considered errors (RMS, ERH, ERZ) and the corresponding number of earthquakes. For 50% of earthquakes, RMS is less than 0.3 s, and for ~ 90% of events does not exceed 0.5 s, which corresponds to an error in determining the coordinates of hypocenters up to 3000 m.The horizontal error (ERH) for 83% of earthquakes is less than 3 km, and in depth (ERZ) for 86% of events is less than 5 km.

Table 2. Error ranges, number of earthquakes and percentage. Period under review (1996-2020 гг.).

RMS, s						
0-0.1	**0.1-0.2**	**0.2-0.3**	**0.3-0.5**	**0.5-1.0**	**>1**	
150	1592	3389	3696	975	19	
1.52%	16.21%	34.5%	37.63%	9.92%	0.19%	
ERH, km						
0-0.5	**0.5-1**	**1-2**	**2-3**	**3-5**	**5-10**	**>10**
1322	2863	2681	1347	1019	518	0.72
13.46%	29.15%	27.29%	13.3%	13.71%	5.2%	2.74%
ERZ, km						
0-1	**1-2**	**2-3**	**3-5**	**5-10**	**>10**	
1908	3312	1585	1647	1057	312	
19.42%	33.72%	16.13%	16.77%	10.76%	3.17%	

In fig. 4 shows the epicentral position of earthquakes that occurred in 1996-2020. and against their background, the epicentral position of events with a level of RMS ≤ 0.3 s is marked in inverse color (Fig. 4a); ERH ≤ 2.5 km (Fig.4b); ERZ ≤ 2 km (Fig.4c). The main part of the events, the position of which is marked in inverse colors, is located on the territory of the BGPG, which is marked on the maps with a turquoise rectangle.

Table 3. Boundary values of the catalogue parameters according to the KNET network data.

Parameters	Date	φ,°n.l.	λ,°e.l.	*H*, km	*K*	*N*o	RMS, s	ERH, km	ERZ, km
Min	06.01.1994	37.35	68.67	0	4.77	3	0	0	0
Max	31.12.2020	45.77	82.24	465	14.83	20	9.78	141.3	99.9

Note. φ – latitude; λ – longitude; *H* – depth; *K* – class; *N*o- number of components; RMS – residuals of the seismic wave arrival times at the station; ERH – horizontal error; ERZ – vertical error.

Table 3 shows the boundary values of all parameters of the catalogue. Events for which any RMS, ERH or ERZ parameter exceeds the allowed values are included in the catalogue, and a note to this event indicates that the error for some parameter has an invalid value. The maximum depth of events indicated in the table (465 km) and the maximum error in its determination (99 km) indicate the presence of individual such events in the catalogue. According to [10], events in the study area can occur up to a depth of 30 km.

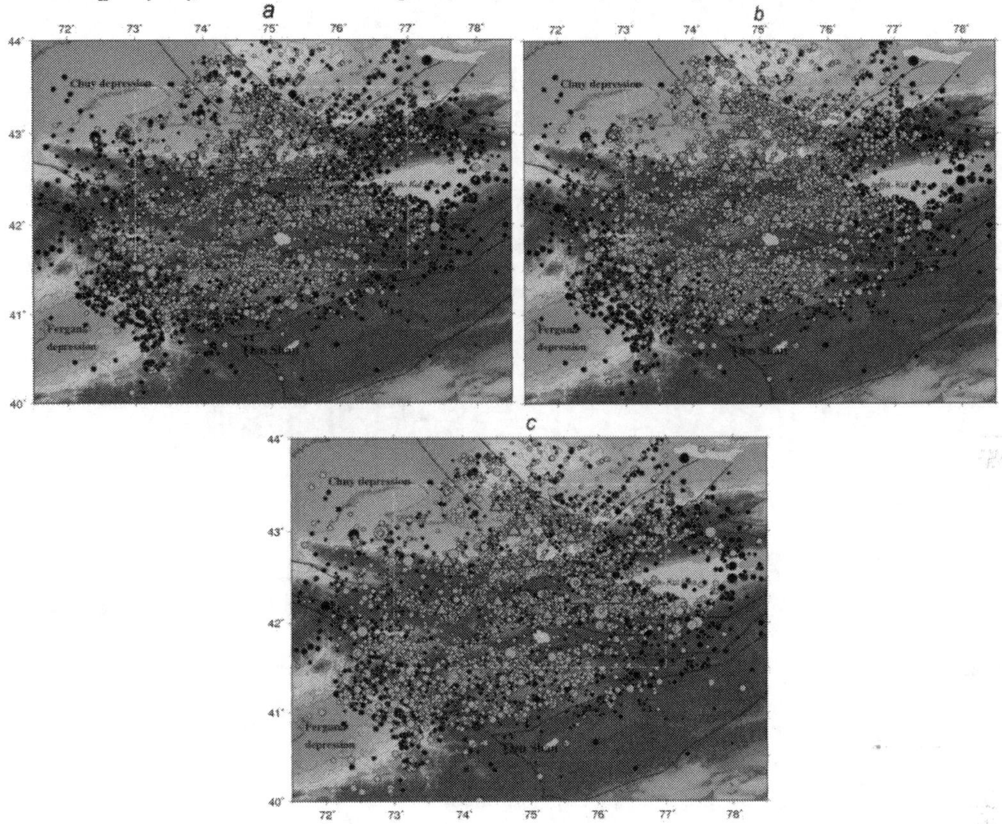

Fig. 4. Epicentral location of earthquakes with RMS ≤ 0.5 s (*a*), ERH ≤ 1.5 km (*b*), and ERZ ≤ 2 km (*c*). The turquoise rectangle marks the territory of the BGPG.

In [11], a comparison was made of the parameters of earthquake hypocenters from the catalogue obtained from the data of the KNET network (RS RAS) with the catalogue of the Institute of Seismology of the National Academy of Sciences of the Kyrgyz Republic (KIS). We selected 970 local events that took place in the Central Tien Shan in 1996–2001. The comparison was carried out for three parameters of earthquakes: latitude, longitude and class. It was determined that the average value of the difference in latitude coordinates is ~ 2 km, in longitude ~200 m and in energy class ~0.09. The noted discrepancies in the parameters under study can be due to the following reasons: different velocity models of the environment used in the CIS and RS RAS, different programs for calculating hypocenters and the handler skill. Insignificant discrepancies in the specified parameters make it possible to consider the catalogue obtained at the RS RAS on the basis of the KNET seismological network as a conditional one.

Statistical characteristics of the catalogue. Some quantitative estimates of the catalogue parameters are shown in Fig. 5. Events in time are distributed unevenly, the most seismically active are 1996, 1999, and 2017 (see Fig. 5b). The main part of the catalogue is made up of weak events $K = 6–8$ (see Fig. 5a), which occurred at a depth of 0–15 km (see

Fig. 5c). Most of the events are determined by 12-16 components (the total number of registered *P* and *S*-waves), the maximum possible number of components for a given network configuration can be equal to 20 (10 stations, *P* and *S*-waves) (Fig.5d).

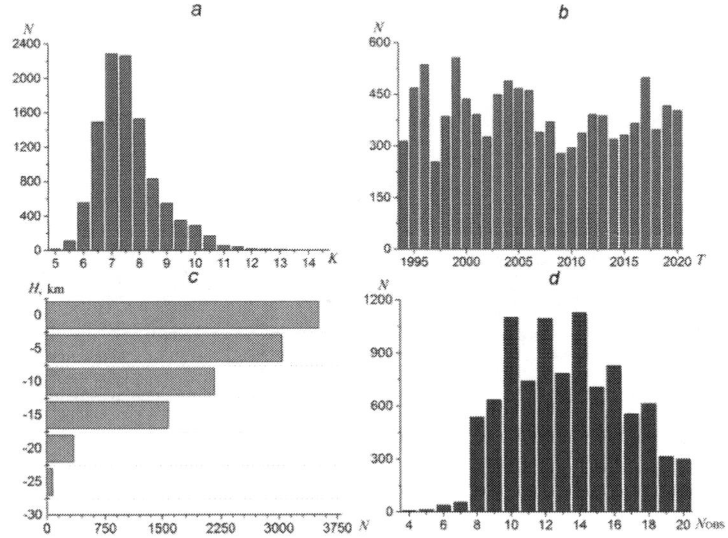

Fig. 5. Histograms of some characteristics of the KNET catalogue (1994–2020): *a* - class; *b* - time; *c* - depth; *d* - the number of phases (*P*- and *S*-waves).

Quantitative distribution of earthquakes by energy classes. Above, in Fig. 5b, the distribution of the number of earthquakes by years is presented, according to which the most seismically active years can be noted. Similar distributions were constructed for separate samples of events that occurred during the network operation in real time (1999-2020) and formed depending on the earthquake class (Fig. 6).

The most active year for weak events is 1999 (see Fig. 6, $6 \leq K < 7$), where their number exceeds 200 earthquakes. In other years, their value varies from 30 to 90 earthquakes per year. For weak events with $7 \leq K < 8$, the maximum number occurs on the period 2003-2006. (more than 200 earthquakes per year). In other years, their number varies from 150 to 200. For events with $8 \leq K < 9$ and $9 \leq K < 10$, the histograms have a similar distribution pattern. There are two periods when the level of these events falls to the minimum value - 2008-2011 and 2013-2014. The maximum number of earthquakes in the considered samples occurred in 2017.The maximum number of earthquakes with $10 \leq K < 11$ occurred in 2006 (24 events) and 2017. (25 events), the minimum in 2010 (less than 5 events). On average, earthquakes from $11 \leq K < 12$ occur 2-6 times a year. The maximum number of such events occured in 2003 (9 events) and 2020 (14 events). The maximum number of earthquakes from $12 \leq K < 13$ occurred in 2005 and 2007. (5 events each). In some years, earthquakes of this class do not occur, and in other years their number varies within 1-2 earthquakes per year. Less often, events occur with $13 \leq K < 14$. In 2004 and 2015, there were 2 earthquakes each, and in the period from 2006 to 2009, there was one earthquake each. In the considered distributions, there is no positive or negative trend in the change in the number of events with a given energy characteristic.

Depth distribution of seismicity. In fig. 5c shows a histogram of the distribution of earthquakes by depth, and Fig. 7 shows their areal distribution at the same depths as on the histogram; earthquakes located below 15 km were considered in one sample.

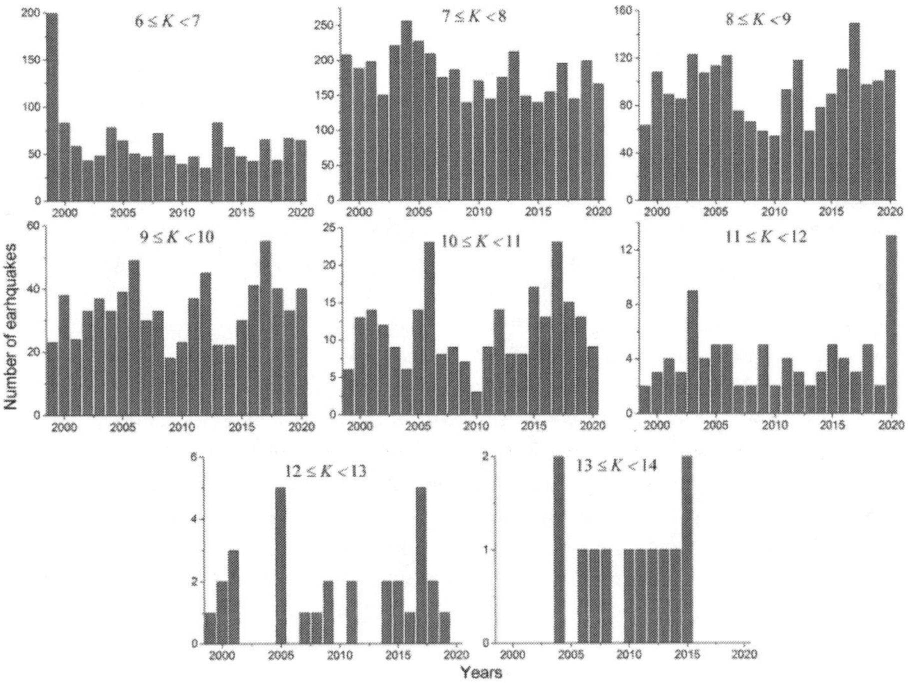

Fig. 6. Distribution of the number of earthquakes by years for the samples formed depending on the class of the earthquake (shown in the graphs). The period under review is 1999-2020.

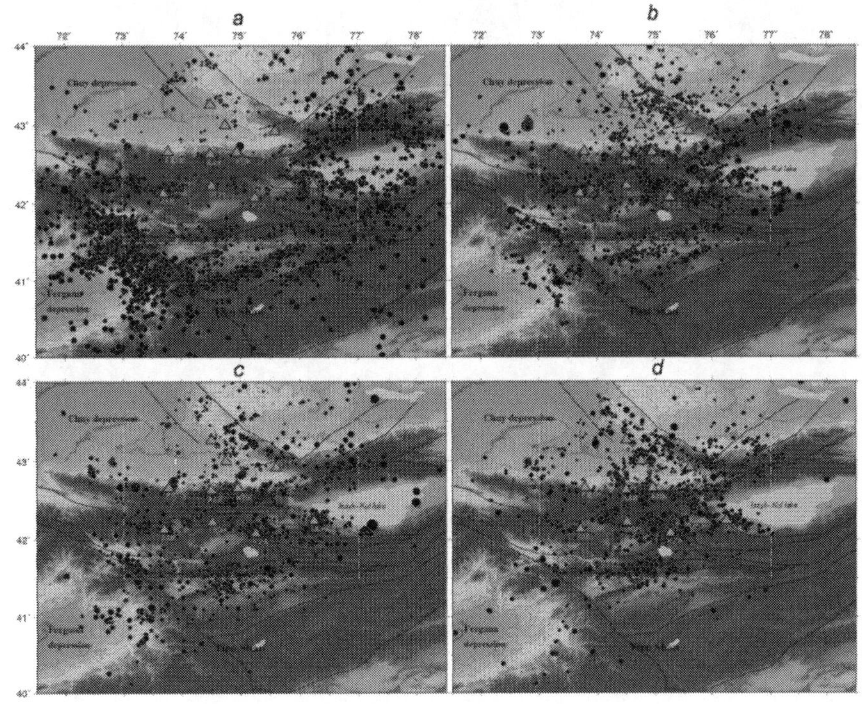

Fig. 7. Epicentral location of earthquakes from the catalogue according to the KNET network data at different depths: *a* - 0-5 km (3613 events); *b* –5-10 km (2960); *c* - 10-15 km (2167); *d* - more than 15 km (1865). The triangles indicate the stations of the KNET network.

At a depth of 0-5 km (see Fig. 7*a*), most of the events are located outside the KNET stations and are located along the sides of the Issyk-Kul depression and Talas-Fergana fault. A small number of events were noted on the territory of the location of the stations of the KNET network. This configuration of near-surface seismicity is most likely associated with artifacts of the computational process. For depths of 5-10 (see Fig. 7b) and 10-15 km (see Fig. 7*c*), the seismicity distribution pattern, which is mainly concentrated within the network and along its margins, changes little, apart from the density of events. At a depth of more than 15 km, most of the events are located on the territory bounded by the coordinates of the network edge stations (North Tien Shan self-generating zone), along the Chu-Ili mountains (see Fig.7*d*) and on the Moldo Too ridge. According to the results obtained, ~50% of earthquakes fall to depths of 5-15 km.

Representative sample of the earthquake catalogue. After removing aftershocks from the earthquake catalogue, a representative sample of the catalogue was calculated. For the calculation, the class of earthquakes and the magnitude were considered, recalculated from the class according to the well-known expression linking magnitude and energy class [12]: $K = 1.8 \cdot M + 4$. A representative sample was determined: for the entire catalogue; for a catalogue without aftershocks and a catalogue without aftershocks bounded by BGPG coordinates. To plot the frequency of occurrence, the number of earthquakes that fell within a certain range of magnitudes (classes) was calculated; the step of changing the range was 0.1. Also, the cumulative Gutenberg - Richter distribution was constructed according to the linear part of which the representative sample and the parameters of the equation describing the approximation of the linear portion of the distribution were determined.

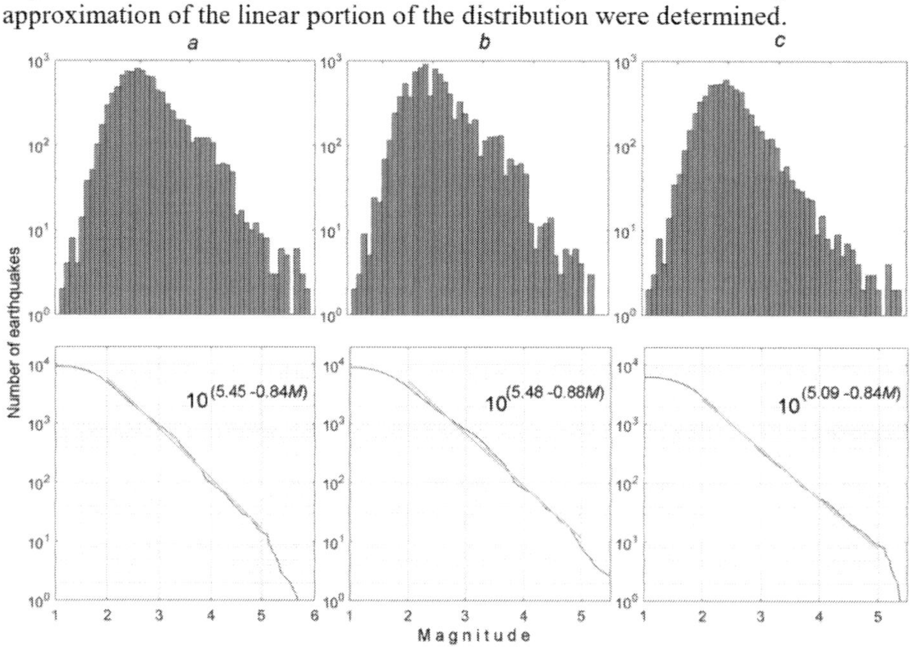

Fig. 8. The repeatability graph and the cumulative Gutenberg - Richter distribution calculated from the magnitude for the complete catalogue using the KNET data (*a*), for the KNET catalogue after removing the aftershocks (*b*) and the catalogue that includes earthquakes that occurred in the BGPG territory (*c*).

In fig. 8 shows the frequency plots (top row) and cumulative Gutenberg - Richter distributions (bottom row) calculated by magnitude, where the equations describing the linear part of the cumulative distribution are also marked. The characteristics of the Gutenberg - Richter distribution obtained for the classes and magnitudes are presented in Table. 4.

The boundaries of the linear part of the Gutenberg - Richter distribution, presented in table. 4 show that the representative part of the catalogue, limited by the BGPG coordinates, begins with earthquakes with $M \geq 1.7$, and for the complete catalogue, events starting from $M \geq 1.8$ are representative. The seismological network KNET is located inside the territory of the BGPG, and allows full registration of events with $M \geq 1.7$, with distance from the boundaries of the BGPG, the number of registered weak earthquakes decreases. Therefore, when solving some problems on the basis of a catalogue based on the data of the KNET network, it is advisable to restrict ourselves to the territory of the BGPG.

Table 4. List of catalogues, characteristics of the Gutenberg - Richter distribution and the boundaries of the linear part of the distribution (1996-2020)

Catalogue KNET	Distribution parameters				Distribution Linear Part			
	K		*M*					
	a	*b*	*a*	*b*	K_{min}	K_{max}	M_{min}	M_{max}
A	7.344	0.472	5.451	0.846	7.3	13.2	1.8	5.1
B	7.548	0.502	5.488	0.885	7.5	12.8	1.9	4.9
C	6.992	0.470	5.097	0.841	7.0	13.2	1.7	5.1

Note. A – earthquake catalogue, B – earthquake catalogue without aftershocks, C – catalogue of earthquakes without aftershocks, limited by BGPG coordinates.

The parameters of the linear part of the Gutenberg - Richter distribution - *a*-value and *b*-value - refer to the most important quantitative characteristics of the seismic regime. The slope of the graph expresses the ratio between the number of strong and weak seismic events, or (already in physical interpretation) the ratio between the number of large and small fractures in the geological environment. The graph level characterizes seismic activity - the total intensity of seismic manifestations in the considered seismically active area. These parameters form the basis of the quantitative seismic zoning method. In [3], it is indicated that for seismically active regions, in most cases, the *b*-value inclination angle takes on a value of 0.9. For the catalogue according to the KNET network, this parameter has a value of 0.84; after removing aftershocks (weak events), its value changes to 0.88. The *b*-value can be used to assess the geodynamic processes of the seismically active zone.

Conclusion

The Kyrgyz seismological network KNET, established in 1991, allows registering earthquakes occurring in the territory of the Northern and Central Tien Shan. During the operation of the network, more than 10,000 earthquakes were registered, which occurred from 1994 to 2020. A statistical analysis of various parameters of the catalogue obtained from the data of the KNET network has been carried out. For 50% of earthquakes in the catalogue, the discrepancy in determining the arrival times of *P*- and *S*-wave fronts (RMS) is less than 0.3 s, and for 86% of events it does not exceed 0.5 s. The horizontal error (ERH) for 80% of events is less than 3 km, and for 83.8% of events, the error in depth (ERZ) is less than 5 km. In total, 130,265 times of arrival of *P*- and *S*-waves were recorded during the study period: 70423 along the *P*-wave and 59842 along the S-wave, which can be used to solve various seismological problems. The quantitative temporal distributions of earthquakes, depending on the class of the event, showed that there is no positive or negative trend in the change in the number of events with a given energy characteristic. The distribution of earthquakes by depths: 0-5 km, 5-10 km, 10-15 km and more than 15 km showed that the most reliable depths are determined for earthquakes with a depth of more than 5 km. After removing aftershocks, a representative sample of the catalogue was determined. The catalogue is most fully represented by earthquakes with $M \geq 1.8$. The catalogue, limited by the BGPG coordinates, expands this sample with weaker events with $M \geq 1.7$. The slope of the Gutenberg - Richter distribution calculated from the catalogue is 0.88. The results obtained in the work showed that the catalogue based on the KNET data is most efficiently used for the territory of the Bishkek geodynamic proving ground.

Acknowledgements

The research was partially carried out within the framework of the state assignment of Research Station of the Russian Academy of Sciences in Bishkek city (topics № AAAA-A19-119020190064-9 and № AAAA-A19-119020190066-3).

References

1. N.A. Sycheva, L.M Bogomolov., S.I. Kuzikov Computational technologies in seismological research (on the example of KNET, Northern Tian Shan) (IMGG FEB RAS, Yuzhno-Sakhalinsk, 2020)
2. B. Gutenberg, C.F. Richter, Bulletin of the Seismological Society of America, **34** (1944)
3. K. Kasahara, Earthquake mechanics (Cambridge Univ. Press, 1981)
4. G.A. Sobolev, A.V. Ponomarev, Earthquake Physics and Precursors (Nauka, Moskow, 2003)
5. G.M. Molchan, O.E. Dmitrieva, Computational seismology, **24** (1991)
6. B.R. Lienert, E. Berg, L.N.F razer, Hypocenter: an earthquake location method using centered, scaled, and adaptively damped least squares, **76**, 3 (1986)
7. S.W. Roecker, T.M. Sabitova, L.P. Vinnik, Journal Geophys. Res., **98**, B9 (1993)
8. N.A. Sycheva, S.I. Kuzikov, Geophysical Research, **13**, 2 (2012)
9. I.V Sychev., I. Koulakov, N.A. Sycheva, A. Koptev, I. Medved, S.E. Khrepy, N. Al-Arifi, Journal Geophysical Research Solid Earth, **123**, 2 (2018)
10. F.N. Yudakhin, Geophysical fields, depth structure and seismicity of Tian Shan (Ilim, Frunze, 1983)
11. N.A. Sycheva Dis.... edging. physical. - a mat. Sciences (Moscow, 2005)
12. Yu.V. Riznichenko, Problemy seysmologii (Nauka, Moscow, 1985) (in Russian)

Generation of equations for shell models of turbulence in the Maple system

Liubov Feshchenko[1,*] and *Gleb* Vodinchar[1,**]

[1]Institute of Cosmophysical Research and Radio Wave Propagation FEB RAS, Paratunka, Russia

Abstract. The paper describes a technology for the automated compilation of equations for shell models of turbulence in the computer algebra system Maple. A general form of equations for the coefficients of nonlinear interactions is given, which will ensure that the required combination of quadratic invariants and power-law solutions is fulfilled in the model. Described the codes for the Maple system allowing to generate and solve systems of equations for the coefficients. The proposed technology allows you to quickly and accurately generate classes of shell models with the desired properties.

1 Introduction

The main idea of shell models of turbulence is to construct a chain of ordinary differential equations describing the energy distribution over the scale spectrum in developed turbulence.

The scheme for constructing cascade models is as follows. The axis of wavenumbers is divided into exponentially expanding zones (shells). A sequence of wave numbers arises $k_n = q^n k_0$, where for a suitable choice of the spatial scale $k_0 = 1$. The denominator of the progression q can be chosen arbitrarily large enough, most often in computational practice $q = 2$ is taken[1–3].

The dynamics of each field X of a turbulent system is described by the collective variables $X_n(t)$. The absolute values of $X_n(t)$ are interpreted as measures of pulsations of the field X in the ranges of scales with wave numbers from $[k_n; k_{n+1})$. For these variables, a system of equations is drawn up that is qualitatively similar to the initial equations of the turbulent system. Since the equations of hydrodynamics and magnetohydrodynamics are quadratically nonlinear, we are talking about a quadratically nonlinear dynamic system. The coefficients at the nonlinear terms of such a system determine in the model the energy transfer along the inertial interval of scales in developed turbulence. Complex or real variables are used when constructing shell models . In this work, we will only consider real ones.

It is known that, in the ideal limit, there are conserved quadratic quantities in the equations of hydrodynamics and magnetohydrodynamics. Therefore, it is necessary that the equations of cascade models also possess quadratic invariants, which are analogs of these quantities.

Actually, the main mathematical problem in deriving the equations of cascade models is to calculate the coefficients of nonlinear terms that would ensure the existence of certain quadratic invariants.

*e-mail: feshenko.lk@yandex.ru
**e-mail: gvodinchar@ikir.ru

In this paper, we describe the developed technology for the automated construction of cascade models, in which the calculation of the coefficients of nonlinear interactions is implemented using one of the computer algebra systems (CAS) - - the Maple system. Note that Maple is used not for studying ready-made shell models, but for the automated derivation of the equations of these models themselves. Previously, the authors of this paper considered the use of CAS for the automated derivation of shell models of MHD turbulence [4, 5]. In this work, we will consider various turbulent systems – «pure» hydrodynamics, magnetohydrodynamics, convective system for conducting and non-conducting fluid.

All the details of the mathematical calculations can be found in the work of the authors [6].

2 General form of the studied models

Let U_n, Θ_n and B_n be collective variables for the fields of velocity, temperature and magnetic induction, respectively. In the most general form, the considered shell models can be written in the following form.

- Hydrodynamics:

$$\frac{dU_n}{dt} = \sum_{i,j} S_{nij} U_i U_j - \mathsf{R}^{-1} k_n^2 U_n + f_n(t), \tag{1}$$

where R – is the Reynolds number, $f_n(t)$ are functions that simulate the supply of energy to a turbulent system from the outside. Usually, this supply is performed only on the main scale of the system (macroscale) $n = 0$.

- Convection for non-conducting fluid:

$$\begin{aligned} \frac{dU_n}{dt} &= \sum_{i,j} S_{nij} U_i U_j - \mathsf{R}^{-1} k_n^2 U_n + \Theta_n, \\ \frac{d\Theta_n}{dt} &= \sum_{i,j} H_{nij} U_i \Theta_j - k_n^2 (\mathsf{PrR})^{-1}\, \Theta_n + f_n(t), \end{aligned} \tag{2}$$

where Pr – is the Prandtl number.

- Magnetohydrodynamics:

$$\begin{aligned} \frac{dU_n}{dt} &= \sum_{i,j} S_{nij} U_i U_j + \sum_{ij} L_{nij} B_i B_j - \mathsf{R}^{-1} k_n^2 U_n + f_n(t), \\ \frac{dB_n}{dt} &= \sum_{i,j} W_{nij} U_i B_j - \mathsf{R}_\mathsf{m}^{-1} k_n^2 B_n, \end{aligned} \tag{3}$$

where $\mathsf{R_m}$ – is the magnetic Reynolds number.

- Magnetoconvection:

$$\begin{aligned} \frac{dU_n}{dt} &= \sum_{i,j} S_{nij} U_i U_j + \sum_{ij} L_{nij} B_i B_j - \mathsf{R}^{-1} k_n^2 U_n + \Theta_n, \\ \frac{d\Theta_n}{dt} &= \sum_{i,j} H_{nij} U_i \Theta_j - k_n^2 (\mathsf{PrR})^{-1} \Theta_n + f_n(t), \\ \frac{dB_n}{dt} &= \sum_{i,j} W_{nij} U_i B_j - \mathsf{R}_\mathsf{m}^{-1} k_n^2 B_n. \end{aligned} \tag{4}$$

The indices n, i, and j in the equations can take any integer values. Various cascade models differ in the coefficients of nonlinear interactions S_{nij}, H_{nij}, S_{nij}, L_{nij}, W_{nij}, which ensure the existence of different quadratic invariants.

In the quadratic terms present in the general forms of shell models (1 -4), the coefficients of these forms are determined by three indices, the first of which corresponds to the number of the equation, that is, the number of the selected scale shell, and the other two – the numbers of the shells interacting with the selected one, i.e. there is some inhomogeneity in the scales space.

However, in reality, there is scale homogeneity for nonlinear energy transfer within the inertial interval. In this case, the interaction of the i-th and j-th shells with the n-th should depend not on the i and j themselves, but only on the distance from them to the n-th shell in space scale. In addition, all nonlinear terms of the exact equations of hydrodynamics, magnetohydrodynamics, and convection contain the nabla operator. An analogue of this operator in the n-th scale shell is the wave number $k_n = q^n$.

Therefore, nonlinear terms for the general form (1-4) of cascade models are usually written in a different way:

$$\begin{aligned}
&S_{nij}U_iU_j = q^n s_{i'j'} U_{n+i'} U_{n+j'},\\
&H_{nij}U_i\Theta_j = q^n h_{i'j'} U_{n+i'} \Theta_{n+j'},\\
&L_{nij}B_iB_j = q^n l_{i'j'} B_{n+i'} B_{n+j'},\\
&W_{nij}U_iB_j = q^n w_{i'j'} U_{n+i'} B_{n+j'},\\
&i' = i - n, \quad j' = j - n.
\end{aligned} \tag{5}$$

The left-hand of the notation turns out to be more convenient for obtaining equations for the coefficients that ensure the fulfillment of the conservation laws, so we took it as the initial one. The right-hand version of the record corresponds to the concept of scale homogeneity within the inertial interval and qualitatively reflects the differential structure of the nonlinear terms of equations. Note also that the matrices of the coefficients $s_{i'j'}$ and $l_{i'j'}$ are naturally assumed to be symmetric.

3 Constraints on the coefficients of nonlinear interactions

In the ideal limit, i.e. when $\mathrm{R} = \mathrm{Pr} = \mathrm{R_m} = \infty$, and in the absence of external forces, the equations of hydrodynamics and magnetohydrodynamics satisfy a certain set of conservation laws. These conservation laws differ in the case of «pure» hydrodynamics and in the case of magnetohydrodynamics, and also depend on the dimension of the physical space in which the turbulent system [1, 3] is considered. The conserved quantities are some quadratic forms of turbulent fields. Therefore, it is required that the shell models, also possess some quadratic invariants of collective variables. These quadratic forms should be formal analogs of the exact conserved quantities [3, 6].

The invariants used in this work are given in the Table 1.

The invariance of each of those given in Table 1 of quadratic forms imposes constraints on the coefficients of nonlinear interactions in the form of homogeneous linear equations.

Formally, these equations, as well as the desired coefficients themselves, are infinitely many, however, restrictions are always introduced on the range of interactions in the space

Table 1. Quadratic invariants for shell models

Two-dimensional flows	Three-dimensional flows
Hydrodinamics	
Enstrophy $\Omega = \sum_n (q^n U_n)^2$	Helicity $H_V = \sum_n (-1)^n U_n (q^n U_n)$
Kinetic energy $E_V = \sum_n U_n^2$	Kinetic energy $E_V = \int \mathbf{v}^2 dV \sim \sum_n U_n^2$
Magnetohydrodinamics	
Total energy $E = \sum_n \left(U_n^2 + B_n^2\right)$	Total energy $E = \sum_n \left(U_n^2 + B_n^2\right)$
Square potential $A = \sum_n (q^{-n} B_n)^2$	Magnetic helicity $H_B = \sum_n (-1)^n (q^{-n} B_n) B_n$
Cross helicity $H_C = \sum_n U_n B_n$	Cross helicity $H_C = \sum_n U_n B_n$
Temperature invariant	
Energy of temperature pulsations $E_T = \sum_n \Theta_n^2$	Energy of temperature pulsations $E_T = \sum_n \Theta_n^2$

of scales. This leads to a finite truncation of both the number of nonzero interaction coefficients and the number of equations for them, since the rest of the equations degenerate into identities. The specific form of the equations depends on the conservation laws used.

Table 2 presents the equations obtained by us for the coefficients of the matrices of nonlinear interactions, grouped by different types of models, in the case of representation of nonlinear terms by the right-hand sides of formulas (5). The derivation of these equations can be seen in the work of the authors [6].

Let us consider one more restriction that naturally follows from wave interactions and is not associated with quadratic conservation laws. This limitation is fundamental in the construction of non-local in scale models, when formally there are interactions of far from each other scales.

From the representation of the equations of motion of the medium in the space of wave vectors, it follows that those and only those waves participate in the nonlinear interaction, from the wave vectors of which it is possible to make a triangle [1, 3]. Then it turns out that for the interaction of shells with numbers n, $n+i$, $n+j$, the pair of indices (i, j) must lie in some region, the specific form of which depends on q, but the general the shape is similar to the region F shown in Fig. 1. The figure 1 itself shows the shape of the domain F for $q\ geq2$. As q decreases, the region expands. The scheme for calculating such a region of pairs of indices of interacting shells is described, for example, in the works [2, 3]. In what follows, we will consider precisely the case $q \geq 2$.

To ensure the fulfillment of this constraint, it is necessary to require that all coefficients s_{ij}, l_{ij}, w_{ij}, h_{ij}, whose indices do not fall into the region F, would be zero.

From Table 2 and the requirement that the indices fall into F, it can be seen that the construction of cascade models of all types under consideration is reduced to the compilation and

Table 2. Equations for nonlinear interactions coeficients

1. Two-dimensional flows
1.1. Hydrodynamics
Kinetic energy conservation: $\left(s_{i,j} + s_{j,i}\right) + q^{i}\left(s_{-i,j-i} + s_{j-i,-i}\right) + q^{j}\left(s_{i-j,-j} + s_{-j,i-j}\right) = 0.$ Enstrophy conservation: $\left(s_{i,j} + s_{j,i}\right) + q^{3i}\left(s_{-i,j-i} + s_{j-i,-i}\right) + q^{3j}\left(s_{i-j,-j} + s_{-j,i-j}\right) = 0.$ Symmetries: $s_{ij} - s_{ji} = 0.$
1.2. Magnetohydrodynamics
Total energy conservation: $\left(s_{i,j} + s_{j,i}\right) + q^{i}\left(s_{-i,j-i} + s_{j-i,-i}\right) + q^{j}\left(s_{i-j,-j} + s_{-j,i-j}\right) = 0,$ $l_{i,j} + l_{j,i} + q^{j} w_{-j,i-j} + q^{i} w_{-i,j-i} = 0.$ Cross helicity conservation: $\left(l_{i,j} + l_{j,i}\right) + q^{i}\left(l_{-i,j-i} + l_{j-i,-i}\right) + q^{j}\left(l_{i-j,-j} + l_{-j,i-j}\right) = 0,$ $w_{i,j} + q^{i} w_{-i,j-i} + q^{j} s_{i-j,-j} + q^{j} s_{-j,i-j} = 0.$ Square potential conservetion: $w_{i,j} + q^{-j} w_{i-j,-j} = 0$ Symmetries: $s_{ij} - s_{ji} = 0, \quad l_{ij} - l_{ji} = 0.$
2. Three-dimensional flows
2.1. Hydrodynamics
Kinetic energy conservation: $\left(s_{i,j} + s_{j,i}\right) + q^{i}\left(s_{-i,j-i} + s_{j-i,-i}\right) + q^{j}\left(s_{i-j,-j} + s_{-j,i-j}\right) = 0.$ Helicity conservation: $\left(s_{i,j} + s_{j,i}\right) + (-1)^{i} q^{2i}\left(s_{-i,j-i} + s_{j-i,-i}\right) + (-1)^{j} q^{2j}\left(s_{i-j,-j} + s_{-j,i-j}\right) = 0.$ Symmetries: $s_{ij} - s_{ji} = 0.$
2.2. Magnetohydrodynamics
Total energy conservation: $\left(s_{i,j} + s_{j,i}\right) + q^{i}\left(s_{-i,j-i} + s_{j-i,-i}\right) + q^{j}\left(s_{i-j,-j} + s_{-j,i-j}\right) = 0,$ $l_{i,j} + l_{j,i} + q^{j} w_{-j,i-j} + q^{i} w_{-i,j-i} = 0.$ Cross helicity conservation: $\left(l_{i,j} + l_{j,i}\right) + q^{i}\left(l_{-i,j-i} + l_{j-i,-i}\right) + q^{j}\left(l_{i-j,-j} + l_{-j,i-j}\right) = 0,$ $w_{i,j} + q^{i} w_{-i,j-i} + q^{j} s_{i-j,-j} + q^{j} s_{-j,i-j} = 0.$ Magnetic helicity conservation: $w_{i,j} + (-1)^{j} w_{i-j,-j} = 0.$ Symmetries: $s_{ij} - s_{ji} = 0, \quad l_{ij} - l_{ji} = 0.$
3. Convective system (additional condition for temperature)
Energy of temperature pulsations conservation: $h_{i,j} + q^{j} h_{i-j,-j} = 0.$

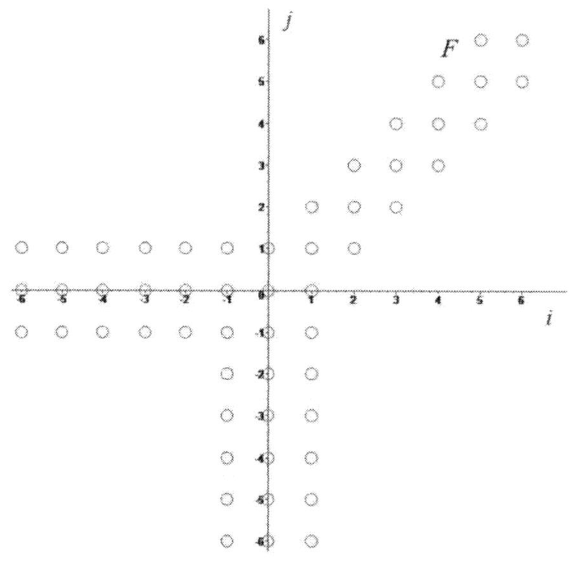

Figure 1. Area F of pairs of indices of interacting shells for $q \geq 2$.

subsequent exact solution of homogeneous systems of linear equations of large dimension. Accuracy is necessary if we want to ensure the conservation laws exactly.

Since the system is homogeneous, situations are possible either only of the zero solution, or of an entire linear variety of solutions. If the solution is only zero, then a shell model with the required set of invariants and a given scale nonlocality simply does not exist. If there are infinitely many solutions, then they define a whole parametric class of models.

4 Scheme for obtaining equations for coefficients in Maple

Let the range of nonlinear interaction in the space of scales (scale nonlocality) be bounded from above by an integer $P \geq 1$. In a real turbulent physical system, such a limitation always exists. This finite scale nonlocality determines the real range of i and j. All coefficients s_{ij}, l_{ij} and w_{ij}, at least one of the indices of which is greater than P in absolute value, will be zero. Since all equations for the interaction coefficients are linear homogeneous, the equations, all the terms of which contain an index whose modulus is greater than P, turn out to be identities. It turns out that the introduction of restrictions on the range of interactions distinguishes finite subsystems in the infinite systems of equations obtained above. It is these subsystems that we will form and solve using Maple tools.

Note also that each equation in Table 2 is defined by a pair of indices (i, j), and this equation contains only coefficients with indices $(\pm i, \pm j)$, $(\pm i, \pm(ij))$ and $(\pm j, \pm(ij))$. There are no other combinations. Therefore, only those equations for which the indices (i, j) will not necessarily degenerate into identities, satisfy the condition:

$$(|i| \leq P \wedge |j| \leq P) \vee (|i| \leq P \wedge |i - j| \leq P) \vee (|j| \leq P \wedge |i - j| \leq P)\,. \qquad (6)$$

They lie in the star-shaped region D shown in Fig. 2. Note that in previous works we used a wider area in the shape of a cross [4, 5]. Now it has been reduced. In what follows, we need the predicate $d(i, j)$, which selects the region D and is defined by the expression (6). It is also clear that if we choose the pair $(i, j) \in D$, then $|i| \leq 2P$, $|j| \leq 2P$ and $|ij| \leq 2P$.

Then the sets of the calculated coefficients are naturally represented in Maple, using two-dimensional arrays, the indices of which vary from $-2P$ to $2P$.

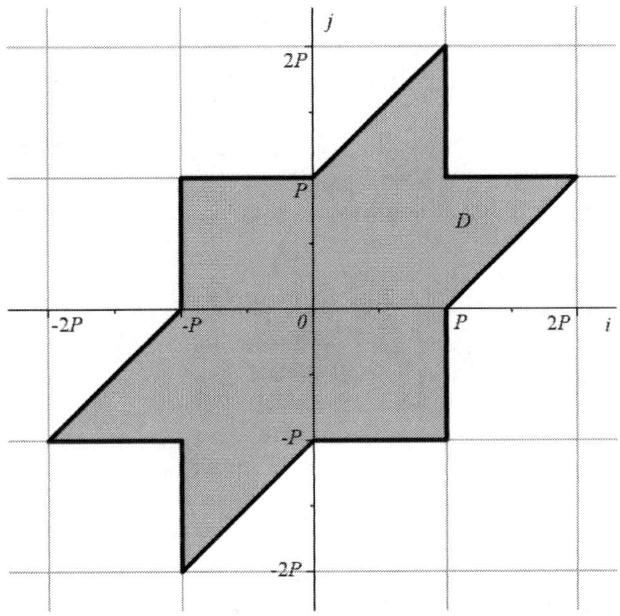

Figure 2. Area D of pairs of indices corresponding to equations that not degenerate into identities in general case.

Now we introduce the predicate $f(i, j)$, which selects the region F of pairs of indices of interacting shells (Fig. 1). It's clear that

$$f(i, j) = (|i| \leq 1 \wedge j \leq 1) \vee (|j| \leq 1 \wedge i \leq 1) \vee (i \geq 1 \wedge j \geq 1 \wedge |i - j| \leq 1). \tag{7}$$

Then all the elements of the arrays, the indices of which satisfy the condition:

$$\neg f(i, j) \vee |i| > P \vee |i| > P, \tag{8}$$

should be zeros. The rest of the elements must be assigned symbolic values, which will then be the components of the vector of unknowns in the system of equations being compiled. Here is the corresponding Maple code snippet.

```

f:=(i,j)->( abs(i)<=1 and j<=1 ) or ( abs(j)<=1 and i<=1 ) or
          ( i>=1 and j>=1 and abs(i-j)<=1 );
k:=1:
for i from -2*P to 2*P do for j from -2*P to 2*P do
  if ( not(f(i,j)) or abs(i)>P or abs(j)>P ) then
    s[i,j]:=0: l[i,j]:=0: w[i,j]:=0:
  else
```

```
       s[i,j]:=x[k]: k:=k+1:
10     l[i,j]:=x[k]: k:=k+1:
       w[i,j]:=x[k]: k:=k+1:
12    end if:
   end do: end do:
14 NumVar:=k-1:
```

As a result of executing this code, the unknown coefficients of the vector $\mathbf{x} = (x_1, x_2, \ldots, x_{NumVar})^T$ will be assigned as symbolic values.

Next, it is necessary to compose expressions for the left side of the equation from Table 2 for the three-dimensional MHD flow for each pair of indices from the region D. In this case, it is advisable to control whether this expression turns out to be zero, since then the equation degenerates into identity. The corresponding Maple code looks like this:

```
40 d:=(i,j)->( abs(i)<=P and abs(j)<=P ) or
             ( abs(i)<=P and abs(i-j)<=P ) or
42           ( abs(j)<=P and abs(i-j)<=P );
   m:=1:
44 for i from -2*P to 2*P do for j from -2*P to 2*P do if d(i,j) then
     expr:=(s[i,j]+s[j,i])+q^i*(s[-i,j-i]+s[j-i,-i])+
46   q^j*(s[-j,i-j]+s[i-j,-j]):
     if (expr<>0) then g[m]:=expr: m:=m+1: end if:
48   expr:=l[i,j]+l[j,i]+q^j*w[-j,i-j]+q^i*w[-i,j-i]:
     if (expr<>0) then g[m]:=expr: m:=m+1: end if:
50   expr:=(l[i,j]+l[j,i])+q^i*(l[-i,j-i]+l[j-i,-i])+
     q^j*(l[-j,i-j]+l[i-j,-j]):
52   if (expr<>0) then g[m]:=expr: m:=m+1: end if:
     expr:=w[i,j]+q^i*w[-i,j-i]+q^j*s[i-j,-j]+
54   q^j*s[-j,i-j]:
     if (expr<>0) then g[m]:=expr: m:=m+1: end if:
56   expr:=w[i,j]+(-1)^j*w[i-j,-j]:
     if (expr<>0) then g[m]:=expr: m:=m+1: end if:
58   expr:=s[i,j]-s[j,i]:
     if (expr<>0) then g[m]:=expr: m:=m+1: end if:
60   expr:=l[i,j]-l[j,i]:
     if (expr<>0) then g[m]:=expr: m:=m+1: end if:
62 end if: end do: end do:
   NumEq:=m-1:
```

As a result of executing this code, expressions of the left sides of the equations for the coefficients will be generated and written as symbolic values into the components of the vector $\mathbf{g} = \left(g_1, g_2, \ldots, g_{NumEq}\right)^T$. Lines 45-49 construct equations for conserving the total energy, in lines 50-55 – for cross-helicity, in lines 56-57 – for magnetic helicity, in lines 58-61 – the symmetry of the coefficient matrices s_{ij} and l_{ij}.

Now we form the main matrix A of our linear system and solve it:

```
70 A:=Matrix(NumEq,NumVar):
   0_:=Vector(NumEq):
72 for i from 1 to NumEq do for j from 1 to NumVar do
     A[i,j]:=coeff(g[i],x[j]):
```

```
74  end do: end do:
    y:=LinearSolve(A,0_):
```

Application of the described scheme and code in the considered case of a three-dimensional MHD flow for $P = 2$ gave the following results (we present only nonzero coefficients):

$$
\begin{aligned}
&s_{-2,-1} = s_{-1,-2} = t_1, \quad s_{-1,1} = s_{1,-1} = q(q-1)t_1, \quad s_{1,2} = s_{2,1} = -q^3 t_1,\\
&l_{-2,-1} = l_{-1,-2} = -\frac{q+1}{2q}(2qt_1 + t_2), \quad l_{-1,1} = l_{1,-1} = \frac{q^2-1}{2q}t_2,\\
&l_{1,2} = l_{2,1} = q^2(q+1)t_1 + \frac{q+1}{2}t_2,\\
&w_{-2,-1} = w_{-1,1} = -\frac{2q^2 t_1 + t_2}{q}, \quad w_{-1,-2} = -w_{1,2} = -t_2,\\
&w_{1,-1} = w_{2,1} = 2q^2 t_1 + qt_2.
\end{aligned} \tag{9}
$$

The result is a 2-parameter class of cascade models for any $q \geq 2$.

Of course, for $P = 2$ it is possible to obtain expressions for these coefficients manually, without making general equations from Table 2, but simply by reducing terms into derivatives of quadratic forms. However, the advantage of the proposed scheme is that it works for any values of the scale nonlocality P, and allows avoiding errors in algebraic calculations.

A well-known property of developed turbulence are power-law spectral laws within the inertial interval. Since collective variables are interpreted as measures of pulsations of the corresponding scales, the manifestation of the spectral laws in shell models will be stationary power-law solutions.

The system of equations for the model coefficients formed and solved in Maple can be supplemented to ensure the existence of stationary power-law solutions. Let us show how this can be done using the example of the model considered above for a three-dimensional MHD flow.

Let in the model (3) $\mathsf{R} = \mathsf{R_m} = \infty$ and $f_n = 0$. We substitute stationary solutions of the form $u_n = u_0 k_n^\alpha = u_0 q^{n\alpha}$ and $B_n = B_0 k_n^\alpha = B_0 q^{n\alpha}$. Since by rescaling the variables of velocity and magnetic field it is possible to achieve that $u_0 = B_0$, we obtain two homogeneous linear equations for the coefficients of the model:

$$
\sum_{i,j} s_{ij} q^{\alpha(i+j)} + \sum_{i,j} l_{ij} q^{\alpha(i+j)} = 0, \quad \sum_{i,j} w_{ij} q^{\alpha(i+j)} = 0. \tag{10}
$$

It is these equations that must be added to the system for the interaction coefficients. In the above Maple code, replace line 62 with the following snippet:

```
62  g[m]:=sum(sum((s['i','j']+l['i','j'])*q^(alpha*('i'+'j')),
                  'i'=-P..P),'j'=-P..P):
64  g[m+1]:=sum(sum((w['i','j'])*q^(alpha*('i'+'j')),'i'=-P..P),
    'j'=-P..P):
66  NumEq:=m+1:
```

We can add several groups of equations of the (10) type with different α, if we want to ensure the existence of several stationary power-law solutions. It is clear that the addition of such equations can reduce, but not necessarily reduce, the number of independent variables in the overall solution for the interaction coefficients. In particular, a situation is possible when their addition will lead to a system with only zero solutions. This suggests that there is no cascade model with such a set of invariants and spectral laws.

The pulsation energies of the velocity and magnetic field in the n-th scale shell are determined, respectively, by the expressions u_n^2 and B_n^2, which for stationary power-law solutions will obviously be proportional to $q^{2\alpha n}$. If, however, the spectral law $E(k) \sim k^\lambda$ holds for the pulsation energy $E(k)$ in a turbulent system, then the energy in the entire n-th scale shell $E_n \sim q^{(\lambda+1)n}$ [3]. Therefore, the power-law solution with exponent $\alpha = (\lambda + 1)/2$ will correspond to the spectral law for energy with exponent λ in the cascade model.

Then, for example, the Kolmogorov spectrum for energy with $\lambda = -5/3$ will correspond to $\alpha = -1/3$. It turned out that adding equations (10) with $\alpha = -1/3$ does not change the eqref solve solution. Therefore, we can say that the formulas eqref solve define a class of cascade models in which there is a stationary solution corresponding to the Kolmogorov spectrum.

Of course, having ensured the existence of stationary power-law solutions, we cannot guarantee their stability. However, it is no longer possible to elucidate stability in a general formulaic form. It is necessary to fix the numerical values of the parameters and carry out computational experiments with a specific model. Moreover, it is well known that, in shell models, power-law solutions are often unstable and the spectrum is determined by averaging over time

Such calculations are no longer related to the subject of this work. However, in principle, the proposed method allows generating models with power-law solutions.

5 Conclusions

To study turbulence using shell models, software tools are very useful, which would allow you to quickly generate classes of models that satisfy one or another set of conservation laws. The choice of a specific model within these classes can then be specified using additional physical considerations, for example, the existence of a given probability distribution for the interaction of certain shells.

The article describes the developed technique for constructing cascade models, in which the compilation of a system of equations for the coefficients of nonlinear interactions in the model and its exact solution is implemented using Maple systems

The technique allows one to vary the dimensions of nonlocality of nonlinear interaction in the space of scales and expressions for cascade analogs of conservation laws and spectral laws. This technique and the programs that implement it solve the problems of fast and error-free generation of entire classes of cascade models.

Acknowledgments

The work was carried out within the framework of the subject research AAAA-A21-121011290003-0 «Physical processes in the system of near space and geospheres under solar and lithospheric influences»
.

References

[1] P. Ditlevsen, *Turbulence and Shell Models* (Cambridge Univ. Press, New-York, 2011).

[2] F. Plunian, R. Stepanov, New Journal of Physics **9**, 294 (2007).

[3] F. Plunian, R.A. Stepanov, P.G. Frick, Physics Reports **523**, 1–60 (2013).

[4] G. Vodinchar, L. Feshchenko, E3S Web of Conf., **127**, 02004 (2019). https://doi.org/10.1051/e3sconf/201912702004

[5] G. Vodinchar, L. Feshchenko, E3S Web of Conf., **196**, 02008 (2020). https://doi.org/10.1051/e3sconf/202019602008
[6] G. Vodinchar, L. Feshchenko, Computational Technologies, **26(5)**, 65-80 (2021). DOI:10.25743/ICT.2021.26.5.006. (In Russ.)

Investigation of acoustic fields in the arctic zone with uneven ice cover

Vladimir Korochentsev[1,*], *Chen* Wenjian[2], *Victor* Petrosyants[1], *Tatiana* Lobova[3], *Julia* Shpak[1].

[1]FEFU, Polytechnical Institute, Department of Electronics, Telecommunications and Instrumentation, Russia
[2]HIU, College of underwater acoustic engineering, China
[3]MSUN named after Admiral G.I. Nevelskoy, Russia

Abstract. A mathematical model for elastic wave propagation in an ice cover with uneven relief (hummock) has been developed. The theoretical model is based on the application of "directed" Green's functions. We obtained numerical results for different distances between radiating and receiving antennas installed inside the ice layer and in water medium. An information-measuring system was created to investigate elastic acoustic waves along ice surface based on electo-hydraulic generator. Experiments of high-frequency acoustic signal propagation from electro-hydraulic generator in water-ice-air system were carried out. We illustrated the model validity for the investigation of hydroacoustic wave propagation in real ice conditions.

1 Mathematical model

We consider the following problem. We are to estimate the field of acoustic wave point sources placed in the ice layer of uneven form (Fig. 1). The following medium characteristics are known: ice, water, air densities and phase velocities of acoustic wave propagation in them; power applied to the radiating antenna; signal frequency; mutual arrangement of sources.

As an initial equation we chose the inhomogeneous Helmholtz equation with Dirac delta-function in the right part [1-3]:

$$\Delta P + k^2 P = \delta(\theta, \theta_0, r, r_0), \quad (1)$$

where k - is the wave number;
$\delta(\theta, \theta_0, r, r_0)$ - is the Dirac delta function;
θ_0, r_0 - are source coordinates;
θ, r - are estimation point coordinates.

*Corresponding author: ga_i_uzt@mail.ru

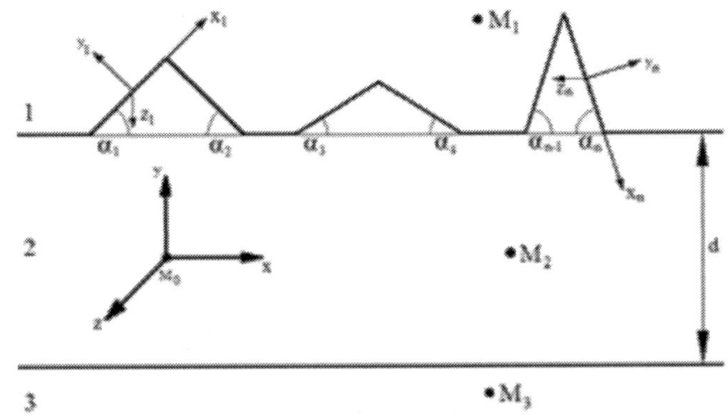

Fig. 1. Geometry of the problem under consideration: M_0 – radiation source; 1 – air half-space, 2 – ice layer, 3 – water half-space; d – ice layer thickness.

We introduce generalized angular coordinates for $\varphi = 0$:

$$U_{1n} = k_0 \cdot \sin(\alpha \pm \alpha_n);$$
$$U_{2n} = k_0 \cdot \cos(\alpha \pm \alpha_n);$$
$$k_0 = 2\pi/\lambda.$$

Point M can be located in the air, ice layer or water. To estimate by "directed" Green's functions, we consider the effect of each plane of ice relief on the resulting field. Radiation point source is located at M_0 point with the coordinates x_0 and y_0. The resulting field in the selected point M with the coordinates x and y is equal to the sum of "directed" Green's functions $P_0^n(x_{0n}, y_{0n})$, estimated from all the planes of ice relief.

"Directed" Green's function of the waves radiated by a point source is written as

$$P_0^n(x) = \frac{i}{2\pi}\int_{U_{0\,min}^n}^{U_{0\,max}^n} \frac{F_0^n(U_{0n})}{\sqrt{k_0^2 - U_{0n}^2}} e^{\{i(x-x_0)\sqrt{k^2 - U_{0n}^2} + [y-y_0]U_{0n}\}} dU_{0n}, \quad (2)$$

where $F_0^n(U_{0n})$ is the diagram function:

$$F_0^n(U_{0n}) = \begin{cases} 1 \text{ if } U_{0\,min}^n \leq U_{0n} < U_{0\,max}^n \\ 0 \text{ in other regions} \end{cases}. \quad (3)$$

"Directed" Green's function of the reflected waves has the from

$$P_l^n(x) = \frac{i \cdot V}{2\pi}\int_{U_{l\,min}^n}^{U_{l\,max}^n} \frac{F_l^n(U_{ln})}{\sqrt{k_0^2 - U_{ln}^2}} e^{\{-i(x-x_0)\sqrt{k^2 - U_{ln}^2} + [y-y_0]U_{0n}\}} dU_{ln}, \quad (4)$$

where V is the interface reflection coefficient determined by the formulas:

$V = \frac{Z_2 - Z_1}{Z_2 + Z_1}$ - for the reflection from upper interface;
$V = \frac{Z_2 - Z_3}{Z_2 + Z_3}$ - for the reflection from lower interface;

Z_1, Z_2, Z_3 – air, ice and water impedans.

"Directed" Greens' function for the waves, which transmitted through the interface, is

$$P_l^n(x)=\frac{i\cdot W}{2\pi}\int_{U_{l\,min}^n}^{U_{l\,max}^n}\frac{F_l^n(U_{ln})}{\sqrt{k_0^2-U_{ln}^2}}e^{\{-i(x-x_0)\sqrt{k^2-U_{ln}^2}+[y-y_0]U_{0n}\}}dU_{ln}\,, \tag{5}$$

where W is the coefficient of transmission though the interface:

$$W = 1 - V.$$

Resulting field amplitude at M is determined by the sum of "directed" Green's functions

$$P_\Sigma(x)=\sum_{n=1}^{N}[P_0^n(x)+P_l^n(x)]. \tag{6}$$

As an example, applying formula (4) we estimate the field from one point source for the frequency of 3 kHz between the radiation point M_0 and M_3, located in water under the ice layer at the distance up to 100 meters.

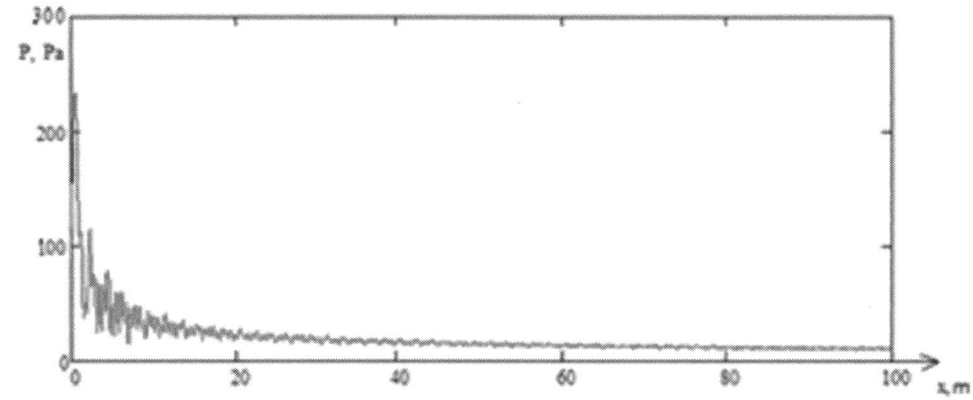

Fig. 2. Distribution of point source field amplitude for the frequency of 3000 Hz.

2 Experimental equipment

The experimental equipment is illustrated in Fig. 3.

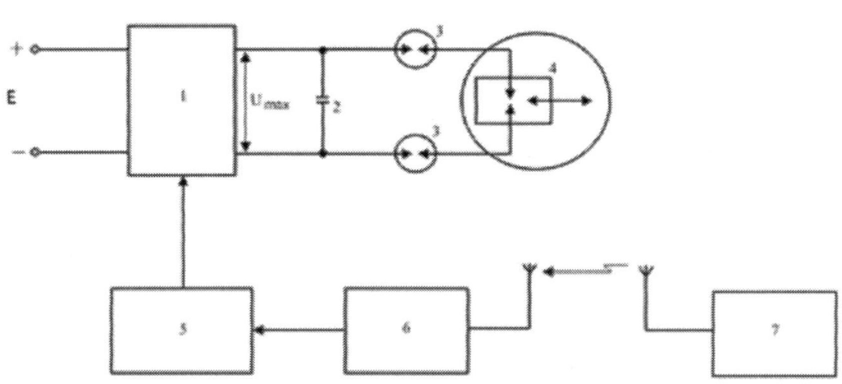

Fig. 3. Block scheme of the setup with shock excitation: 1 – step up voltage converter; 2 – storage capacitor; 3 – high-voltage dischargers; 4 – hydroacoustic radiator; 5 – control system; 6 – Bluetooth-module; 7 – control signal transmitter.

The voltage converter 1 is powered from an accumulator of 24 V. The maximum voltage at the converter output U_{max} may reach 30 kV. The control system 5, based on the microcontroller, allows us to adjust charge amplitude of the storage capacitor 2 and discharge pulse frequency. Two dischargers are applied in the discharge circuit. The second discharger is required to increase the discharge pulse steepness for the shock effect to be maximal. The hydroacoustic radiator with shock excitation 4 consists of a high-voltage discharger and an antenna. Two types of antennas were used as an antenna. They had the forms of parallel planes and of a cylinder.

The setup is controlled remotely by a Bluetooth-module 6. A mobile phone is used as a control signal transmitter 7. The setup control software allows us to adjust the discharge pulse amplitude and repetition frequency.

Physical configuration of the experimental setup is illustrated in Fig. 4. The left part (Fig. 4, a) shows the electrohydrodynamic shock generator without a cover and the right part presents the cylindrical hydroacoustic radiator (Fig.4, b).

a *b*

Fig. 4. Experimental equipment.

The investigations with the experimental setup were carried out in a pool of the depth of 1.5 m with the diameter of 3 m. As a hydroacoustic signal sensor we used a hydrophone with piesoceramic primary transducer having the sensitivity of 140 μV/Pa and operating frequency range up to 20 kHz. To observe the recorded hydroacoustic signals, a digital edge-triggered oscilloscope produced by Keysight, model DSOX 1204A/G, was applied.

The hydroacoustic radiator and the hydrophone were installed at the distance of 2.45 m from each other. The radiator and the hydrophone were arranged so as to have enough time to receive the investigated signal before reflected signal arrival. The depth was 1 m. The radiator was mounted on a rotator.

Fig. 5 shows the oscillogram of the received signal. The signals were received for the following operating mode of the setup: voltage on the storage capacitor was 7 kV, repetition frequency was 2 Hz and the radiator rotation angle was 45 decrease relatively the hydrophone.

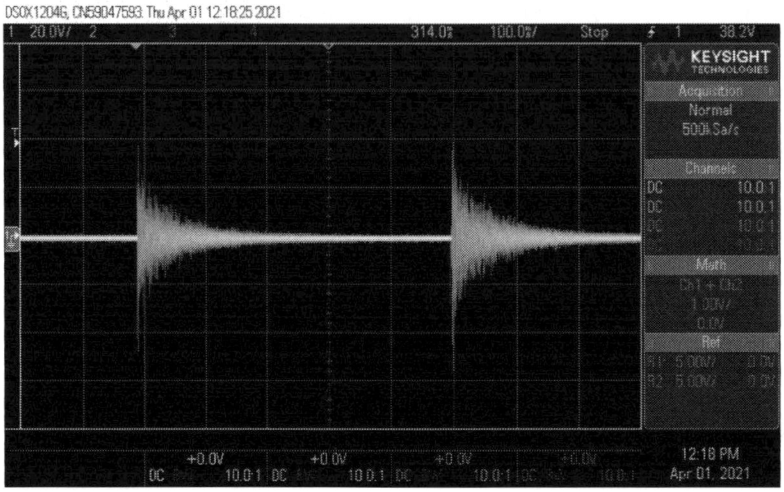

Fig. 5. Received signal oscillogram.

The measurement results for several rotation angles of the radiator are illustrated in Table 1.

Table 1. The measurement results for several rotation angles of the radiator

Radiator rotation angle relatively the receiver, φ	Experimental effective sound pressure, Pexp		Estimated effective sound pressure, Pest	
	kPa	p.u.	kPa	p.u.
0 degrees	252	2,9	234	2,7
45 degrees	177	2,1	175	2,0
90 degrees	85,8	1	86,1	1

3 Experimental investigation of acoustic wave propagation in the presence of ice hummock

An information-measuring system, illustrated in Fig. 6 [4] was used for the experimental investigation of hydroacoustic wave propagation along ice cover with hummocks.

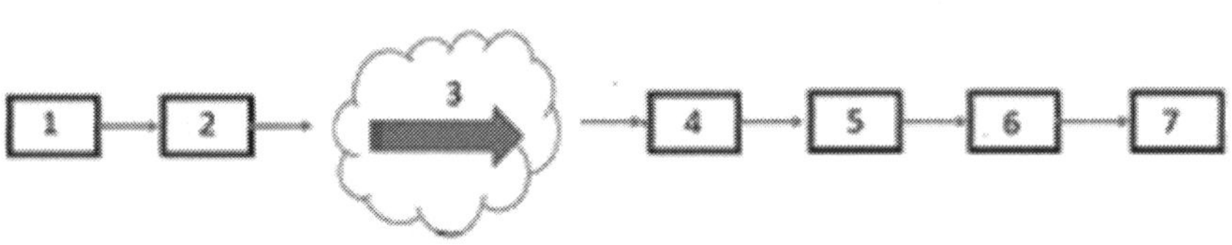

Fig. 6. Functional diagram of the information-measuring system: 1 – hydroacoustic setup with shock excitation (Fig. 4); 2 – hydroacoustic radiator; 3 – hydroacoustic wave propagation channel; 4 – hydrophone; 5 – hydrophone amplifier; 6 – analogue-to-digital converter; 7 – PC.

Fig.7 shows the scheme of the hydroacoustic radiator and the hydrophone installation relatively the ice cover. The hydroacoustic radiator 2 is submerged into a hole, made in the ice, at the required depth. Hydrophone 4 is submerged into another hole at the required distance x_i from the first one.

A hydroacoustic pulse is sent by the hydroacoustic radiator 2, received by the hydrophone 4, amplified by the hydrophone amplifier 5, converted into a digital code in the unit 6 and sent to a PC 7 for processing and information display.

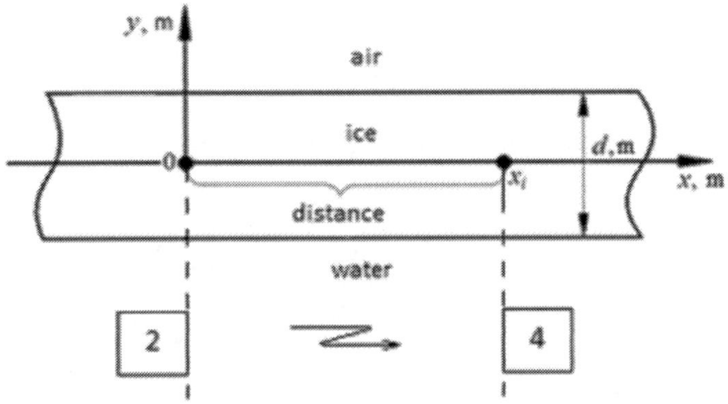

Fig.7. Scheme of radiator and receiver installation.

Fig. 8 shows the results for the experimental investigation of the dependence of sound pressure (Pa), generated by cylindrical antenna in the direction of radiation axis direction, on the distance (m). Five independent measurement series are presented.

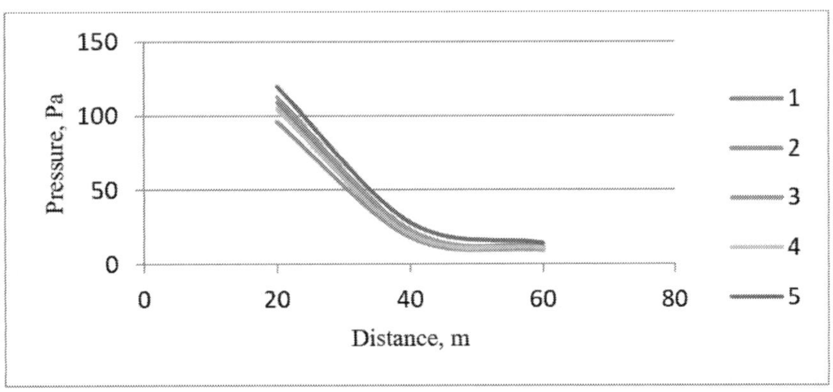

Fig.8. Curves of pressure dependence on distance.

4 Conclusions

The proposed mathematical model of elastic wave propagation may be applied for modeling of hydroacoustic wave propagation from earthquake sources under real ice conditions.

The experimental investigations confirmed high efficiency of the generator with electro-hydraulic shock to form hydroacoustic high-power signals in the systems for signal transmission for extra-long distances.

References

1. V.I. Korochentsev, Wave problems of the theory of directed and focus antennas: monography, (Dal'nauka, Vladivostok, 1998).
2. A.V. Korchaka, A.A. Em, V.I. Korochentsev, The Far Eastern Federal University: School of Engineering Bulletin, **38**, 1 (2019) (in Russian).
3. V. Korochentsev, A. Em, Y. Shpak, W. Xue, G. Shabanov, E3S Web Of Conf., **127** (2019).
4. W. Chen, V. Chernenko, V. Petrosyants, V. Grishchenko, E3S Web Conf. **127** (2019).

Investigation of electromagnetic fields in the arctic zone with uneven ice cover

Vladimir Korochentsev[1,*], *Vey* Syue[2], *Sergey* Gorovoy[1], *Vasily* Chernenko[1], *Artem* Em[1], *Timur* Tagaev[1].

[1]FEFU, Engineering school, Department of instrumentation, Russia
[2]HIU, College of underwater acoustic engineering, China

Abstract. Numerical investigation of electromagnetic waves propagating near an ice cover with hummocks of different height has been carried out. Half-wave vibrators, melted into ice at the depth of about 20 sm, were used as an antenna. Transmitted signal frequencies were from 10 to 30 MHz. A mathematical model for amplitude spatial distribution of electromagnetic wave source, placed inside an ice cover with hummocks, was developed. The results of numerical study show that, when antenna is inside hummocks, the signal is amplified. Experimental investigations show good agreement with the theoretical model.

1 Mathematical model

We consider the following problem. We are to estimate the field of electromagnetic wave point sources placed in the ice layer of uneven form (Fig. 1) [1-3]. The following medium characteristics are known: magnetic and dielectric permeability of mediums; power applied to the radiating antenna; signal frequency.

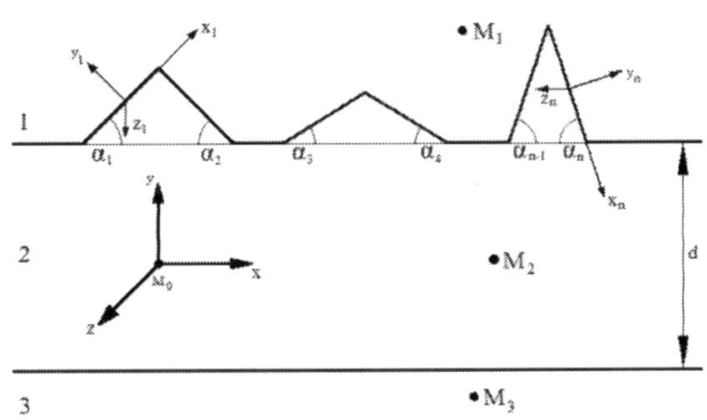

Fig. 1. Geometry of the problem under consideration:
M_0 – radiation source; 1 – air half-space, 2 – ice layer, 3 – water half-space; d – ice layer thickness

*Corresponding author: ga_i_uzt@mail.ru

We introduce generalized angular coordinates for $\varphi = 0$:

$$U_{1n} = k_0 \cdot sin(\alpha \pm \alpha_n);$$
$$U_{2n} = k_0 \cdot cos(\alpha \pm \alpha_n);$$
$$k_0 = 2\pi/\lambda.$$

Point M can be located in the air, ice layer or water. To estimate by "directed" Green's functions, we consider the effect of each plane of ice relief on the resulting field. Radiation point source is located at M_0 point with the coordinates x_0 and y_0. The resulting field at the selected point M with the coordinates x and y is equal to the sum of "directed" Green's functions $E_0^{(n)}(x_{0n},y_{0n})$, estimated from all the planes of ice relief.

"Directed" Green's function of the waves radiated by a point source is written as

$$E_0^n(x)=\frac{i}{2\pi}\int_{U_{0\,min}^{(n)}}^{U_{0\,max}^{(n)}}\frac{F_0^{(n)}(U_0)_n}{\sqrt{k_0^2-U_{0n}^2}}e^{\{i(x-x_0)\sqrt{k^2-U_{0n}^2}+[y-y_0]U_{0n}\}}dU_{0n}\ ,$$

where $F_0^n(U_{0n})$ is the diagram function:

$$F_0^n(U_{0n}) = \begin{cases} 1 \text{ if } U_{0\,min}^n \le U_{0n} < U_{0\,max}^n \\ 0 \text{ in other regions} \end{cases} . \tag{1}$$

"Directed" Green's function of the reflected waves has the from

$$E_l^n(x)=\frac{i\cdot V}{2\pi}\int_{U_{l\,min}^{(n)}}^{U_{l\,max}^{(n)}}\frac{F_l^{(n)}(U_l)_n}{\sqrt{k_0^2-U_{ln}^2}}e^{\{-i(X-X_0)\sqrt{k^2-U_{ln}^2}+[Y-Y_0]U_{0n}\}}dU_{ln}, \tag{2}$$

where V is the interface reflection coefficient determined by the formulas:

$$V = \frac{Z_2-Z_1}{Z_2+Z_1} \text{ for the reflection from upper interface;}$$

$$V = \frac{Z_2-Z_3}{Z_2+Z_3} \text{ for the reflection from lower interface;}$$

Z_1, Z_2, Z_3 – air, ice and water impedans.

"Directed" Greens' function for the waves, which transmitted through the interface, is

$$E_l^n(x)=\frac{i\cdot W}{2\pi}\int_{U_{l\,min}^{(n)}}^{U_{l\,max}^{(n)}}\frac{F_l^{(n)}(U_l)_n}{\sqrt{k_0^2-U_{ln}^2}}e^{\{-i(X-X_0)\sqrt{k^2-U_{ln}^2}+[Y-Y_0]U_{0n}\}}dU_{ln}, \tag{3}$$

where W is the coefficient of transmission though the interface:

$$W = 1 - V.$$

Resulting field amplitude at M is determined by the sum of "directed" Green's functions

$$E_\Sigma(x)=\sum_{n=1}^{N}[E_0^n(x)+E_l^n(x)]\ . \tag{4}$$

As an example, applying formula (4) we estimate the field from one point source for the several frequencies.

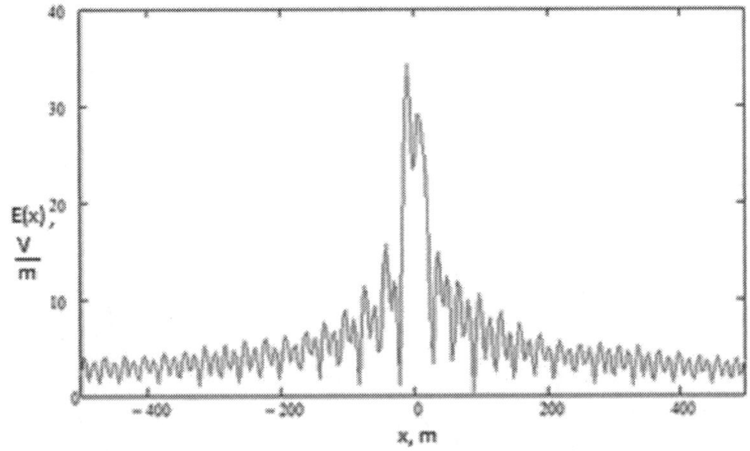

Fig. 2. Amplitude distribution of point source field for the frequency of 10 MHz

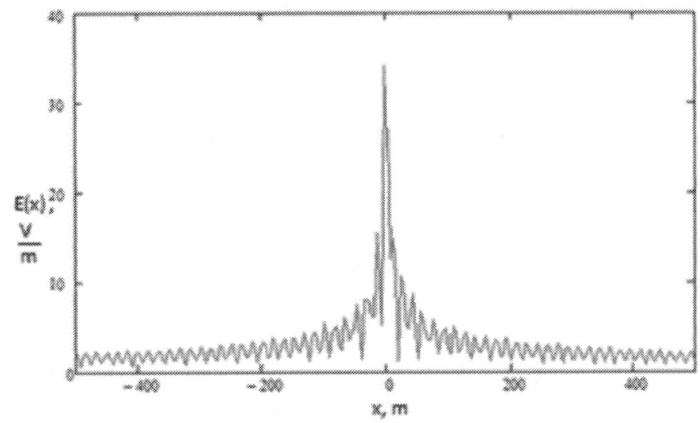

Fig. 3. Amplitude distribution of point source field for the frequency of 15 MHz

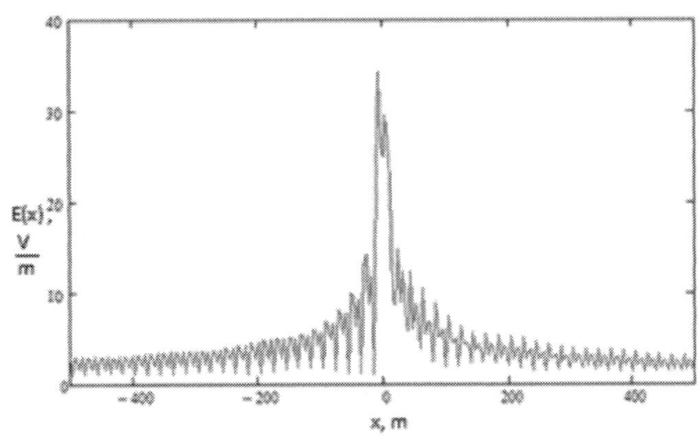

Fig. 4. Amplitude distribution of point source field for the frequency of 28.5 MHz

The calculation time on a medium-powered computer is not more than 2 minutes.

2 Aims and tasks of the experimental investigations

The aim of the experiment is to measure amplitudes of the signals with the frequencies of 10 MHz, 15 MHz, 28,5 MHz in the ice layer.

In the course of the experiment, the antennas were arranged as follows (Fig. 5)

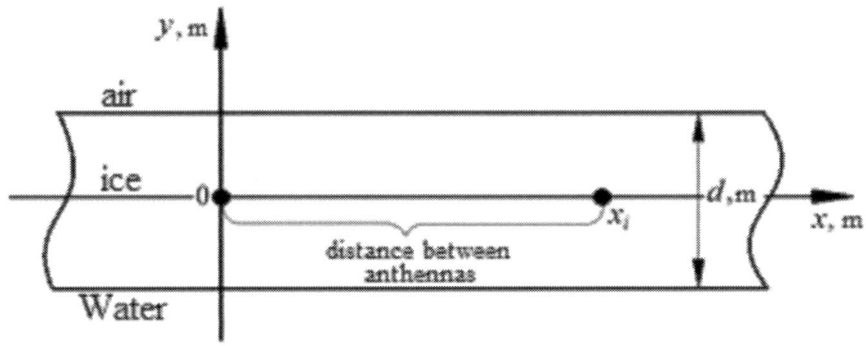

Fig. 5. Antennas arrangement

In Fig. 5, x_i is the receiving antenna coordinate, radiating antenna is installed at point 0 (origin of coordinates), d is the ice layer thickness.

The measuring bench (Fig. 6) consists of a radiating (blocks 1–3) and receiving (blocks 4–6) paths.

The measuring bench is operating as follows:

1. The radiostation (block 1) forms an electric signal with required parameters.

2. A signal from the radiostation (block 1) goes to a tuner (block 2), the device for matching the radiostation and the antennas. From the tuner (block 2) output, the signal is sent to the radiating antenna (block 3).

3. The signal is received by the antenna (block 4), then it goes to the receiving radiostation (block 6) through the tuner (block 5). The power applied to the radiating antenna is 30-35 W.

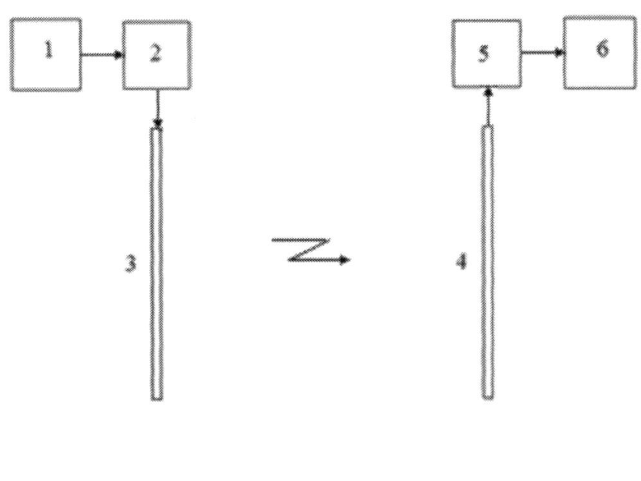

Fig. 6. Installation diagram

Wave propagation is affected by medium boundaries, ice layer thickness and distribution of solid and fluid phases in it.

The investigations were carried out in Novik bay (Fig. 7) located on the territory of Russkiy Island.

Fig.7. Laying of antenna into the ice

3 Results of experimental investigations

It is clear from Fig. 2 – Fig. 4 and Fig. 6 – Fig. 8 that the dependences of signal amplitudes obtained in the result of numerical and experimental investigations coincide.

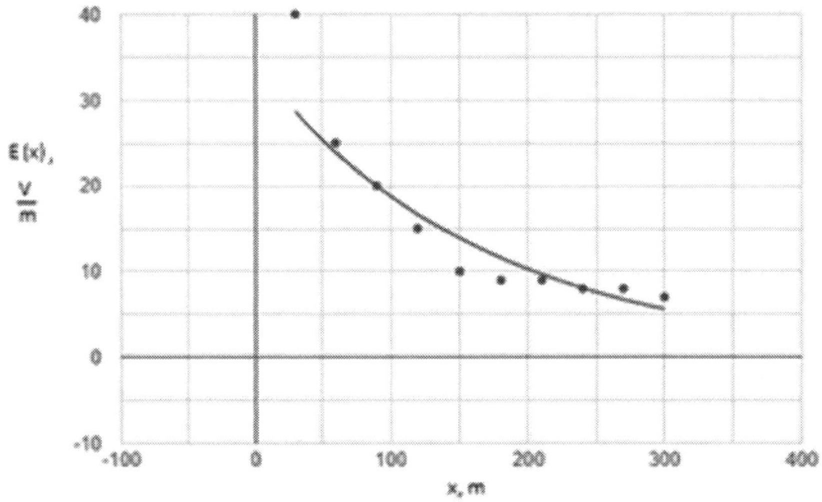

Fig. 8. Dependence of signal amplitude on the distance between the antennas. Signal frequency is 10 MHz

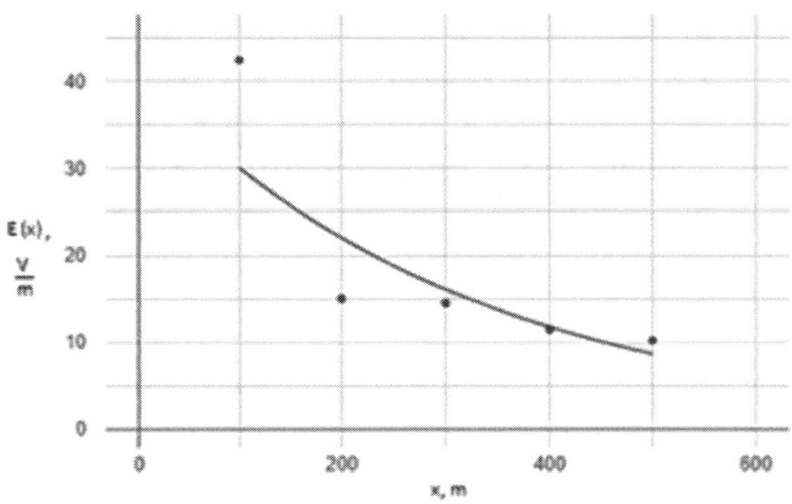

Fig. 9. Dependence of signal amplitude on the distance between the antennas. Signal frequency is 15 MHz

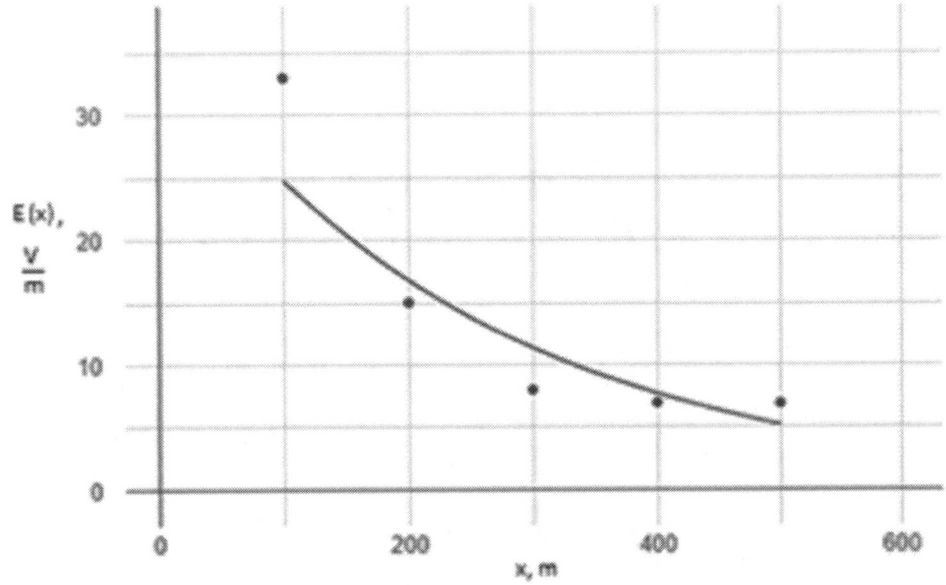

Fig. 10. Dependence of signal amplitude on the distance between the antennas. Signal frequency is 28,5 MHz

4 Conclusions

We have developed a valid mathematical model to investigate electromagnetic wave fields in the air and in the ice layer. The experimental investigation of electromagnetic wave propagation in the ice layer with hummocks show good agreement with theoretical calculations.

References

1. V.I. Korochentsev, Wave problems of the theory of directed and focus antennas (Dal'nauka, Vladivostok, 1998) (in Russian).
2. A.V. Korchaka, A.A. Em, V.I. Korochentsev, The Far Eastern Federal University: School of Engineering Bulletin, **38**, 1 (2019) (in Russian).
3. V. Korochentsev, A. Em, Y. Shpak, W. Xue, G. Shabanov, E3S Web Of Conf., **127** (2019).

Monitoring of Ionosphere and Magnetosphere Plasma and High Energy Charge Particle Fluxes in Multi-satellite Measurements in Wide Range of Altitudes

Vladislav Osedlo[1], *Oleg* Abanin[1], *Natalia* Vlasova[1], *Vladimir* Kalegaev[1,2], *Nikolay* Pavlov[1], *Vasilii* Petrov[1], *Mikhail* Podzolko[1], *Ilia* Rubinstein[1], *Sergey* Svertilov[1,2*], *Vladimir* Tulupov[1] and *Andrey* Shemukhin[1]

[1] M.V. Lomonosov Moscow State University, D.V. Skobeltsyn Institute of Nuclear Physics, 119991, Leninskie Gory 1(2), GSP-1, Moscow, Russia
[2] M.V. Lomonosov Moscow State University, Physics Department, 119991, Leninskie Gory, Moscow, Russia

Abstract. A system for monitoring the radiation parameters of near-Earth space is described. This system is based on the multi-satellite measurements made on spacecraft Meteor, Electro, Arktika launched into orbits with a wide range of altitudes. The main instrument for space radiation monitoring is spectrometer of electrons and protons SKIF. Such instruments operate in all spacecraft of mentioned above series. The results of observations of different events connected with solar and geomagnetic activity in 2017 and 2021 years are presented and discussed.

1 Introduction

Near-Earth space (NES) contains charged particles of ionospheric and magnetospheric plasma, energetic particles of the Earth's radiation belts (ERB), solar and galactic cosmic rays (SCR and GCR) in a wide range of energies [1]. They can lead to damage and failure of electronic equipment installed on spacecraft (SC), can cause disruption of short-wave communications in high-latitude regions, lead to failures in navigation systems and reduce the accuracy of global navigation and positioning systems [2].

Within the framework of multi-satellite measurements on spacecraft launched into orbits with a wide range of altitudes (Meteor - 830 km, Arktika - an elliptical orbit with apogee and perigee heights of ~39 thousands km and ~600 km, Electro - geostationary in the plane equator - 36 thousands km), for several years at the Skobel'tsyn Institute of Nuclear Physics of Moscow State University (SINP MSU), monitor observations of ionospheric and magnetospheric plasma and energetic electron and proton fluxes have been carried out and continue (spacecraft Arktika-M1). The picture becomes more complete with the involvement of information from identical measuring instruments. Such an instrument, developed and manufactured at the SINP MSU, is a space radiation spectrometer SKIF for

* Corresponding author: sis@coronas.ru

measuring fluxes, energy spectra, temporal variations and angular distribution of ERB protons and electrons, SCR particles from solar flares under their combined effect.

Obtaining a large volume of regular monitoring data on the radiation situation in the NES on the basis of multi-tiered measurements allows studying its short- and medium-term dynamics, its relationship with solar activity, is a necessary condition for refining empirical and for creating prognostic dynamic models of space weather, for formation a three-dimensional real time dynamic picture of radiation distribution in the ionosphere and the inner magnetosphere.

In this article, a description of the SKIF equipment will be given and examples of the use of complex multi-satellite measurements for the analysis and diagnostics of the radiation state of the NCP will be given.

2 Observations

2.1 SKIF instrument

The SKIF instrument, developed and created at SINP MSU is intended for monitoring the radiation environment in NES and provides detection of charged particle fluxes (electrons and protons) in a wide energy range from 0.05 keV to more than 160 MeV. The SKIF instrument consists of telescopes based on several semiconductor detectors and one scintillation, gas-discharge counters and spectrometers of low-energy electrons and protons based on electrostatic analyzers. The spectrometer of energetic particle, so-called SEP unit of the SKIF instrument provides measurements of the spectra of protons and electrons in the energy ranges of 2 - 160 MeV, 0.15 - 10 MeV, respectively. The SEP unit logic forms several energy intervals for particles of each type, i.e. 6 intervals for electrons and 12 for protons. Measurements of the spectra of low-energy electrons and protons in the energy range 0.05 - 20 keV are performed by electrostatic analyzers (ESA) of electrons and protons (units SGMTD-E and SGMTD-P). The entire recorded energy range 0.05 - 20 keV ESA of both electrons and protons is divided into 12 intervals. The SKIF instrument includes two identical units duplicating each other's work, but installed in mutually normal directions to study the pitch-angle distribution of particles. The SKIF instrument also includes two identical gas-discharge detector units, in which the SBM20-1 detectors are used. Its readings are used as reference values of fluxes in the upper energy ranges. The detectors are located behind the 2-mm protection of the instrument front wall, normally to the axes of view of the remaining detecting units.

At present, the SKIF instrument operates on three types of Russian SC, i.e. Meteor, Arktika and Electro series. Meteor SC is a polar sun-synchronous satellite with a circular orbit at an altitude of ~ 830 km with an orbital period of ~ 110 min. SC Arktika has an elliptical orbit with apogee and perigee heights of ~ 39 thousand km and ~ 600 km, respectively, and an orbital period of ~ 8 hours. SC Electro is a geostationary satellite located in the plane of the geographic equator at an altitude of ~ 36 000 km. This constellation of satellites makes it possible to effectively monitor radiation conditions in near-Earth space.

2.2 Experimental data

The Center for Operational Space Monitoring Data (COSMD) [http://smdc.sinp.msu.ru] has been established at the SINP MSU. The COSMD computer complex carries out storage and processing of experimental data, monitoring and forecasting of space weather parameters

[4]. The figures below were constructed and can be reproduced on the pages of the COSMD website [http://swx.sinp.msu.ru/tools/ida.php?gcm=1].

For a comparative analysis of the proton and electron flux dynamics according to the data from the SKIF instrument, we also used the experimental data on the electron fluxes from the GOES satellite [http://goes.gsfc.nasa.gov/], on the solar wind (SW) velocity and the interplanetary magnetic field (IMF) with the ACE spacecraft [http://www.srl.caltech.edu/ACE/]. The GOES satellite is a geostationary satellite the ACE spacecraft is located at the libration point L1 at a distance of ~ 1.5 million km from the Earth. Geomagnetic index data were obtained from World Data Center C2 for Geomagnetism, Kyoto [http://wdc.kugi.kyoto-u.ac.jp/].

3 Results of observations

3.1 Geomagnetic and radiation conditions in NES during May 19 – 28, 2021

During the period May 19-28, 2021, significant variations of the NES radiation associated with solar activity were detected. Experimental data from three spacecraft, i.e. Meteor, Electro and Arktika located at different orbits, as well as data on the solar wind and geomagnetic activity parameters, allow reliable diagnostics of the state of not only the Earth's magnetosphere, but also of the interplanetary medium.

The geomagnetic environment in OKP on May 19-28, 2021 can be characterized as moderately disturbed. Two magnetic storms were observed with a minimum D_{st} variation of -34 nT at 20.05 and -28 nT at 27.05 (Fig. 1). The storms were caused by the arrival of interplanetary structures to the Earth, characterized by an increase in the SW velocity and the southern component of the IMF. As a result of a weak storm on 20.05, an increase in the flux of relativistic electrons occurred, which can be seen from geostationary satellites (Electro-L2 and GOES-16) data, see Fig. 1, as well as in polar orbit (SC Meteor-M2), see Fig. 2. The second storm on May 27 led to a drop in the flux of relativistic electrons. Dynamics of the relativistic electron fluxes at the maximum of the outer radiation belt according to data from a polar orbit is shown in Fig. 3.

For radiation monitoring, together with the ERB electron component, the proton component is also an important factor. The radiation belt structure when the Arktika-M1 satellite passes through it with the indication of the L-parameter values can be seen in Fig. 4. Differences in the flux magnitudes are determined by the particle flux pitch-angle distribution.

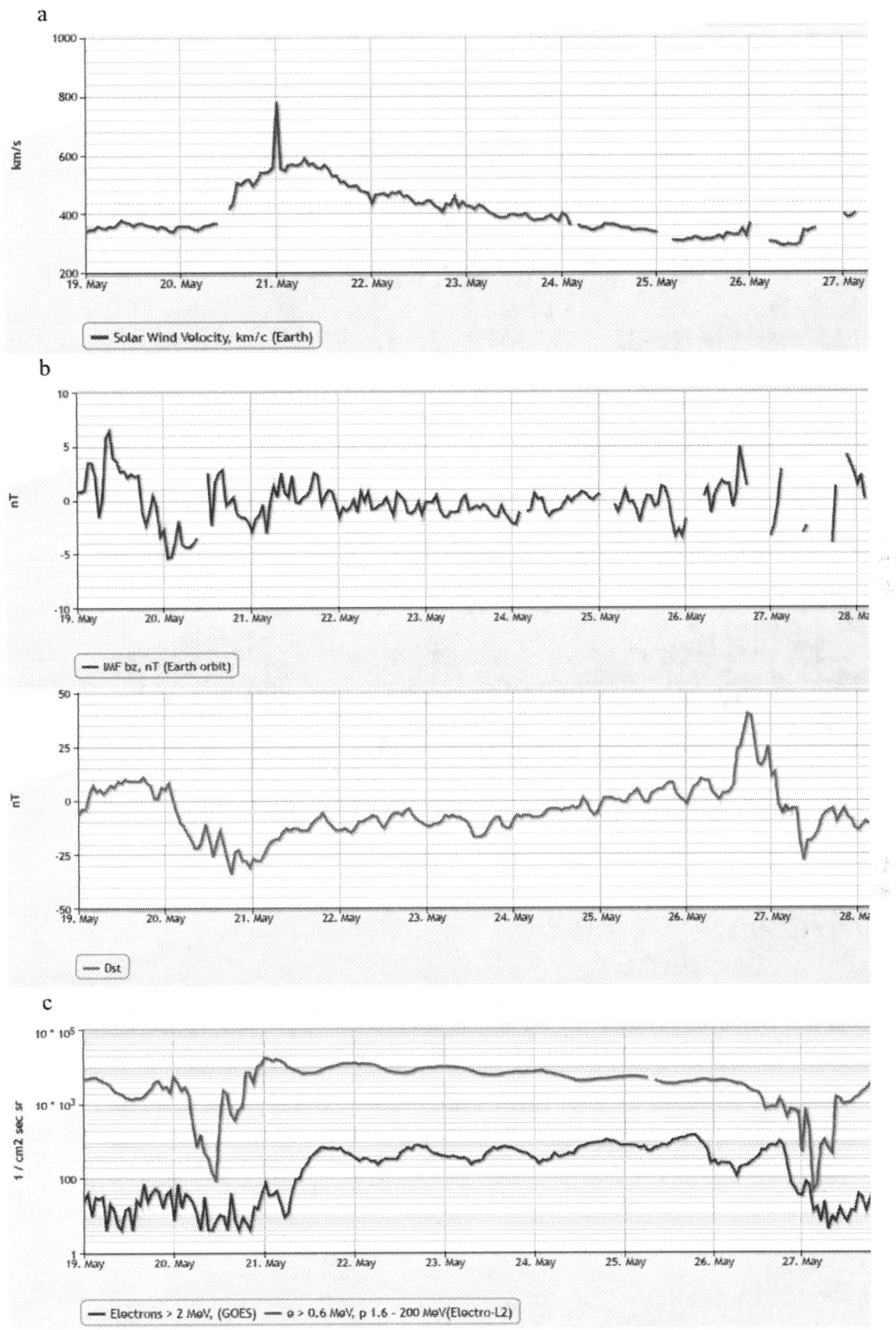

Fig. 1. Time profiles 19-28.05.2021 of SW velocity (a), interplanetary magnetic field (IMF) B_z-component (b, upper panel), D_{st}-variation (b, low panel), ERB electron fluxes at geostationary orbit (c).

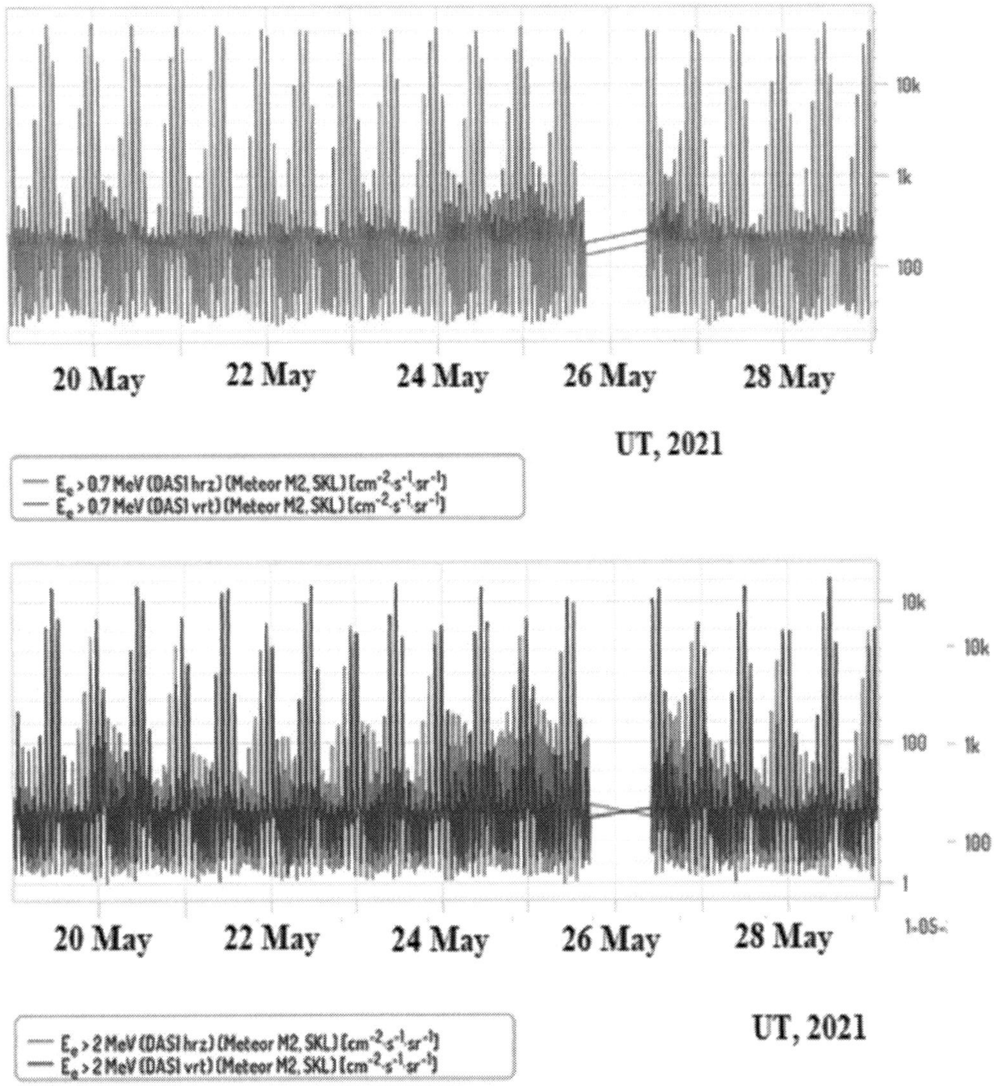

Fig. 2. Time profiles of the ERB electron fluxes in the ranges >0.7 MeV (upper panel) and >2 MeV (bottom panel) obtained from polar orbit 19-28.05.2021.

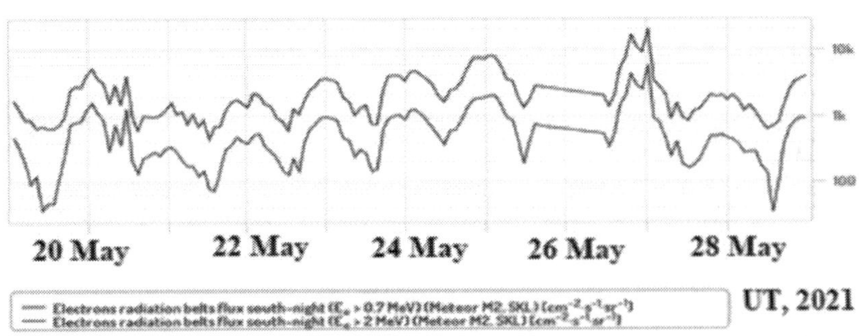

Fig. 3. Time profiles of electron fluxes at ERB maximum obtained from polar orbit data 19-28.05.2021.

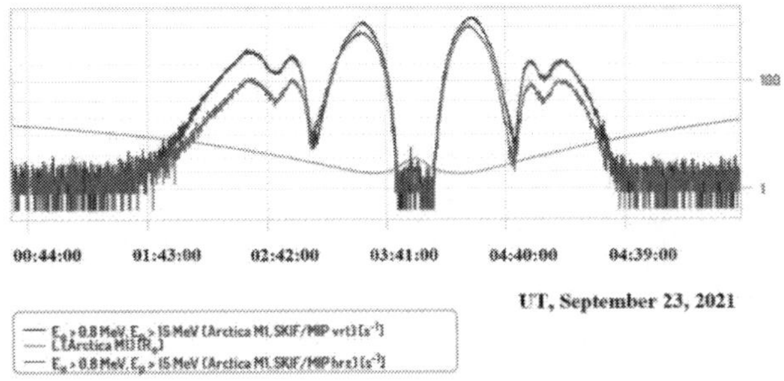

Fig. 4. Space-time profile of the ERB (E_e > 0.8 MeV, E_p > 15MeV on the Арктика-M1 SC orbit 23.05.2021 in 00-06 UT.

3.2 Solar proton events during September 04 – 10, 2017

One of the tasks of the experiment with SKIF instrument is the detection of energetic charged particles during solar proton events (SPE). On September 04-10, 2017, three series of powerful flares occurred on the Sun. In the first series on September 4, at 20:30 UT, a flare of the maximum M5.5 class was detected, in the second, on September 6, at 12 UT, an X9.3 class flare, and in the third, on September 10, an X10 class flare. The latest flare is the most powerful X-ray flare that has occurred on the Sun in the past 12 years.

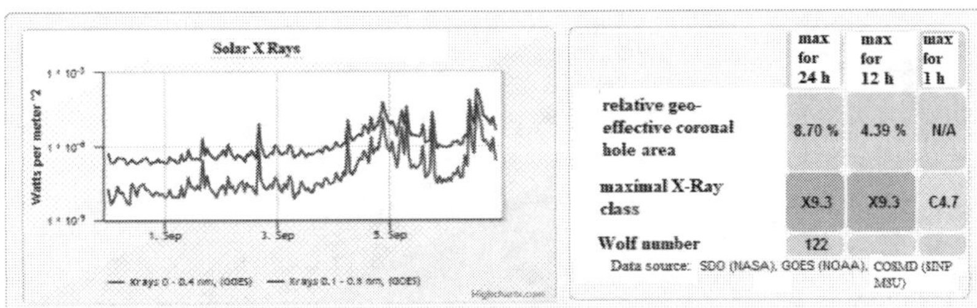

Fig. 5. Presentation of solar activity data on COSMD website for September 06, 2017.

The website of the Center for Operative Space Monitoring Data (COSMD) provides daily data on solar activity observations [http://swx.sinp.msu.ru/apps/geospace_now.php?gcm=1]. The power of X-rays from the Sun is shown according to the data of the NOAA GOES SC data, and the area of coronal holes is presented according to the SDO SC data. The data on the X-ray emission of the Sun in September 2017 were obtained from the GOES-15 spacecraft (see Fig. 5).

The dynamics of the solar X-ray power and the SW velocity, as well as the solar proton fluxes according to the geostationary satellite Electro - L2 and GOES - 15 data is shown in Fig. 6. The result of the first series of flares was a solar protons flux increase on September 04-08, 10-16, 2017, registered in the NASA SPE catalog [https://umbra.nascom.nasa.gov/SEP/seps.html] as one event, while in the catalog, presented on the COSMD website [http://swx.sinp.msu.ru/apps/sep_events_cat/index.php?gcm=1&lang=ru], this period is considered as several separate SPE, as it is associated with various solar phenomena [5]. It can also be seen that the solar proton flux propagated in the disturbed interplanetary medium (Fig. 6b). The complex time profile of particle fluxes is associated with these factors (Figs. 6c and 6d). The increase in solar proton fluxes on September 10 is one of the strongest SPEs of the 24 solar activity cycle, the second GLE and the last event of this cycle.

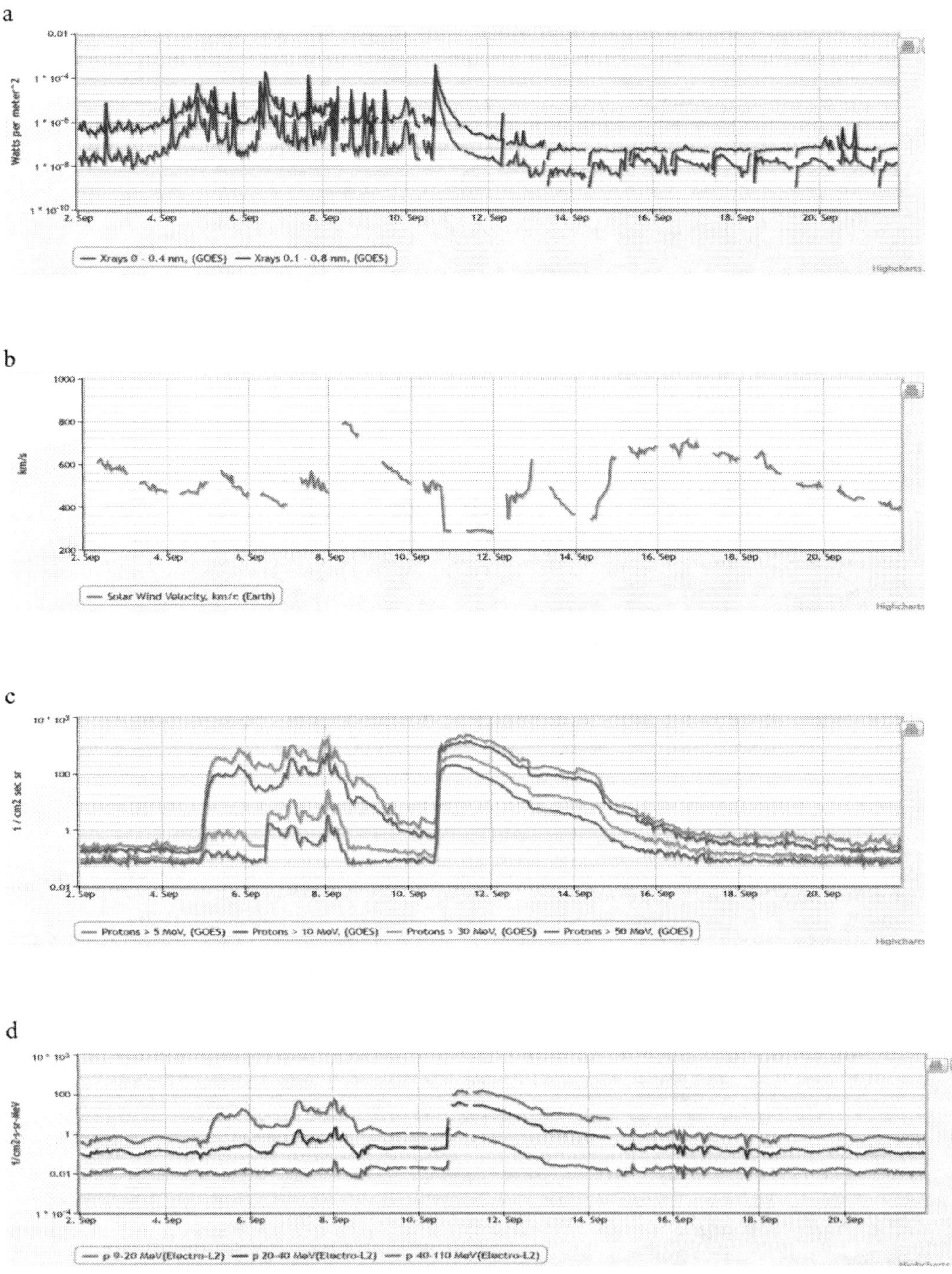

Fig. 6. Time profiles of solar X-ray power (a), SW velocity (b) and solar proton fluxes on the geo-stationary orbit (c, d) 02.09 - 20.09.2017.

3.3 Solar proton events during May 28, 2021

The solar proton event on May 28, 2021 is one of the first SPEs of the 25th solar activity cycle. The event is weak, despite this, the flux of solar protons was detected on all spacecraft on which the SKIF instrument was installed.

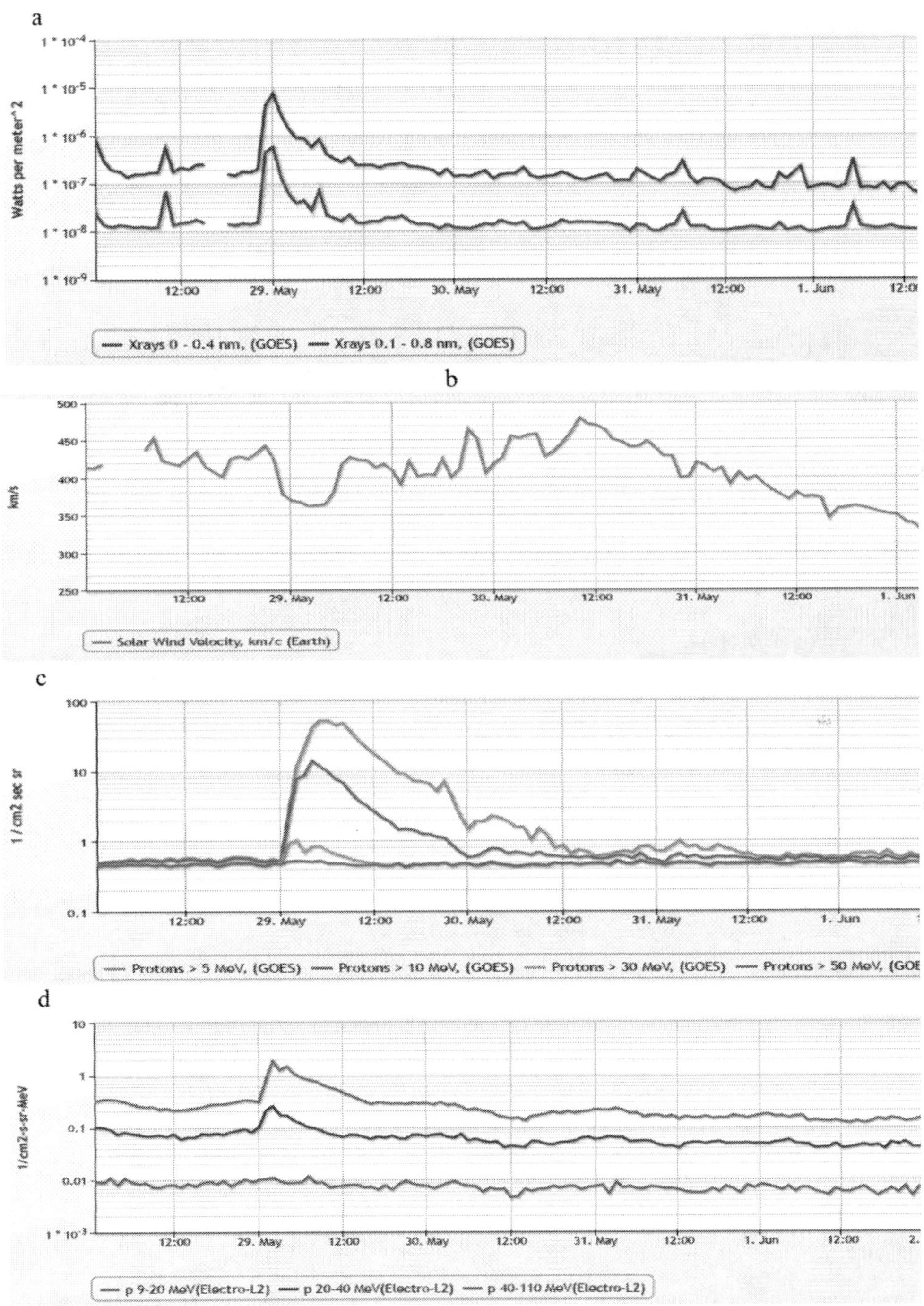

Fig. 7. Time profiles of solar X-rays (a), SW velocity (b) and solar proton fluxes at geostationary orbit (c, d) during 28.05 - 02.06 2021.

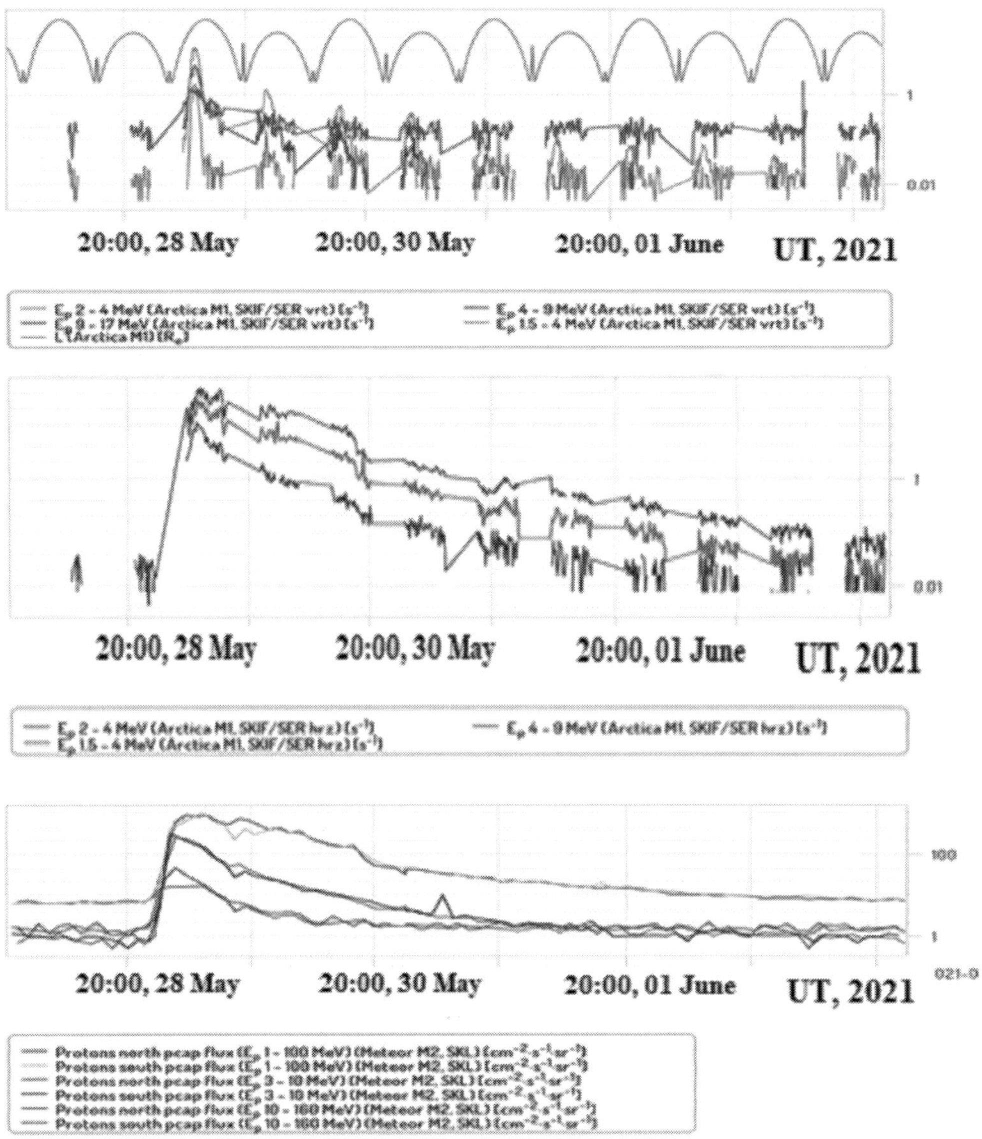

Fig. 8. Time profiles of solar proton fluxes during 28.05 - 02.06 2021.

An increase in solar proton fluxes on May 28, 2021 is associated with an X-ray class C9.5 flare on the Sun (see Fig. 7) at 22:19 UT with helio coordinates N18W67 in the active region (AR) 12824. The flare occurred at the western longitude of the solar disk, and therefore the time profiles of solar proton fluxes have a rather steep leading edge (Figs. 7c and 7d and Fig. 8). Solar protons propagated in the interplanetary medium inside a high-speed solar wind flow (Fig. 7b); therefore, the time profile of the fluxes is rather smooth. The SPE is weak, i.e. the maximum flux of protons with energies > 10 MeV was (according to the GOES satellite data, Fig. 7c) was ~ 13.9 pfu, where 1 pfu = 1 particle/($cm^2 \cdot s \cdot sr$). The time profiles of solar protons detected inside the magnetosphere in the region of closed geomagnetic field lines (data from the Electro and GOES-16 satellites) is shown in Figs. 7c and 7d. The time profiles according to the Meteor - M2 satellite data in the Earth polar cap areas and the Arktika - M1 satellite data on the magnetic field force open lines and partially outside the magnetosphere are shown in Fig. 8. The data from polar low-orbit SC make it possible to substantially supplement the data about SCR events received from geostationary satellites. The proton flux profiles according to the Meteor - M2 satellite data obtained separately over the Earth northern and southern polar caps are shown in bottom panel of Fig 8. The asymmetry of the particle arrive is practically not detected in this event.

4 Conclusion

The first results of energetic charge particle flux variations in NES from multi-satellite observations at different altitudes are presented. Such observations give a relatively complete picture of electron and proton flux dynamics at different areas, including high-latitude crossings of ERBs at altitudes about 800 km (Meteor SC), the edge of outer electron belt at geo-stationary orbit (36 000 km, Electro-L SC) and polar cap at 600 km altitude as well as the exit to the interplanetary space (39 000 km, Arktika – M1 SC). The results of complex measurements are presented for different kinds of NES radiation environment disturbances, such as geomagnetic storms caused by solar activity (events 19-28.05.2021), and solar proton events 04-10.09.2017, 28.05.2021. These observations indicate on that all kinds of geo and helio- activity lead to significant variations of energetic charge particle flux variations and should be monitored for radiation condition forecasting.

References

1. M.I. Panasyuk, et al., Model cosmosa. T.1. Physicheskie uslovija v kosmicheskom prostranstve (Russ.) (KDU, Moskva, 2007)
2. O.P. Verkhoglyadova, C.D. Bussy-Virat, A. Caspi, et al., Space Weather, **18** (2020)
3. M.I. Panasyuk M.I., M.V. Podzolko, V.V. Kalegaev, et. al., Moscow University Physics Bulletin **73**, 687 (2018)
4. V. Kalegaev, M. Panasyuk, I. Myagkova, et al., J. Space Weather and Space Climate, **9** (2019)
5. G.A. Bazilevskaja, E.I. Daybog, Yu.I. Logachev, et al., Geomagnetizm i aeronomia (Russ.) **61**, 8 (2021).

Monitoring of Physical Processes in Upper Atmosphere, Ionosphere and Magnetosphere in Ionosphere Space Missions

Anatoly Petrukovich[1], *Mikhail* Mogilevskii[1], *Igor* Kozlov[1], *Sergey* Pulinetz[1], *Yurii* Dobrolenskii[1], *Konstantin* Anufreichik[1], *Dmitrii* Chugunin[1], *Vladimir* Nazarov[1], *Aleksandr* Abbakumov[1], *Vladimir* Kuznetsov[2], *Stanislav* Andrievskii[2], *Yurii* Mikhailov[2], *Mikhail* Panasyuk[3,4], *Sergey* Svertilov[3,4*], *Ivan* Yashin[3], *Vitaly* Bogomolov[3,4], *Anatoly* Iyudin[3], *Vladimir* Kalegaev[3,4], *Vladislav* Osedlo[3], *Vasilii* Petrov[3], *Andrey* Shemukhin[3], *Aleksey* Repin[5], *Nadezhda* Kotanaeva[5], *Pavel* Moiseev[6], *Sergey* Poljakov[7], *Yurii* Shlugaev[7], *Vasilii* Asmus[8], *Sergey* Uspenskii[8], *Leonid* Makridenko[9], *Aleksandr* Gorbunov[9] and *Vladimir* Kozhevnikov[9]

[1] Institute of Space Research, Russian Academy of Science, 117997, Profsoyuznaya st. 84/32., Moscow, Russia
[2] N.V. Pushkov Institute of Terrestrial Magnetism, Ionosphere and Radio-wave Propagation, Russian Academy of Science, 108840, Kaluzhskii road 4, Troitsk, Moscow, Russia
[3] M.V. Lomonosov Moscow State University, D.V. Skobeltsyn Institute of Nuclear Physics, 119991, Leninskie Gory 1(2), GSP-1, Moscow, Russia
[4] M.V. Lomonosov Moscow State University, Physics Department, 119991, Leninskie Gory, Moscow, Russia
[5] E.K. Fedorov Institute of Applied Geophysics, 129128, Rostokinskaja st. 9, Moscow, Russia
[6] Astron Electronics, 302019, Veselaya st. 2, Orel, Russia
[7] Nizhnii Novgorod State University, Radio-physical Institute, 603950, Bolshaya Pecherskaja st. 25, Nizhnii Novgorod, Russia
[8] Scientific and Research Center of Space Hydro-meteorology Planeta, 123242, Bolshoy Predtechenskii alley 7, Moscow, Russia
[9] VNIIEM Corporation, 107078, Horomny tupik 4, 1, Moscow, Russia

Abstract. The Ionosphere missions are the part of Ionosond-2025 space project, which main scientific objectives are monitoring of physical processes in the Earth upper atmosphere, ionosphere and magnetosphere, as well as of solar activity. Within the framework of the Ionozond-2025 project, it is planned to launch four spacecraft Ionosphere and one satellite Zond. The Zond satellite is planned to be launched in 2025. The main task of Zond mission is patrol of solar activity. The launch of the first pair of spacecraft is planned in early 2022, the second pair - late 2022 or early 2023. In case of successful implementation of the program of experiments on the Ionosphere satellites, control of the physical parameters of electromagnetic fields and corpuscular radiation in the near-Earth space will be provided, new information will be obtained on the geophysical processes occurring in the magnetosphere, ionosphere and upper atmosphere in their connection with solar activity.

* Corresponding author: sis@coronas.ru

1 Introduction

As it is well-known, the ionosphere is the upper layer of the atmosphere, located above about 85 km, the matter of which is strongly ionized mainly by ultraviolet and X-ray radiation from the Sun. The ionosphere plays an important role in the propagation of radio waves, therefore its monitoring is essential to ensure the quality of radio communications. In addition, the ionosphere can be considered as a natural laboratory, in which the processes of interaction of plasma and waves can be studied directly by the in situ measurements. This causes a great interest in ionospheric research, which has been intensively carried out over the past decades both by ground, air (rockets, high-altitude balloons) and orbital instruments.

Despite that to date, extensive experimental material has been accumulated about the ionosphere and theoretical models have been developed that make it possible to describe its states in significant details, there are still urgent research problems associated with the need to study those regions of the ionosphere in which our knowledge (and models) do not allow us to confidently predict the parameters of the ionospheric plasma. These areas include primarily the auroral and equatorial regions.

The peculiarities of the auroral region are that the classical mechanisms of plasma formation, based on the effects associated with exposure to solar radiation, do not work in it. Therefore, it is assumed that the key role in its formation (especially during the polar night) magnetosphere tail particle precipitation are played. As a result, the F layer is strongly weakened (or disappears) and the E layer plays a decisive role. Therefore, the problem of the parameters of the auroral ionosphere becomes "volumetric", the solution of which depends on the state of the magnetosphere and the features of precipitation. As a result, the "quality" of the ionosphere also changes, i.e. it becomes highly heterogeneous and non-stationary.

The equatorial region also differs from the mid-latitude region due to the so-called floating "bubbles" that violate the generally accepted structure.

A separate range of tasks in the study of non-stationary processes in the ionosphere is based on the use of artificial disturbances of the ionosphere by impact of high-frequency (HF) electromagnetic radiation generated by heating stands (Sura, Tromsö, HAARP and others), low-frequency (LF) radiation from powerful ground-based transmitters , as well as disturbances arising from the influence of electric circuit harmonics. These studies have a "double" meaning. On the one hand, it is controlled effect on the ionospheric plasma, which should make it possible to simulate the phenomena occurring in nature, but with known exposure parameters. On the other hand, there is number of physical problems for which "unlimited" conditions for carrying out measurements are important (the dimensions of the measuring instruments are much smaller than all the characteristic dimensions in plasma) and such measurements can be implemented in a kind of "space laboratory", which in fact, the ionosphere is.

To solve these problems, an integrated approach is required based on the use of instruments that provides detection of different types of radiation, as well as spatially separated measurements based on a satellite system. This approach lies at the heart of the Ionosphere space mission goals.

2 Concept of space missions Ionosphere

The main tasks of the Ionosphere space missions include:

- monitoring of the ionosphere;
- monitoring of the magnetosphere;
- measurement of wave activity (electromagnetic, acoustic waves);

- diagnostics of corpuscular ionizing radiation;
- diagnostics of the Atmosphere ozone (determination of the planetary ozone distribution).

Ionosphere satellites should be launched into near-Earth orbits with a set of instruments to monitor the environmental helio-geophysical parameters in the upper atmosphere, ionosphere and near-earth space. Spacecraft "Ionosphere" should operate in a near-circular sun-synchronous orbits in two orbital planes (2 spacecraft in each plane). The spacecraft in each plane should be separated by an angle of 180±30°. Orbital parameters of the first pair of spacecraft: altitude ~820 km; orbital period ~101 min; inclination ~98°; eccentricity ~0.001; the position of the orbital plane relative to the mean Sun right ascension ~135°; the local time of the orbital ascending node is ~21 h. The orbital parameters of the second pair of spacecraft differ in the position of the orbital plane relative to the mean Sun right ascension - ~46° and the local time of the orbital ascending node - ~15 h. The launch of the first pair of spacecraft is planned in early 2022, the second pair - late 2022 or early 2023.

Scientific payload includes instruments for radio tomography of the upper atmosphere and ionosphere:

- ionozond for the range 0.1 - 20 MHz with a 137 MHz transmitter (LAERT instrument);
- Langmuir probe and ionospheric plasma energy spectrometer (ESIP instrument);
- GPS - GLONASS meter of total electron content (TES instruments);
- VLF wave complex of instruments;
- onboard radio transmitters at frequencies of 150, 400 MHz (equipment Mayak 150/400);
- hard X-ray and gamma ray spectrometer (SG/1 instrument);
- spectrometer of plasma and energetic charged particles (SPER/1 instrument);
- detector of galactic cosmic rays (GALS/1 instrument);
- ozonometer;
- onboard complex for payload control and scientific data collection (BKUSNI);
- system for transmitting output data from the antenna-feeder device of the target information radio link (AFU RLCI).

Thus, scientific payload includes five groups of instruments: instruments for ionosphere plasma parameters monitoring, instruments for electromagnetic wave monitoring, instruments for magnetosphere particle and space weather monitoring, ozonometer and instruments for data collecting and transmitting. The payload composition of all four Ionosphere spacecraft is the same, except that the GALS/1 instruments and the ozonometer will only be installed on the second pair of spacecraft.

The key instrument of the ionospheric group of equipment is the ionospheric locator, i.e. the LAERT ionozond, which allows measuring the ionospheric plasma density profiles. The ionozond measurements can be used to reconstruct the height distribution of the electron concentration up to the height of the F2 layer maximum. These measurements are remote, and their interpretation can be carried out both on board the spacecraft and after receiving telemetry (TM) information in ground-based laboratories. The second approach is most used for complex distributions of electron concentration and in non-stationary geomagnetic conditions.

Another approach to determining the electron concentration at significant distances from the measurement site is based on the use of signals from navigation satellites such as GPS/GLONASS. This method makes it possible to determine the concentration of electrons in the column along the signal propagation path from the source (GPS/GLONASS satellite) to the receiver of the Ionosphere spacecraft.

Accurate measurements of the electron concentration and temperature of particles at the spacecraft location point carried out using a Langmuir probe and an ionospheric energy spectrometer.

To determine small-scale ionospheric irregularities, their dynamics and build a three-dimensional picture of the plasma distribution, it will be carried out using onboard transmitters "Mayak". The signals from this device will be received on the Earth's surface and processed by the tomographic method.

To determine the distribution of ozone, a simple ozonometer will be used to determine the density of the gas below the satellite.

The SPER/1 instrument is designed to determine the epithermal plasma and energetic charged particle fluxes parameters, which is especially important in the auroral region, where the properties of the ionosphere are largely determined by streams of precipitating particles.

The GALS/1 instrument is designed to measure the fluxes of galactic cosmic rays.

The SG/1 gamma spectrometer is designed to detect the bremsstrahlung of energetic electrons generated in high-altitude electromagnetic discharges, as well as to monitor solar activity in the hard X- and gamma ray ranges and to measure atmospheric albedo fluxes.

The organization of the payload operation on spacecraft board is provided with the help of the acquisition and control device, i.e. BKUSNI. Target data will be transmitted via a special telemetry system.

3 Monitoring of ionosphere by plasma density measurements

3.1 Measurements of electron concentration global distribution by sounding method with the use of LAERT instrument

The LAERT instruments is designed for global operative monitoring of the Earth's ionosphere by the method of pulsed multi-frequency radio sounding in the frequency range from 0.1 to 20 MHz. The LAERT ionozond provides an operative determination of the critical frequency of the ionosphere along the satellite orbit, the calculation of vertical profiles of the electron concentration, the determination of the parameters of the ionospheric plasma at the altitude of the spacecraft orbit using the plasma resonance frequencies. According to these parameters, it is possible to construct a 3-dimensional global distribution of electron concentration in the Earth's ionosphere, to determine the boundaries of the main formations in the ionosphere (equatorial anomaly, main ionospheric trough, auroral oval and polar cap), which affect the propagation of radio waves in various frequency ranges. The ionozond also has the ability to synchronize with ground-based ionospheric stations for conducting sessions of trans-ionospheric and reverse ionospheric sounding.

The LAERT ionozond includes:

- transmitter;
- onboard receiving device;
- signal processing unit;
- transmitting antenna;
- receiving antenna;
- preamplifiers.

The transmitter is designed to emit radio pulses in the frequency range 0.1 - 20 MHz. It includes: broadband amplifier, antenna switch, broadband matching transformer, cable set.

Main parameters of the transmitter:

- the range of radiated frequencies is 0.1–20 MHz;
- input impedance 50 Ohm; - output impedance 2x300 Ohm;
- input signal power 0 dBm - power at the antenna input 280 W.

The onboard receiver provides two-channel reception of signals emitted by the transmitter of the satellite ionozond and reflected from the ionosphere; shaping the excitation signal of the transmitting device; formation of a highly stable frequency signal for a central processing unit (CPU).

The on-board receiving device has the following technical parameters:

- the range of received frequencies is from 0.1 to 20 MHz;
- bandwidth (-3 dB) - 15 kHz;
- sensitivity (with a signal-to-noise ratio of 10 dB) not less than 5 μV;
- the frequency range of the excitation signal of the power amplifier is from 0.1 to 20 MHz.

During the instrument operation, the number of sounding frequencies and the frequency step are determined by the programmed operating mode. The sounding of the ionosphere is carried out by radio pulses with duration of 100 μs with a repetition rate of 60 Hz. The standard time for obtaining an ionogram at a frequency of sending probing pulses of 60 Hz and 400 probing frequencies is 6.7 seconds. The session time is reduced to 5 seconds with the adaptive selection of the frequency grid of sounding. For special measurement modes, a longer ionogram recording period can be provided.

The periodicity of sounding sessions is 10, 30, 60 seconds (set by the commands of the command and control unit (SCU)), while the sounding mode with a frequency of 10 seconds is considered the main one.

There are three main modes of ionozond operation:

- full sweep 0.1 - 20 MHz;
- adaptive mode, the ionozond itself determines the critical frequency according to a special algorithm, and in accordance with this, the maximum sweep frequency is set;
- special mode, it is determined by the programmed, it can be work at fixed frequencies, work for reception without a transmitter, work in synchronization mode with a ground ionozond, work in trans-ionospheric sounding mode, etc.

3.2 Measurement of the ionosphere electron concentration altitude distribution by the method of radio occultation estimation of the amplitude and phase delays of the signals from the global navigation systems GPS/GLONASS with the use of the PES instrument

With the use of PES instrument, the following main tasks should be solved:

- global monitoring of the structure and dynamics of the ionosphere;
- detection of anomalous phenomena and characteristic signs of changes in the ionosphere, caused by the processes of preparation of earthquakes and man-made influences;
- development of technical means and methods for monitoring natural and man-made disasters from the spacecraft.

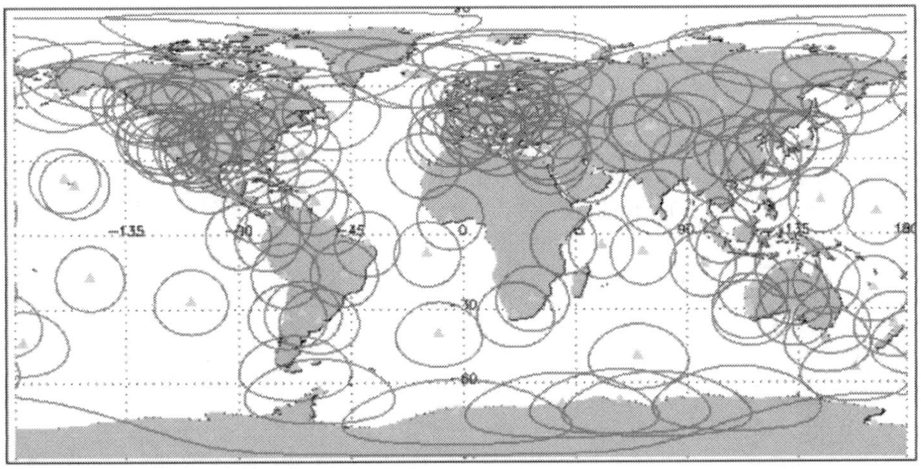

Fig. 1. Global network of ground GPS receivers. The vision areas at 400 km altitudes at the elevation angle 10^0 are shown.

As it is well-known, higher requirements for positioning accuracy have led to the creation of new generation navigation satellite systems GPS-NAVSTAR (NAVSTAR - Navigational Satellite Time and Ranging) and GLONASS (Russian Global Navigation Satellite System). The satellites of these systems have a circular semi-synchronous orbit with the Earth's rotation with an altitude of about 20 thousand km (orbital period 12 hours) and are placed in the indicated orbits in three planes spaced 120 degrees apart. There are 4-5 satellites in the radio visibility zone of each point on Earth at the same time. The duration of observation of one satellite reaches six hours and, therefore, provides much more information on the spatial and temporal changes in the parameters of the ionosphere.

Currently, the global network of ground-based GPS receivers (see Fig. 1) supplying observational data to regional centers, accessible through public Internet sites, numbers about 1000 points

Continuous high-precision phase-difference measurements at GPS frequencies (1.2 / 1.6 GHz) behind all satellites in the field of view at each receiving point of the global network allow obtaining information on daily variations in the total electron abundance in the ionosphere above the point of observation. The processing of measurement data from a network of ~ 100 GPS receivers (GDGPS system of global differential corrections) makes it possible to obtain practically on-line distribution of TEC (Total Electron Content) in the near-Earth ionospheric plasma.

Ionospheric correction is transmitted to users of single frequency GPS receivers in order to achieve similar accuracy to users of dual frequency receivers. The GDGPS system reconstructs in real time a global map of ionospheric electronic content (currently the map is updated every 5 minutes). In addition, according to the data of the global network of GPS receiving points, it is possible to study the morphology and dynamics of medium- and large-scale inhomogeneous structures, which are manifested in variations in the total electron content [1].

Recently, special GPS receivers have been produced that allow measuring the amplitude of navigation signals 1.2/1.6 GHz with a sampling frequency (up to 50 Hz) required for recording radio measurements. With the use of such receivers, strong scintillations of GPS signals were recorded not only in the equatorial zone, but also at mid-latitudes during a moderate magnetic storm [2].

Receiving radio signals from GPS/GLONASS global navigation systems using equipment installed on LEO satellites makes it possible to implement measurement conditions in which a change in the electron concentration integral along the path of radio wave propagation is due to a change in the path length and the addition (or exclusion) of a new high-altitude ionospheric layer. These conditions arise during "radio ascents" and "radio calls" of satellites of navigation systems with respect to the low-orbit spacecraft and the Earth, when it is possible to measure the time delay and phase of GPS/GLONASS signals when the lowest point crosses the ionospheric altitude signal propagation path in the interval from ≈50 km to the orbital altitude of a low-orbit spacecraft.

The geometry of radio occultation measurements is shown schematically in Fig. 2.

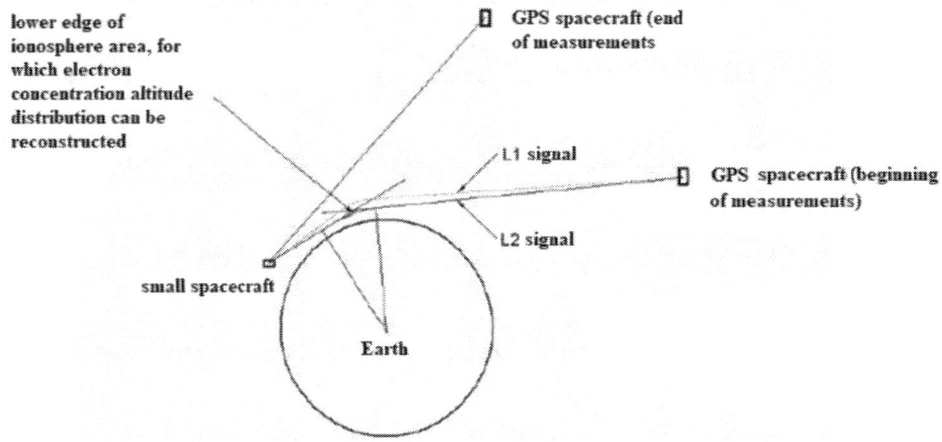

Fig. 2. The scheme of radio occultation measurements, on which there are presented the mutual position of low orbital small spacecraft and satellite of GPS navigation system.

It follows from Fig. 2 that for the conditions of the daytime mid-latitude undisturbed ionosphere for an altitude interval of 100-1000 km, the electron concentration values calculated from the data of radio occultation measurements practically coincide with the model values.

When signals are received on board the spacecraft from satellites of the GLONASS and GPS navigation systems, it is simultaneously possible (with continuous operation of the TES equipment) to measure more than a thousand altitude profiles of the electron concentration of the Earth's ionosphere during the day, i.e., to carry out a global survey of the ionosphere. The measurement program allows you to control the state of the ionosphere over specified regions, including seismically active ones.

Signals from GPS/GLONASS satellites are received by the antenna, filtered and amplified by an antenna amplifier, which is integrated into the antenna housing. Signals of two navigation frequencies come from the antenna to the input of the navigation receiver. In the navigation receiver, the frequency conversion of the input signals and distribution over the tracking channels are performed. Tracking is controlled by the processor. After capturing the satellite channel, the received signal is demodulated in order to extract from it a modulating message about the code phase, carrier phase, satellite clock and navigation message.

The navigation message contains information about the current state of the satellite transmitting the signal, its ephemeris, the state of the GPS/GLONASS systems, the time and almanac of the movement of the satellites. The carrier phase and code phase measurements are sent to the input/output (I/O) controller. The I/O controller provides temporary storage of measurement results and automatic control of the navigation receiver operating mode, in accordance with the measurement program. The controller sets the operating mode of the navigation receiver after the supply voltage is applied, receives and executes control commands and transmits the measurement results via the serial data interface.

Switching the operating modes of the TES equipment is carried out in accordance with a given program, taking into account the spatial position of the spacecraft relative to the satellites of the GPS/GLONASS navigation systems, its geographic position and current time. The measurement results are tested for compliance with the established criteria in order to transfer only useful data to the satellite board systems.

The antenna of the PES equipment is active and provides reception of signals in two frequency ranges 1217 - 1265 MHz (L1) and 1565 - 1615 MHz (L2).

The navigation receiver contains a high-frequency linear and digital part. The high-frequency linear part of the receiver operates in the input frequency ranges L1, L2 and is designed to transfer the spectra of the received GPS signals to the video frequency range.

3.3 Radio-interferometric methods of diagnostics of the ionosphere with the use of Mayak 150/400 instruments

Radio interference methods were developed by the Soviet radio-physical school, created by academicians L.I. Mandelstam and N.D. Papaleksi [3, 4]. The use of these methods for measuring the regular and inhomogeneous structure of the electron concentration in the near Earth surface and space plasma using geophysical rockets, satellites and interplanetary spacecraft has shown their effectiveness and formed an independent radio-physical research direction, i.e. radio beacon methods for diagnosing the environment. The name of these methods is associated with the name of the first coherent transmitters of VHF radio waves "Mayak", which were installed on research rockets. Since 1957 coherent transmitters "Mayak" in various modifications are widely used in space experiments on LEO spacecraft, on geostationary satellites and on interplanetary spacecraft.

The consistent development of satellite radio beacon methods ultimately led to a technological breakthrough, i.e. to the creation of space systems for high-precision temporal and spatial positioning - first the first-generation low-orbit radio navigation systems, and then high-orbit GPS and GLONASS systems.

The physical foundations of the radio beacon method are quite simple and clear [5, 6]. The beacon radio transmission method is based on the fact that when radio waves propagate in a collisional anisotropic partially ionized inhomogeneous plasma, which is the Earth's ionosphere, radio waves interact with the propagation medium due to known effects, such as refraction, dispersion, absorption, scattering on fluctuations of electron density, etc.

As a result of this interaction, the radio waves received on Earth contain information about the parameters of the sub-satellite region, incl. information about the regular, wave and stochastic structure.

The organization and geometry of radio beacon measurements determines the set of environmental parameters determined from ground-based registration data. The most used options for using radio beacon methods (depending on which facility the radio beacon is installed on):

- rocket experiments - provide high-precision measurements of the height profile of the electron concentration from the base of the ionosphere to the height of the rocket ascent; the concentration profile is determined both in the ascent and in the descent of the rocket [7];
- low-orbit satellites - provide measurement of the space-time structure of the sub-satellite ionosphere (total electron content, TEC), horizontal TEC gradients, as well as spectra of ionospheric irregularities (with synchronous measurements using receiving antennas with a small separation base of ~ 500-1000 m) [8];
- geostationary satellites - high-precision measurements of TEC and diurnal variations of the sub-satellite thickness of the ionosphere and magnetosphere along the fixed direction

"SC - receiver"; measurements of this kind make it possible to diagnose the impact of powerful ground-based HF stands on the near-earth plasma [9];

- radio occultation measurements - the geometry of "radio sunrise" and "radio entry" is used for radio beacon studies of the ionospheres of planets of the solar system (Venus, Mars) using interplanetary spacecraft [10], as well as for measuring the altitude profiles of the Earth's atmosphere and ionosphere - when the reception on low-orbit satellites of radio beacon signals from the spacecraft of the GPS/GLONASS navigation constellation [11].

The most widespread are radio beacon measurements for the diagnosis of the ionosphere in natural conditions and under anthropogenic impacts using low-orbit satellite radio navigation systems (SRNS), which are equipped with transmitters emitting two coherent signals with frequencies of 150 and 400 MHz. On the network of ground receiving stations, the amplitudes and Doppler shift of the received radio navigation signals are measured, time stamps and navigation messages are allocated, and the regular (total electron content in the sub-satellite region of the ionosphere) and statistical scintillation characteristics characterizing the space-time parameters of inhomogeneities in the sub-satellite region of the ionosphere are calculated.

Radio-beacon measurements on LEO spacecraft are usually organized according to the following scheme:

- a radio beacon (transmitter) is installed on board the spacecraft;
- the radiation pattern of radio beacon signals in the alignment of angles ± 60 ° from the direction to the nadir ensures their reception on the Earth within the line of sight of the spacecraft;
- the radio beacon is switched on either in continuous radiation mode, or operates according to a cyclogram, which provides the ability to receive radio beacon signals at "dedicated" ground receiving points, equipped with the necessary complex of antenna and specialized receiving equipment;
- with appropriate organization of radio beacon measurements on a "chain" of ~ (4-6) ground receiving points, a radio-tomographic method of data processing can be implemented, which gives the most complete reconstruction of the spatial structure of the ionosphere over a given region.

Radio-tomography of the ionosphere based on LEO signals using 150/400 MHz radio beacons makes it possible to reconstruct the two-dimensional distribution of electron concentration in the sub-satellite ionosphere [12 - 14]. The applied phase-difference measurement method has no drawback associated with the problem of determining the unknown constant $2\pi n$ of the total phase of the signal. In contrast to phase tomography, the phase-difference method is more sensitive to relatively small inhomogeneities of the electron density, which make a small contribution to the total phase, but more significant to its derivative. The reconstructions of the electron density field, performed using the phase tomography method, do not show details less than a few hundred km in size, while phase-difference tomography allows the reconstruction of ionospheric structures ~ 100 km in size and the value of electron density fluctuations of ~ 5% of the maximum concentration.

Currently, tomographic studies of the ionosphere are carried out on four chains of ground-based receivers, i.e. in Russia, Finland, Taiwan and the USA (Alaska).

Radio-tomographic data processing makes it possible to reconstruct the large-scale inhomogeneous structure of the sub-satellite region of the ionosphere. To measure the parameters of small-scale ionospheric irregularities, the time series of amplitude or phase beacon registrations should be subjected to a standard FFT (Fast Fourier Trans-form) or MEM (Maximum Entropy Method) processing procedure; as a result, power spectra are obtained, respectively, of amplitude or phase fluctuations depending on the frequency of fluctuations. For low-orbit spacecraft (with a circular orbit of ~1000 km), the radio beam scans the F-region (~300 km) at a speed of 2.2 km/s. Therefore, when processing a section of the register program with a duration of 10 s and a sampling frequency of 125 Hz, the spectral range of wavelengths from 22 km to 18 m (frequency fluctuation range from 10-1 to 102 Hz) will be covered [15].

The airborne radio beacon used for remote sensing of the ionosphere by radio scanning methods is a multichannel (usually 2 channels) transmitter that emits on the harmonics of the reference generator, thereby ensuring the coherence of signals.

Let the operating frequencies f_1 and f_2 of a coherent radio beacon

$f_1 = m_1 f_0\ ; f_2 = m_2 f_0$,

where f_0 is the frequency of the transmitter reference oscillator (quartz frequency stabilization is sufficient); m_1 and m_2 are integers defining the ratings of the radiated frequencies.

In this case, the phase difference of the received frequencies, reduced to the frequency f_1

$\Delta\Phi = \Phi_1 - (m_1/m_2)\ \Phi_2$

does not depend on the "vacuum" component of the Doppler frequency shift of a moving transmitter, but is determined entirely by the influence of the ionosphere, that is, in the case of a coherent transmitter, the "vacuum" component of the Doppler frequency shift, which is several orders of magnitude higher than the "ionospheric" component, can be automatically excluded in the receiving device.

The MAYAK-150/400 instruments consist of two antennas and an electronics unit (the transmitter itself).

The antennas of the Mayak-150/400 transmitter are crossed half-wave vibrators, the directional patterns of which are oriented in the nadir.

3.4 Measurement of ionospheric plasma parameters along the spacecraft orbit with the use of ESIP instrument

Measurements of the main parameters of the ionospheric plasma along the spacecraft orbit are necessary both for studying the structure and dynamics of the ionosphere as a whole, and for studying individual physical processes in the ionospheric plasma and global monitoring of the ionosphere. For these purposes, the ESIP (Energy-Spectrometer of Ionospheric Plasma) equipment is used, in which several types of sensors are used to register the parameters of the ionospheric plasma. According to the principle of operation, they belong to two groups: analyzers with inhibitory potential (AIP) and Langmuir probes. Thus, the ESIP equipment includes the following measuring sensors:

- ion flux sensor (IFS);
- ion spectrum sensor (ISS);
- drift meter (DM);
- 4 potential sensors (PS);
- sensor for fast measurement of electronic temperature (ETS);
- flat Langmuir probe (FLP).

For all sensors, the directly measured quantity is the incoming particle flux. In the case of ions, the flux is given by the expression:

$F_i = F_i(N_i, T_i, M_i, \alpha, \beta, V_{SC}, V_D, U_{SC}, U_{AI}, U_D)$,

where N_i, T_i, M_i are the concentration, temperature and mass of particles, α, β are the angular coordinates of the incident ion flux, V_{SC}, V_D are the vectors of the spacecraft velocity and ionospheric plasma drift, U_{SC} is the spacecraft body potential, U_{AI} is the potential difference at the AIP electrodes , UD is the potential of the sensor body.

In this expression, the device controls (sets) only the potentials of the electrodes. The spacecraft speed is considered to be known. All other parameters are unknown. Since the thermal velocity of ions, as a rule, is significantly lower than the velocity of the spacecraft, the ion flux to any sensor can be represented as a parallel beam of ions incident on the sensor with the velocity $V = - V_{sc} U_{AI}$.

In the case of electrons, the situation is reversed - the thermal velocity of electrons is significantly higher than the velocity of the spacecraft, so the flow of electrons to the sensor is isotropic.

Ionic parameters (concentration of N_i, temperature T_i, masses of the main ions M_i, energy distribution, magnitude and direction of drift velocity V_D) are measured using analyzers with the braking potential of the AIP or otherwise ion traps (Faraday cups). They are an inlet aperture, a flat multigrid design, and a collector. They are used to measure the dependence of the ion flux incident on the sensor aperture (ion current to the collector) depending on the potentials on the sensor electrodes. On this basis, by setting different potentials on the electrodes, sensor options are created for measuring various plasma parameters.

In the ESIP instrument based on AIP, three types of sensors are implemented, i.e. IFS, ISS and DM.

IFS is the Ion Flux Sensor, it is an AIP with a potential distribution on the electrodes that allows measuring the ion flux (concentration) of N_i ions. Used for fast (up to 1000 times per second) continuous N_i changes. The geophysical parameter determined with this sensor is the total ionic concentration of N_i ions in the range $10^2 - 10^7$ cm^{-3}.

ISS is the Ionic Spectrum Sensor, it is an ATP with a potential distribution on the electrodes, providing measurement of the integral energy spectrum of ions (IES) in the energy range $E \leq 15$ eV. To measure the energy spectrum, a varying voltage must be applied to the analyzing grid within the limits sufficient to decelerate the heaviest incident ions. The range of measured values of N_i and T_i is $10^2 - 5{\cdot}10^6$ cm^{-3} and $3{\cdot}10^2 - 10^4$ K, respectively. Measurements are carried out 1 time per second.

DM is the drift-meter, it is built on the basis of a IFS sensor with a "split" collector. This sensor allows you to measure the direction of the drift velocity V_D, i.e. the vertical and horizontal components of V_D on the angles of arrival. Measurements are carried out 1 time per second. According to the readings of the angles of arrival, the spacecraft orientation relative to its velocity vector can be determined within $\pm 45°$ in pitch and $\pm 45°$ in yaw with an accuracy of $\pm 1°$. This calculation can be done by the built-in microprocessor and displayed on board in real time.

For maximum efficiency of the device operation, the spacecraft must be oriented along the vector of the incident plasma flow (along the spacecraft velocity vector).

Measurement of the angles of arrival and the energy spectrum of ions makes it possible to determine the full vector of the ionospheric plasma drift velocity, the knowledge of which in combination with the calculated value (or measured in the presence of a magnetometer on board) of the magnetic field will make it possible to calculate the full vector of the constant electric field in the ionosphere.

The parameters of the electronic component of the ionospheric plasma (N_e concentration and T_e temperature) are measured using various Langmuir probes.

The electronic temperature sensor ETS is a modified Langmuir probe and measures the electron temperature (T_e) by the method of high-frequency electron temperature probing. In addition, the potential of the spacecraft body is determined from its readings. The high frequency electron temperature probe is a flat four-electrode differential symmetric design. Each of the 4 collectors is surrounded by a guard electrode, which lowers the parasitic capacitance and increases the measurement frequency. In addition, to increase the resistance to external interference, the electrodes are cross-connected in pairs. A sinusoidal modulated voltage (no more than 500 mV) with a frequency of less than 50 kHz is supplied to the working collector according to a special program. Due to the nonlinearity of the current-voltage characteristic "probe-plasma", the floating potential of the collector changes, the value of which is determined by the temperature T_e.

If a sinusoidal voltage with a given amplitude A and frequency f is applied to the collector of the T_e sensor, then the potential of the collector relative to the plasma will change by an value that is a known function of the electron temperature and the amplitude of the applied sinusoidal voltage: $dU = dU(T_e, A)$.

$dU = (kT_e/e)\times\ln I_0(eA/kT_e)$,

where e is the electron charge, k - the Boltzmann constant, T_e is the electron temperature, I_0 is the modified Bessel function of the first type of zero order. Since the amplitude A is known, from this formula we can calculate T_e from the inverse relation $T_e = T_e(dU, A)$.

The FLP sensor (flat Langmuir probe) uses the method of the classical Langmuir probe and measures the energy distribution of electrons, from which the Ne concentration and the electron temperature Te are determined during ground processing.

The range of kT_e values measured by the ETS and FLP sensors is 0.1–5.0 eV. The measurement frequency for the ETS sensor is not less than 1 kHz, for the FLP sensor it is 3 Hz.

The ESIP equipment sensor unit must be installed on a rod that removes it at a distance of at least 0.5 m from the spacecraft structural elements. In this case, it must be oriented with its axis along the spacecraft velocity vector. In this case, the sensor unit will be in the oncoming plasma flow.

4 Monitoring of electromagnetic waves with the use of VLF – Wave Probe instrument

As it is well-known, outer space is a medium in which electromagnetic radiation propagates and is generated in a very wide frequency range, ranging from permanent magnetic and electric fields, up to X- and gamma rays. The range from constant fields to the first tens of megahertz is specific for near-Earth space (NES); these fields are associated with the plasma shells of the Earth - the magnetosphere and the ionosphere. The restructuring of the structure of plasma shells, the development of dynamic processes in them lead to the appearance of a complex of electromagnetic phenomena recorded both in the NES and on the Earth's surface.

In turn, a number of natural electromagnetic phenomena occurring in the Earth's atmosphere contribute to the electromagnetic fields recorded in the magnetosphere (as a rule, strongly modified by the ionosphere). A classic example is the radiations recorded by satellites that arise during lightning discharges and are called whistling atmospherics.

Organization of long-term global monitoring of electromagnetic radiation in the range from constant fields to tens of kilohertz is one of the methods for controlling events (natural and anthropogenic) in near-earth space.

The main tasks of electromagnetic monitoring are:

- measurements of background electromagnetic radiation of the VLF frequency range in the NES;
- revealing the correlation of the radiation detected on satellites with extreme phenomena and processes on the ground and in the depths of the Earth (typhoons, earthquakes, powerful explosions, energy catastrophes, etc.); anthropogenic electromagnetic impact on the ionosphere and magnetosphere;
- identification of medico-biological problems of the impact of natural and anthropogenic electromagnetic fields of the considered frequency range.

The tasks solved on the basis of measurements of quasi-constant electric fields and the intensity of ULF-VLF radiation in this project include:

- study of the dynamics of ionospheric - magnetospheric plasma under various conditions in the solar wind.
- observation of the effect of electric fields on the processes of precipitation of energetic particles into the ionosphere and other auroral and subauroral phenomena.
•
- observation of the mechanisms of formation of thin electromagnetic and plasma structures in the ionosphere and magnetosphere (electrostatic field jumps, Alfven vortices, etc.)
- observation of electromagnetic and plasma responses in the ionosphere and magnetosphere of the Earth to the formation and development of phenomena in the lower atmosphere and lithosphere associated with intense effects on them of various sources of natural origin (typhoons, earthquakes, etc.).

The use of a new type of spacecraft orbit in the "Ionosphere" project will make it possible to carry out long-term measurements in the selected power tube and, on the basis of these simultaneous measurements on two spacecraft, to separate the spatial and temporal variations of quasi-constant electric fields and the intensity of ULF-VLF radiation, which is fundamentally important for solving the problems formulated above.

Measurements of electric fields on board satellites are mainly carried out using the double probe method. This method is characterized by simplicity of implementation, reliability, versatility, and sufficiently high accuracy. To implement a measurement using this method, it is necessary to measure the potential difference between two points in space. By measuring the potential difference ΔU_x between two bodies spaced apart in space, you can calculate the field strength between them, dividing it by the length of the measurement base d:

$E_x = \Delta U_x/d$.

If to use several pairs of sensors in mutually perpendicular directions, you can measure 3 components of the electric field and also get the field vector. This method can be used to measure quasi-constant and alternating electric fields in the range from zero Hz to several tens of kilohertz.

In this case, the dynamic range of the measured values must be ensured for a quasi-constant field:

- Ex ~ 0.1-300 mV/m,
- Ey~ 0.1-700 mV/m,
- Ez ~ 0.1-300 mV/m.

Dynamic range for variable field ~1 μV/m – 10 mV/m.

The selected dynamic range of changes in the measured values is due to the range of scientific tasks set in the project, as well as the fact that when measuring quasi-constant electric fields in space plasma, together with the investigated fields of physical nature, the induced field $\boldsymbol{E} = \boldsymbol{V}\text{x}\boldsymbol{B}$ is also measured, associated with the movement of probes in the external geomagnetic field. , as well as the corotation field, the values of which are much higher than the investigated parameters. Subsequently, these fields are removed by digital filtering methods.

The required rate of interrogation of the measured values is at least 1 time per second, which is necessary for a spatial resolution of ~ 10 km to study the mechanisms of the formation of fine structures.

The required measurement accuracy of the electric field components for the tasks solved in the project, $\Delta E_{x,y,z} = 2$ mV/m. It is determined by several factors that introduce errors in the measured values for $\Delta E_{x,y,z}$, respectively. In particular, the zero drift at maximum temperature changes gives for $\Delta E_{x,y,z} = \Delta E_y = \Delta E_z = 0.5$ mV/m. With an orientation accuracy of 1° - 1.5° in the roll and pitch angles and 2° - 2.5° in the yaw angle, the maximum error due to the deviation of the satellite coordinate system from the orbital is less than 0.1 mV/m for the X- and Y- components and 0.5 mV/m for the Z- component at maximum roll angles. Errors due to emission currents (mainly photocurrents) arising from the inhomogeneity of the probe surface, ΔE <0.5 mV/m. Errors due to the deviation of the probe shape from spherical do not exceed 0.03 mV/m. Thus, the maximum error in measuring the electric field does not exceed ± 1.5 mV/m for the E_x and E_y components and ± 2 mV/m for the E_z - component of the field.

The VLF - wave probe complex should provide monitoring of low-frequency wave phenomena in near-earth space in the range from fractions of a Hz to 20 kHz. At the same time, the equipment provides for both observation of natural electromagnetic radiation and observation of anthropogenic impact on the surrounding space.

During the experiment, the following observations should be made:

- propagation of whistling atmospherics (SA), which are packets of low-frequency waves with varying frequency (it is of interest to observe partially dispersed (BH) whistling atmospherics that have passed into the ionosphere along a short path from the Earth's surface, the passage of strong BHs is associated with the penetration of electromagnetic fields from high power sources);
- seismic-ionospheric effects in quasi-constant electric and magnetic fields, as well as in the fields of the ULF/ELF range;
- occurrence and movement of typhoons, including more likely to track typhoons in Oceania.

The VLF - wave probe complex is made in the form of 3 electronic units and a system of sensors of electrical and magnetic components, including:

- SHASH 2 instrument, designed for recording and processing a large amount of information in relatively short time intervals;
- ANCh instrument designed for smooth monitoring of spectral components of the electromagnetic field in the daily storage mode of information using a relatively small amount of information;
- IESP-3R electric field measurement unit;
- electric sensors ULCH/CNCH;
- magnetic sensors LEMI-106HS.

To convert the fields into electrical signals, the pair of electric DE sensors and LEMI106HS magnetic sensor are used.

The sensors are mounted on a boom with a minimum distance of any of the sensors from the spacecraft - 2m and at a distance from possible radiation sources in the frequency range of the receiving band of the sensors.

The IESP-3R instrument is an analogue electric field meter in a wide frequency band that measures components of the electric field by the double probe method in a wide frequency band of 0 - 1 MHz.

5 Monitoring of space weather factors with the use of SPER/1, SG/1 and GALS/1 instruments

As noted above, one of the factors determining the state of the ionosphere is charged particle fluxes, mainly electrons and protons of supra-thermal and high energies, falling from the tail of the magnetosphere, radiation belts, as well as energetic particles of solar and galactic cosmic rays. The fluxes of these particles in near-Earth space are mainly due to the factors of geomagnetic and solar activity, the so-called space weather. Therefore, a necessary condition for monitoring the state of the ionosphere is measuring variations of charged particle fluxes in near-Earth space in connection with Geo and Solar activity.

a

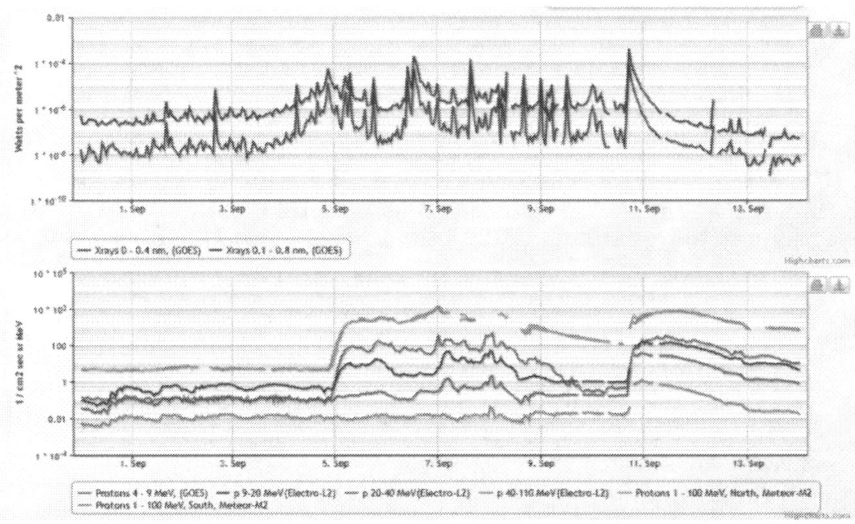

b

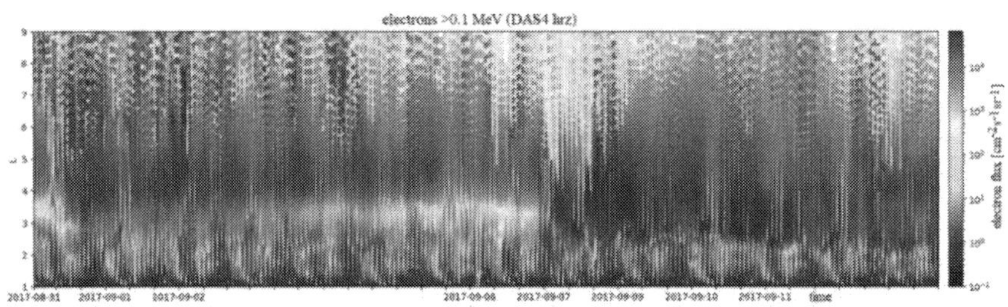

c

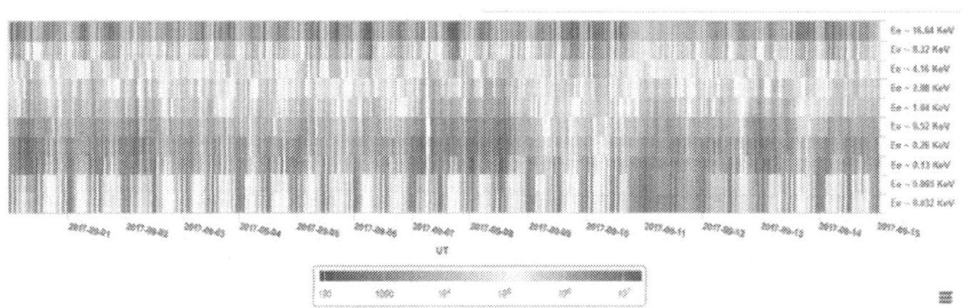

d

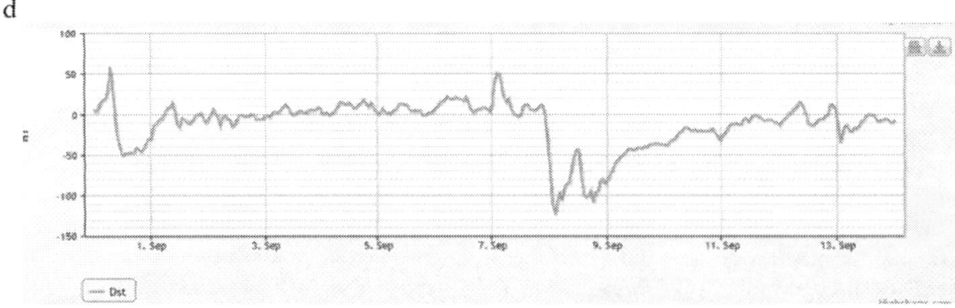

Fig. 3. Event detected 1 -15 September, 2017. Time profile of the intensity of X-rays from the sun according to GOES data (a, upper panel), SCR fluxes in the GSO and in the polar cap according to GOES data, Electro L2, SKIF instrument, Meteor, SCR instrument (a, lower panel), electron fluxes in ERBs according to Meteor data, MSGI instrument (b), spectrogram of auroral electron fluxes, Meteor-MSGI (c), time variation of D_{st} (d).

As examples of variations in fluxes of energetic charged particles of different types depending on the level of solar flare activity are shown in Fig. 3. It can be seen from the figure that after an intense solar flare in the second half of the day on September 10, 2017, increases in solar cosmic ray (SCR) fluxes are observed in the polar cap and in the geostationary orbit. At the same time, the fluxes of energetic electrons in the radiation belts and in the auroral region significantly increase after the noticeable D_{st} variation on September 8, which, in turn, may have been inspired by solar flares that took place on September 5-9, which are also associated with increases in SCR fluxes on September 5-9.

5.1 Monitoring of charge particle fluxes with the use of SPER/1 instrument

It should be noted that the measurements of the fluxes of supra-thermal and energetic electrons on the Meteor and Electro-L2 satellites were carried out using the SKIF instrument, which is a prototype of the SPER/1 instrument, which is supposed to be used for similar measurements in the Ionosphere missions.

The SPER/1 instrument is designed to measure fluxes and differential energy spectra of low-energy electrons and protons in the energy range 0.1 - 20 keV, as well as charged particles of medium and high energies, i.e. electrons in the energy range ~ 0.15 - 10 MeV and protons in the energy range ~ 2 - 160 MeV. The integral spectrum of α-particles of MeV energies is recorded separately. The count rates should be measured at least once a second, and the total spectrum acquisition time should be no more than 20 s.

The SPER/1 instrument is structurally a complex of detector assemblies providing detection of charged particle fluxes mentioned above, united by a common power supply and control interface. The detector assemblies include:

- unit of a medium-energy spectrometer (UMES);
- unit of a low-energy spectrometer (ULES);
- unit of a gas-discharge detector (UGD).

The detector assemblies are located in one direction on the front wall of the instrument and during the spacecraft flight must be directed to the nadir. The information from the detector units should go to the data processing unit, where it is processed and accumulated for transmission to the onboard complex for payload control and scientific data collection BKUSNI.

The detecting units are switched on independently of each other by control commands, which gives a certain flexibility in controlling the device. The sampling rates of detectors, telemetry collection parameters and some other device settings can also be changed.

The units of the low-energy spectrometer are a scanning electric separator in the form of axially symmetric spherical surfaces that focus the flows of charged particles on the input window of the secondary electron multiplier. Electron and proton detectors differ only in the polarity of the separator connection and the overclocking potential between the separator and the wind turbine. In the "default" mode, the full scan cycle for energies from 100 eV to 20 keV takes about 10 s, but this setting can be changed by control commands depending on the current task and the total detector load.

The unit of the medium-energy spectrometer is made according to the classical scheme of a semiconductor telescope with three silicon detectors with thicknesses of 40, 500 and 1000 μm, respectively, which, among other things, ensures the continuity of the data obtained with the data from the payload of meteorological spacecraft.

In the gas-discharge detector unit the SBM20-1 detector is used. Its readings are used as reference values of fluxes in the upper energy ranges. The detector is located behind the 2-mm protection of the instrument front wall, normally to the axes of view of the remaining detecting units.

5.2 Monitoring of gamma quantum fluxes with the use of SG/1 instrument

To monitor the gamma rays from the Earth's atmosphere and control the solar flare activity in the hard X- and gamma ray ranges, the SG/1 instrument is used, which measures the fluxes and differential energy spectra of hard X- and gamma rays in the near-Earth space in the energy range 0.02 - 10.0 MeV at an energy resolution of 7.2% on the Cs-137 line (0.662 MeV). The effective area of the instrument for gamma ray detection is ~ 50.0 $\cdot cm^2$. In the normal position, the instrument detector should be directed by the entrance window into the nadir. In this case, a viewing angle will be of at least ~50.0$\cdot cm^2$.

A CsI(Tl) scintillation crystal of 8 cm thick and 8 cm in diameter, viewed by a photomultiplier tube (PMT) of the Hammamatsu R1307 type, should be used as a detector in the SG/1 instrument. The instrument allows separating events associated with the detection of gamma quanta and neutrons in a CsI (Tl) crystal. The main mode of using the instrument is to carry out continuous measurements for a long time. The instrument provides measurement of count rates in 16 channels, differing in the values of energy release in the detector crystal, as well as regular transmission of three 256-channel spectra, collected separately in the low-energy region, in the hard part of the range, as well as energy release spectra corresponding to the detection of fast neutrons. In addition, the instrument has the ability to transmit a part of the primary information containing the amplitude and the exact time for each case of interaction in the detector, i.e. the so-called "event-by-event" recording. Data of this type can be generated both regularly and by the burst trigger formed at the moment the detector count rate increases.

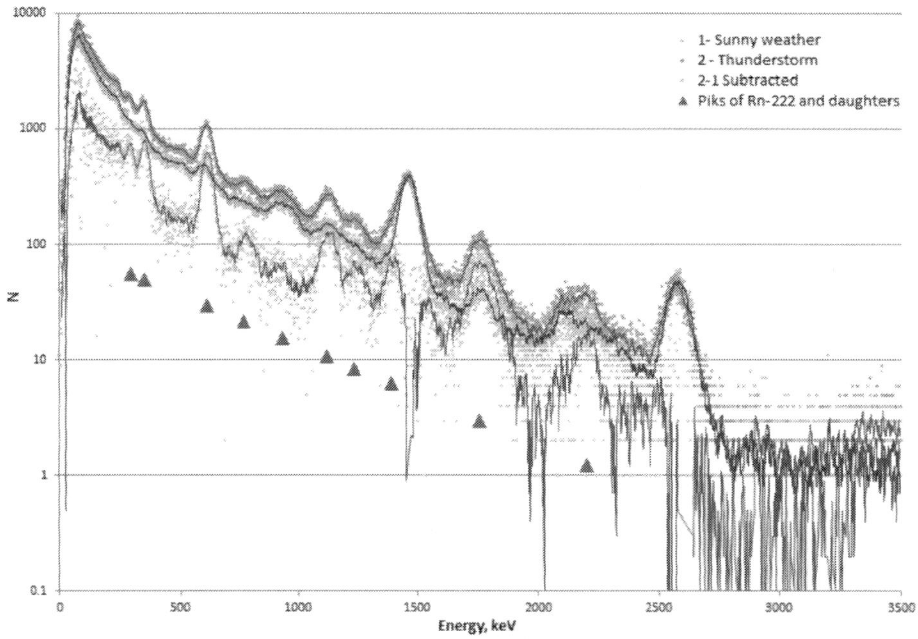

Fig. 4. Energy spectrum of atmospheric gamma rays detected on July 27, 2015 in the Moscow region during a thunderstorm and in clear weather.

Thus, scientific data from the SG/1 instrument are transmitted by information packets ("frames") of two types, i.e. "main regular information package" and "event by-event array". The size of the data field of each of them is 2032 bytes. "event by event array" frames can be generated regularly or by a burst trigger, while the internal structure of the frame does not change, but a different value of the start and end markers is used. Sampling rate is 1 Hz in background and 1 kHz in burst mode. Sweep time, i.e. total time of energy spectrum measurement is 0.1 - 10 s.

The data volume transmitted by the SG/1instrument can be adjusted using control commands that change the frequency of formation of packets of one type or another, the thresholds of the burst trigger, and other adjustable parameters. The daily quota for the SG/1 instrument is 280 MB.

As an illustration of the capabilities of the SG/1 instrument as a gamma spectrometer, we can consider the results of measurements using a laboratory prototype of the instrument, which in composition, including the detector and electronics, is fully equivalent to the flight model. In particular, the spectra of atmospheric gamma ray background at sea level, measured during a thunderstorm and for clear weather are shown in Fig. 4 [16]. The spectra show many lines of gamma rays. The background lines K-40 (E_γ = 1.46 MeV) and Tl-208 (E_γ = 2.6 MeV) are visible with the same intensity during a thunderstorm and in clear weather. Other lines are associated with the isotope of radon Rn-222, which in turn is a daughter isotope of Ra-226. As is known, the lifetime of the isotope Rn-222 3.9 days makes it possible to penetrate a sufficient number of radioactive atoms into the atmosphere. After a series of decays of Rn-222, short-lived isotopes are formed, in particular Bi-214 with a half-life of ~ 20 min, which give the observed lines of gamma rays. Gamma ray lines Rn-222 are visible at all observation intervals, but their amplitude is much greater during thunderstorms than at intervals corresponding to clear weather. The spectrum obtained by subtracting the spectrum measured during clear weather from the spectrum measured during a thunderstorm is also shown in Fig. 4. It shows that all the additional gamma rays detected during a thunderstorm are due precisely to Rn-222 radon and its daughter isotopes.

5.3 Monitoring of galactic and solar cosmic ray fluxes with the use of GALS instrument

Monitoring of fluxes of high-energy particles, i.e. galactic and solar cosmic rays (GCR and SCR) is carried out with use of the GALS/1 instrument, which should measure the flux density of protons with energies of >600 MeV in three integral energy ranges. The instrument is a radiation detector based on a Cherenkov counter. The selection of energy ranges is provided by threshold discrimination of the detector output pulses. The operating mode of the instrument is continuous, while the data is read once per second at a frequency of 1 kHz.

The dependence of the effective detection area on the energy of the isotropic proton flux is such that at a proton energy of more than ~ 800 MeV, it weakly depends on the energy and amounts to ~ 36 cm^2. To ensure a lower energy detection threshold with a sufficient effective area, the problem arises of separating the useful signal against the background of the dark current pulses of the multiplier itself. The requirement for the signal-to-noise ratio is reduced to the requirement that the counting rate of the detector, due to the detection of particles when measuring small (up to 5 particles/cm^{2}s) fluxes of GCR particles, is at least an order of magnitude higher than the counting rate due to dark current pulses. This requirement is met by using a low-noise photomultiplier tube and ensuring its optimal operation, taking into account the sensitivity of the electronic signal processing circuit. Based on the experience of previous similar experiments, when using a low-noise photomultiplier, it can be set at an energy level of 600 MeV. A decrease in the detection threshold can lead to an avalanche-like increase in the contribution of the photomultiplier noise pulses when operating conditions change.

The use of data from the GALS/1 spectrometer throughout the spacecraft's orbit, in particular, when it operates in the zone of the outer radiation belt, can be difficult due to the contribution of background electron radiation to its counts. To take into account the contribution of background radiation and, accordingly, to expand the possibility of using the measurement data of the GALS/1 spectrometer, additional measures must be taken. The level of sensitivity to a signal from 600 MeV protons corresponds to a signal from electrons with an energy of about 1.5 MeV. The protective casing of the sensor prevents the main flux of electrons with energies up to 5 - 6 MeV from entering the radiator, but it cannot completely remove the electronic background. Cutting off the electronic background can be achieved only by lowering the sensitivity of the multiplier signal registration circuit, which will lead to a simultaneous increase in the energy threshold for the detection of space radiation. To take into account the contribution of the electronic background during the semantic processing of information, an additional 2 thresholds for particle registration were introduced into the registration circuit of the photomultiplier signal of the Cherenkov detector due to the introduction of additional threshold discriminators with higher response levels. An aggregate analysis of the counts of simultaneous measurements in all channels will make it possible to exclude the contribution of the background radiation and obtain additional information on the fluxes of relativistic electrons in the NES.

6 Monitoring of atmospheric ozone with the use of the Ozonometer instrument

The measurement of the total content of ozone and other gases in the Earth's atmosphere has been carried out by ground-based and satellite methods for a long time. With the help of ground-based measurements, a large array of data has been obtained, which is used to solve various scientific and applied problems in the physics and chemistry of the Earth's atmosphere. However, such measurements are of a local nature and cannot give a complete picture of the ozone distribution. Since the mid-1960s, satellite methods have begun to play an increasing role in obtaining information on the ozone content, primarily a method based on the interpretation of measurements of reflected and scattered solar radiation in the ultraviolet (UV), visible and infrared (IR) spectral regions. The possibility of obtaining practically global information on the concentration of gaseous components of the atmosphere (including ozone) using satellite methods made these measurements extremely important in the study of various aspects of ozone dynamics, its relationships with various atmospheric parameters, changes in solar activity, in the study of long-term trends and factors, causing them.

Thus, to date, prerequisites have been created for the development and practical use of integrated methods of satellite monitoring of the atmosphere for solving regional and global problems of monitoring atmospheric pollution, interregional and transboundary transport of pollution, assessing the impact of these processes on climate change. Three main methods of remote sensing of the Earth's ozone layer can be distinguished: the emission method based on measurements of the Earth's own radiation and the atmosphere, the backscatter method based on measurements of ultraviolet solar radiation scattered by the atmosphere, and the absorption method associated with measuring the transparency of the atmosphere in the direction of an Earth satellite - a source of radiation (Sun or stars).

Measurement of radiation in these methods is carried out in nadir or at different angles to the nadir. According to the direction of the observation line of the equipment relative to the nadir, the measurements are conventionally divided into limb and nadir measurements.

In the case of the nadir scheme, the outgoing radiation of the atmosphere is measured in the vertical direction or at different angles to the vertical. With limb measurements, the observation line is directed towards the Earth's horizon. Due to the large length of the beam path in the atmosphere, these measurements make it possible to determine, along with ozone, the content of other small gas components in its upper layers.

In the Ozonometer instrument, the backscattering method (nadir observation) is used for determining the ozone content in the atmosphere. Its essence is as follows. The short-wavelength part of solar radiation, as noted earlier, undergoes noticeable scattering by molecules and fluctuations in the atmospheric density. The scattering of radiation obeys Rayleigh's law, according to which the intensity of the scattered radiation is inversely proportional to the wavelength to the fourth power and is distributed in such a way that it is twice as large in the direction of propagation as in the transverse direction. The ozone content even at the maximum of its altitudinal distribution is much less than the concentration of the main components of the atmosphere. Therefore, the absorption of sunlight in the ozone bands occurs gradually, as it "deepens" into the atmosphere. Radiation penetrating the atmosphere causes a noticeable flux of backscattered UV radiation. Of course, the depth of penetration of sunlight into the atmosphere depends on the absorption capacity of ozone at a given wavelength. In the region of the main Hartley absorption band, the sun's rays are completely absorbed in the ozone layer, the intensity of outgoing UV radiation in this part of the spectrum is determined by the solar radiation scattered by the upper atmosphere and absorbed by ozone.

In the region of weaker absorption of the Higgins band, solar radiation with weakening reaches the Earth's surface. In this part of the spectrum, the outgoing radiation is determined by the scattered atmosphere and solar radiation reflected from the Earth's surface and its cloud cover.

Calculations show that in the wavelength range of 0.25 - 0.3 μm, UV radiation from the Sun is absorbed in the ozone layer without reaching the Earth's surface, and each of the wavelengths in this region has a certain scattering depth and a rather sharply pronounced relatively narrow effective scattering layer. The shorter the wavelength, the higher its "effective layer" is. This portion of the spectrum can be used to determine the vertical distribution of ozone.

UV radiation in the spectral range 0.3 - 0.34 μm penetrates into the lower layers of the atmosphere. Scattering at these wavelengths captures practically all dense layers of the atmosphere from the Earth's surface to an altitude of 60 km. This wavelength range is used to determine the total ozone content in the atmosphere.

The total scattered radiation from all layers determines the intensity of UV solar radiation scattered by the Earth's atmosphere. The ultimate goal of the method is the distribution of ozone in the atmosphere according to spectral measurements obtained from the satellite.

Thus, the main task of the experiment with the Ozonometer instrument is the global and continuous measurement of the ozone content in the Earth's atmosphere based on spectroscopic measurements from two strong absorption bands - the Hartley band (ozone absorption band in the range 200 - 300 nm with a maximum of about 255 nm) and the Higgins band (a system of absorption bands with alternating minima and maxima in the range 300 - 360 nm). Absorption in the Chappuis band (band 450 - 650 nm) is weaker, but it can also be used to study the gaseous composition of the atmosphere from satellites, provided that the spectrometer sensitivity is sufficiently high. Also, by the ultraviolet and visible spectrum of the scattered radiation of the Sun, it is possible to determine the concentrations of other important gas components: NO, NO_2, NO_3, SO_2, BrO, H_2CO and others with absorption bands in the studied range. For example, nitrogen oxides NO_X (NO, NO_2, NO_3) play an important role in atmospheric chemistry.

The result of measurements using the Ozonometer instrument will be:

- global maps of the distribution of the total ozone content in the atmosphere over the entire globe with a frequency of 1 day. Measurements are not carried out during the polar night;
- the same maps of nitrogen oxides (NO2);

- maps of excess content of NO3, H2CO, SO2 over the natural level;
- products of the generalization of these measurements (climatology, changes from year to year, long-term trends, etc.).

To solve the set tasks, it is proposed to create a spectrometer for the UV and visible radiation range. The spectrometer assumes measurements in a wide wavelength range from 0.3 to 0.8 μm with a spectral resolution of 0.4 to 0.6 nm. The spectrometer is developed on the basis of a reflective diffraction grating using a symmetric crossed Czerny-Turner scheme. Observations are carried out in nadir. Solar radiation reflected from the Earth's surface, i.e. Twice passed through the atmosphere, it hits the rotating mirror, passes through the entrance telescope, which creates an optical beam of the required diameter, and is fed to the entrance slit of the analyzer with a size of 5 - 10 by 1000 microns. After the slit, the light hits a collimating mirror, which forms a parallel beam. This beam is directed to a diffraction grating (600 lines/mm), the main element of the spectrometer. After the grating, the diffracted light hits the second mirror, which focuses the first-order spectrum in the detector plane. The detector is a Toshiba silicon-based CCD array with a spectral range of 200 - 1100 nm, 3648 pixels, and a pixel size of 8×200 μm.

The design of the developed instrument provides the presence of instantaneous rectangular fields of view with angular dimensions within 1°, elongated in the direction perpendicular to the direction of movement of the vehicle, which at an orbital altitude of 600 km gives a spatial resolution on the earth's surface of $\approx$10 km. Subsequent mapping of the distribution of ozone or other gases will represent this area as a point on the map.

7 Conclusion

Thus, in case of successful implementation of the Ionosphere space project, a multi-satellite constellation will be deployed for the first time, which will allow simultaneous monitoring of various regions of the ionosphere, including near-equatorial and Polar regions. In this case, both active experiments on sounding the ionosphere in order to determine the parameters of the ionospheric plasma, in particular, obtaining a three-dimensional picture of the electron concentration distribution, and passive measurements of radiation scattered in the atmosphere from ground-based and orbital transmitters will be carried out. In addition, comprehensive monitoring of space weather factors will be carried out, including control of the magnetic-wave situation, fluxes of energetic charged particles and gamma-quanta in near-Earth space. A separate range of tasks is associated with monitoring atmospheric ozone in connection with geomagnetic and solar activity.

The results to be obtained during the implementation of the "Ionosphere" mission should provide important scientific information about dynamic processes in the "atmosphere - ionosphere - magnetosphere" system, "wave-particle" interactions, acceleration of charged particles, possible nonlinear effects in the ionosphere and magnetosphere. plasma. Also, the implementation of this project will be of great practical importance in terms of monitoring the geophysical situation in near-earth space, monitoring space weather.

References

1. E.L. Afrajimovich, E.I. Astaf'eva, O.I. Berngardt, V.V. Dem'janov, T.N. Kondakova, O.S. Lesuita, B.G. Shpynev, Izv. Vuzov, Radiophysica (Russ.) **47**, 509 (2004)
2. E.L. Afrajimovich, E.I. Astaf'eva, Solar-Terrestrial physics (Russ.) **3**, 50 (2003)
3. L.I. Mandelshtam, N.D. Papaleksi, JETPh (Russ.) **7**, 659 (1937)
4. V.V. Migulin, Uspekhi Phys. (Russ.) **128**, 667 (1979)

5. Ja.L. Al'pert, Uspekhi Phys (Russ.) **71**, 369 (1960)
6. G.K. Solodovnikov, V.M. Sinelnikov., E.B. Krokhmal'nikov, Distantcionnoe zondirovanie ionosphery Zemli s ispolzovaniem radiomajakov i kosmicheskykh apparatov (Russ.) (Nauka, Moskva, 1988)
7. V.M. Sinelnikov., S.V. Pakhomov, A.P. Glotov, G.P. L'vova, T.L. Gyljaeva, *Fazovy doplerovskii metod izmerenija vysotnogo profilja electronnoy concentracii v nizhnej ionosphere s pomoschiy geophysicheskikh raket*, in Issledovanija structury i volnovykh svojstv prizemnoj plasmy (Russ.), IZMIRAN, Moskva, 40 (1980)
8. L.M. Erykhimov, O.I. Maksimenko, *Issledovanie ionosphernykh neodnorodnostey pri pomoschi ISZ*, in Dreify neodnorodnostey v ionosphere (Russ.), Nauka, Moskva, 41 (1973)
9. L.M. Erykhimov, E.E. Mitjakova, E.N. Mjasnikov, S.V. Poljakov, A.V. Rakhlin, V.M. Sinelnikov, Izv. vuzov. Radiophysica (Russ.) **20**, 1814 (1977)
10. M.A. Kolosov et al, Doklady Akademii nauk SSSR (Russ.) **241**, 555 (1978)
11. Y.-H. Kuo, S. Sokolovsky, R.A. Anthes, F. Vandenberghe, Special Issue of Terr., Atmosph. And Ocean. Sci. **11(1)**, 157 (2000)
12. V.E. Kynitsyn, E.D. Tereschenko, Tomografija ionosphery (Russ.) (Nauka, Moskva, 1991)
13. V.N. Oraevskii, V.E. Kynitsyn, Yu.Ja. Ruzhin, E.S. Andreeva, A.H. Depueva, E.F. Kozlov, O.G. Razinkov, I.I. Shagimuratov, Geomagnetism i aeronomia (Russ.) **35**, 117 (1995)
14. V.E. Kynitsyn, E.D. Tereschenko, E.S. Andreeva, Radiotomografija ionosphery (Russ.) (FIZMATLIT, Moskva, 2007)
15. Yu.G. Ivanov, V.M. Russkin, V.M. Sinelnikov, G.K. Solodovnikov, *Avtomatizirovannaja systema registratcii I obrabotky amplitudno-fazovykh fluktuatcii radiosignalov*, in Volnovye processy v prizemnoy plazme (Russ.), Nauka, Moskva, 61 (1988)
16. V. Bogomolov, A. Chilingarian, G. Garipov, et al., *Results of TGE Study in 0.03-10 MeV Energy Range in Ground Experiments near Moscow and Aragatz*, in Proceedings of International Symposium on Thunderstorms and Elementary Particle Acceleration (TEPA-2016), Cosmic Ray Division, Yerevan Physics Institute, ISPIRE C15-10-02, 3-7 October 2016, Nor-Amberd, Armenia (2016)

Operation of a multi-frequency parametric side-scan sonar in an ice waveguide

Petr Pivnev[1*], *Sergey* Tarasov[1], *Zhu* Jianjun [2], *Vasily* Voronin[1]

[1]Southern Federal University, Taganrog, Russia
[2]Harbin Engineering University, China

Abstract. Hydroacoustic systems for mineral exploration, solving engineering problems and monitoring the ecological state of the world's oceans are currently being intensively developed. However, the practical use of hydroacoustic systems for solving the problems under consideration, operating in the traditional mode, has some significant limitations. These restrictions are largely related to the state of the marine areas in which such work is carried out. Especially little is known about the patterns of propagation and interaction of acoustic waves in marine basins with ice cover. These areas are rich in minerals and intensive shipping is developing in them. Therefore, an important place in acoustic research is occupied by the study of the acoustic properties of the ice cover of the polar regions of the Earth. This is determined by the fact that the ice sheet is a unique constantly collapsing and renewable natural system. In this regard, the conditions for the propagation of acoustic waves are changing. More than 70% of the Arctic basin is covered with ice, the lower boundary of which has significant irregularities with a standard deviation of up to 3 m, so the scattering of acoustic waves at such a boundary is significant and different at different frequencies. The formation of the acoustic field of the hydroacoustic systems used in these conditions is quite complex. Therefore, the task of assessing changes in the characteristics of the field and the use of appropriate hydroacoustic systems for their effective use is urgent.

1 Page layout

One of the tools for studying the bottom relief and bottom sediments is a side-scan sonar (SSS). It allows to us significantly increase the productivity of survey due to the large area of the surveyed space in one measurement cycle. A characteristic feature of the SSS is the shape of the beam pattern: wide in the vertical plane (to provide a wide viewing band of the bottom and the aquatic environment) and narrow in the horizontal plane (to increase the resolution of the locator by angular coordinates). Traditionally, a wide directivity characteristic in SSS is created either by the small wave dimensions of the array in a given plane, or by the using of developed curved array surfaces.

* Corresponding author: pivnev@mail.ru

Array with curved surfaces at low frequencies have too large dimensions. Arrays that form a wide directivity due to small wave sizes will not allow placing a large power on the radiating surface. So, the energy potential are reduces. The creation of small-sized arrays in the low-frequency range is possible by using of radiating parametric arrays, whose operation is based on the nonlinear interaction of acoustic waves in an aquatic environment. The interaction of two or more waves propagating in the medium generates waves with combination frequencies, including in the low-frequency range. To profile the bottom and bottom structures or to search for silted objects, it is necessary to use acoustic signals in the low-frequency range for their good penetration into the bottom sediment thickness. Therefore, the low-frequency part of the spectrum of waves generated by a parametric antenna is of interest.

The efficiency of generating secondary waves in a parametric radiating array depends on the concentration of the energy of the pump waves in space, that is, it depends on the width of the directivity characteristic. An increase in the energy of waves in the low-frequency range is associated with an increase in directivity [1]. In SSS, a wide directivity characteristic in a parametric array can be formed by using several small-sized highly directional pumping arrays. In other words, we can apply the approximation of an arc antenna by plane partial pumping arrays with the addition of a low-frequency field in space [2,3].

The total beamwidth will be determined by the beamwidth of the partial array, the number of individual partial pumping arrays, and the angle between them. This design of the parametric arrays provides a wide field of view, when all elements are switched on simultaneously.

At the same time, it can work as a scanning one, when individual converters with a rather narrow beam width, are switched on alternately. To conduct research with a parametric side-scan sonar, a parametric radiating array with a beam width about 25 degrees in the vertical plane and about 3 degrees in the horizontal plane was developed. The antenna consists of five partial pumping arrays with pumping frequencies of 130 and 150 kHz. This makes it possible to form the necessary beam pattern in the horizontal plane. The energy efficiency of such arrays is ensured by the narrow beam patterns of partial arrays in both planes.

2 The use of multicomponent signals to increase the efficiency of radiation at low frequencies

To increase the efficiency of converting the energy of pumping waves into the energy of a difference frequency wave, it is proposed to use the interaction of waves in a multicomponent pump signal [4,5]. At the same time, a multicomponent signal will be understood, as a pumping signal in which the individual components (waves) are separated by frequency from each other at the same frequency, usually equal to the lower difference frequency of the parametric array.

The amplitude of the spectral low-frequency component of the signal generated in the medium will be equal to:

$$P_{-} = \sum_{m=1}^{n-m} P_m , \qquad (1)$$

Where, P_m is the amplitude of the m-th component of the difference frequency wave. It is equal to:

$$P_m = \sum_{m=1}^{n-m} A_{k,k+m} p_k p_{k+m} , \qquad (2)$$

n - is the number of components; m - is the component number of the difference frequency signal; p_k, p_{k+m} -are the amplitudes of the interacting components of the pump waves.

$$A_{k,k+m} = B_{k,k+m} I_{k,k+m} \tag{3}$$

$B_{k,k+m}$ – coefficient that takes into account the characteristics of interacting waves and the parameters of the propagation medium:

$$B_{k,k+m} = \frac{a^2 \varepsilon m \Omega}{8 c_m^4 \rho} \exp\left(-\frac{z}{L_m}\right), \tag{4}$$

$$I_{k,k+m} = i \cdot \int_0^z \frac{\exp(-\alpha z(1 - i\Delta D))}{\left(1 - i\frac{z-y}{L_{dm}}\right) + y\left(\frac{2iL_{dm}}{l_{dk} \cdot l_{dk+1}} + \frac{z}{l_{dk} \cdot l_{dk+1}}\right)} dy\ , \tag{5}$$

where, $l_{dk} = \frac{a^2 \omega_k}{2c_k}$, $l_{dk+1} = \frac{a^2 \omega_{k+1}}{2c_{k+1}}$ - the diffraction zones of the pump array by the k-th and k+1- st components of the pump waves; $L_{dm} = \frac{a^2 m \Omega}{4 c_m}$ - - the length of the diffraction zone of the difference frequency wave component; $\alpha = \frac{2c_m^3 \rho}{b(m\Omega)^2}$ – the attenuation coefficient of the difference frequency wave component;

c_m, c_k, c_{k+1} – the sound velocity of the *m*-th component of the difference frequency wave and the interacting pump waves;

$\Delta D_m = (k_j - k_{j+1} - K_m) l_{zm}$ - characterizes the change in the phase shift between interacting waves at a distance proportional to the attenuation length of the pump waves and determines the period and amplitude of the oscillations of the formed waves of difference frequency.

Analyzing the above expressions, it can be stated that the amplitude of the lowest harmonic of the difference frequency wave during the interaction of a multicomponent signal increases almost proportionally to the number of interacting components of the difference frequency waves (the number of components minus one). Thus, the range of the parametric side-scan sonar is significantly increased. For example, the studies conducted in [6, 7] showed that the use of a six-component pump signal allowed to reach the depth of sounding of the bottom layers five times more than when using a two-component pump signal.

Figure 1 shows the calculated values of the harmonic amplitudes of the difference frequency signal. They obtained by the interaction of a six-component pump signal for different penetration depths of difference frequency signal into the ground. The calculated values, are shown by points combined by curves to illustrate the dependence at the same penetration depths. Curve 1 corresponds to the depth of the layer h = 1 m, 2 – 2 m, 3 – 3 m, 4 – 5 m. The point indicated by the number 5 represents the value of the amplitude of the difference frequency wave with a two-component pump signal.

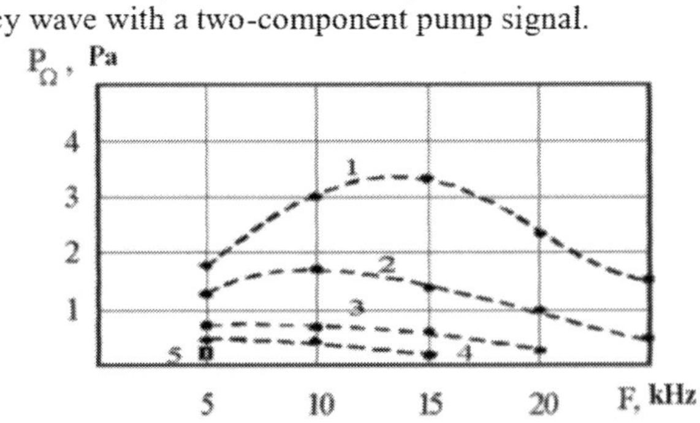

Fig. 1. Amplitudes of harmonic components of the difference frequency signal in the interaction of a six-component pump signal.

The amplitudes of the waves of the second and higher harmonics of the difference frequency also increase significantly. The spectrum of secondary waves is enriched. For sonar, the energy of higher harmonics of difference frequency waves may be used. However, the use of such multicomponent pumping signals requires an increase in the dynamic range of pump wave amplifiers.

3 Features of behaviour of parametric antenna characteristics in an "ice" waveguide

The broad beam pattern of the parametric side-scan sonar in the vertical plane when operating in a waveguide can illuminate a certain area of the water surface. In this case, there is a reflection of both pumping waves from the surface and waves of difference frequency. Let's consider how the combined reflection of these waves affects the field of a reflected wave of difference frequency [8-9]. When interacting waves reflected from the waveguide boundaries, two parametric arrays are formed, as it were. The first is limited by the distance from the pumping array to the reflecting surface. The second array formed by waves reflected from the surface. Moreover, virtual arrays have different interaction characteristics. The second parametric array generates waves of difference frequency due to the interaction of reflected pump waves in a virtual pump array formed as a result of reflection of pump waves from the waveguide boundary. As a result, the signal is formed by summing the difference frequency waves reflected from the waveguide boundary with the difference frequency waves generated after the pump waves are reflected from the waveguide boundaries.

The acoustic resistance of the waveguide boundaries can take different values. These are acoustically rigid boundaries with a reflection coefficient of $V=1$, acoustically soft boundaries with a reflection coefficient of $V= -1$ and boundaries with a complex reflection coefficient. In this case, $V=V_1+iV_2$, where V_1 and V_2 are the real and complex part of the reflection coefficient.

Considering only the axial distribution of difference frequency waves, the solution for difference frequency wave in an infinite space can be written in a simplified form through integral exponential function as:

$$P_{-}(0,z_3) = \frac{ip_{01}p_{02}\varepsilon\Omega^2 a^2}{8c_0^4\rho_0(i+Bz_3)} e^{\frac{d-iz_3}{i+Bz_3}}\left[E_i\left(\frac{Bz_3+d}{i+Bz_3}\right) - E_i\left(\frac{d-iz_3}{i+Bz_3}\right)\right], \quad (6)$$

where $E_i(z)$ – integral exponential function [8, 9].

In the general case, at an arbitrary angle of incidence, the complex amplitude of the difference frequency wave is determined by the addition of the wave formed before the reflecting boundary with the wave appearing after the reflection of the pump waves with the corresponding phases. In the case of a normal beam incident on the waveguide boundary , it is possible to obtain an expression for the axial distribution of the sound pressure of a difference frequency wave in the form:

$$P_{-} = \frac{ip_{01}p_{02}\varepsilon\Omega V L_{д}}{2c_0^3\rho_0(i+Bz_3)} e^{\frac{d-iz_3}{i+Bz_3}}\begin{bmatrix}(1-V)E_i\left(\frac{Bh_3z_3+i(h_3-z_3)+d}{i+Bz_3}\right)+ \\ +E_i\left(\frac{d-iz_3}{i+Bz_3}\right) - VE_i\left(\frac{Bz_3+d}{i+Bz_3}\right)\end{bmatrix}, \quad (7)$$

for $V=1$, i.e. an acoustically rigid boundary, this expression completely coincides with the previous one and the result turns out to be the same as for an infinite medium. Therefore, it can be conclude that the acoustically rigid boundary does not affect the basic parameters of the parametric array.

In the case of reflection of a parametric array beam from an acoustically soft or impedance boundary, the axial and spatial distribution of the difference frequency wave can changes significantly due to changes in the phase ratios of the difference frequency waves formed before and after the interface.

Experimental studies were conducted on the effect of reflecting boundaries on the field of the parametric array.

Figure 2 shows the transverse distributions of the amplitude of the difference frequency wave obtained experimentally and by calculation in the presence of an acoustically soft boundary. The location of the boundary at an angle of 45 degrees demonstrates the phase changes of the reflected waves of the difference frequency, as if it were an impedance boundary. The experimental curves are shown as solid lines, and the calculated ones are dashed. Measurements and calculations were carried out at different distances from the reflecting boundary - 0.04; 0.08; 0.14; 0.2 and 0.28 m. The distance from the pumping array to the boundary was h = 0.28 m, and the angle of incidence was 45 degrees. Experiments and calculations were carried out for a difference frequency of 150 kHz. Pumping waves with an average frequency of 2.06 MHz were emitted by a pump converter with a diameter of 18 mm [1,2,8,9].

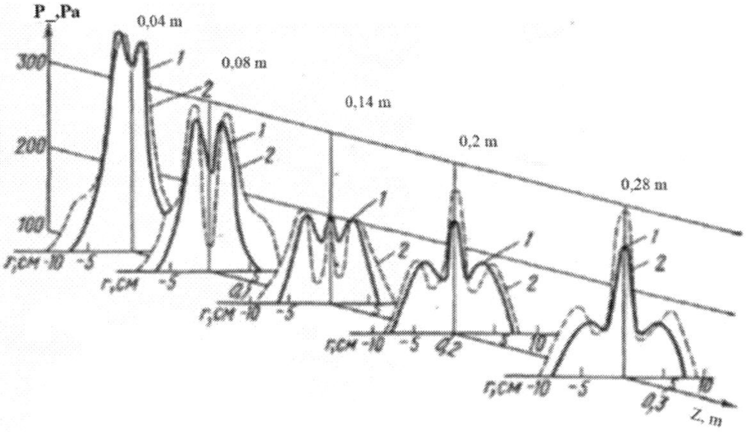

Fig. 2. Experimental (curves 1) and calculated (curves 2) transverse distributions of the difference frequency wave.

The results of the research show that the transverse distributions of waves of difference frequency vary significantly with distance and with a change in the impedance of the reflecting boundary of the waveguide.

Let us briefly consider the scattering and reflection of acoustic waves by the ice boundary of the waveguide [10, 11]. The reflection and scattering of sound by the ice cover depends on the characteristics of the relief of its lower surface and the physical characteristics of the ice. If the sound wave length is much greater than the thickness of the ice cover, then this boundary can be considered as a thin plate lying on an aqueous liquid half-space. Then the reflection coefficient from such a boundary, as shown in [11], is modulo close to unity and the reflection can be considered as from the free surface of the sea, i.e., as from a soft boundary and the arguments about the nature of the field given above are valid. With increasing frequency, the wavelength becomes commensurate with the thickness of the ice cover of the waveguide. At the same time, during reflection, the physical characteristics of the cover begin to significantly affect, in particular, the relief of the lower boundary of the cover, i.e., the upper boundary of the waveguide. If the wavelength becomes commensurate with the terrain irregularities, then they play a major role in the scattering and reflection of waves from the ice cover of the waveguide. The angular dependence of the wave scattering force by the boundary increases significantly. The attenuation of sound in the ice cover also increases with frequency and this affects the amplitude and phase of the sound scattering coefficient by the ice boundary. The same phenomena affect the shape of the scattering indicatrix. It can be approximated by Lambert's law.

4 Conclusions

The paper shows that one of the tools for studying the bottom relief and bottom sediments is a side-scan sonar, which allows to significantly increase the productivity of search operations due to a wide band of the surveyed space. To conduct research with a parametric side-scan sonar, a parametric radiating array with a wide beam pattern in the vertical plane and a narrow one in the horizontal plane was developed. To increase the efficiency of converting the energy of pumping waves into the energy of a difference wave, it is proposed to use the interaction of waves in a multicomponent pump signal. The broad beam pattern of the parametric side-scan sonar in the vertical plane when operating in a waveguide can illuminate a certain area of the surface. At the same time, there is a reflection of both pumping waves from the surface and waves of difference frequency. The effect of the combined reflection of these waves on the field of the reflected wave of difference frequency is considered.

Studies show that when considering the field of a parametric array in a waveguide with an ice boundary, it is necessary to take into account the change in the field depending on the frequency and state of the ice boundary.

References

1. V.K. Novikov, V.I. Timoshenko, Parametric antennas in sonar (Shipbuilding, Leningrad, 1989) (in Russian).
2. V.A. Voronin, S.P. Tarasov, V.I. Timoshenko. Hydroacoustic parametric systems (Rostizdat, Rostov-na-Donu, 2004) (in Russian).
3. S. Tarasov, P. Pivnev, W. Chen, D. Durov, E3S Web of Conf., **127**, (2019) https://doi.org/10.1051/e3sconf /20191270.
4. V.A. Voronin, T.N. Kutsenko, S.P. Tarasov, Izvestiya TRTU, **15**, 1(2000) (in Russian).
5. A.S. Eller , J.A.S.A., **6**, 56 (1974).
6. V.A. Voronin, A.G. Ishutko, T.N. Kutsenko, Izvestiya TRTU, **35**, 6 (2003) (in Russian).
7. P.P. Pivnev, A.V. Voronin, Izvestiya TRTU, **90**, 1 (2009) (in Russian).
8. Karabutova, B.K. Novikov, S.P.Tarasov. Acoustic. Jour., **27**, 3 (1981) (in Russian).
9. L. K. Zarembo, V. I. Timoshenko, Nonlinear acoustics (Publishing House of MSU, Moscow, 1984) (in Russian).
10. V. V. Bogorodsky, V. P. Gavrilo, V. A.Nikitin, Acoustic. Jour., **22**, 2 (1976) (in Russian).
11. G.A. Lebedev, K.K. Sukhorukov. Propagation of electromagnetic and acoustic waves in sea ice (Gidrometizdat, St. Petersburg, 2001) (in Russian).

Registration of synchronous geomagnetic pulsations and proton aurora during the substorm on March 1, 2017

Ilya Varlamov[1,2,*], *Stanislav* Parnikov[2], *Igor* Ievenko[2], *Dmitry* Baishev[2], and *Kazuo* Shiokawa[3]

[1]M.K. Ammosov North-Eastern Federal University, Yakutsk, Russia
[2]Yu.G. Shafer Institute of Cosmophysical Research and Aeronomy SB RAS, Yakut Scientific Centre SB RAS, Yakutsk, Russia
[3]Institute for Space-Earth Environmental Research, Nagoya University, Japan

Abstract. Data of synchronous geomagnetic pulsations and proton aurora registrations were analyzed during the substorm on March 1, 2017 at Zhigansk (L=4.5, induction magnetometer), Maimaga (L=4, all-sky imager and Yakutsk (L = 3.3, induction magnetometer) stations, simultaneously with satellite measurement of EMIC waves. Ground-based registration of proton aurora is very difficult due to the fact that their intensity is much lower than the aurora intensity caused by precipitations of electrons, but in the event of substorm activity at the zenith of Maimaga station, a narrow (1 degree in latitude) proton arc was observed. Irregular pulsations of the diminishing periods (IPDPs) in the range of Pc1 geomagnetic pulsations associated with the injection of energetic protons were recorded simultaneously at Zhigansk and Yakutsk stations. This is the first report when STEVE (Strong Thermal Emission Velocity Enhancement) was observed in the course of a substorm with the onset at 12:45 UT after the decay of Pc1-associated proton arc. It is shown that the proton arc and geomagnetic pulsations are a consequence of ion-cyclotron instability in the area of the outer plasmasphere overlapping by energetic protons.

1 Introduction

In the area of overlapping of the outer plasmasphere (plasmapause) by ring current ions, many processes arise, the results of which, in particular, are electromagnetic ion cyclotron (EMIC) waves, mid-latitude stable auroral red (SAR) arcs and proton auroras, equatorward of the auroral oval. EMIC waves are generated as a result of ionic instabilities arising from mixing of hot and cold plasma. SAR arcs are the result of heating the outer plasmasphere by energetic ring current ions. The flux of superthermal electrons arising due to the Coulomb interaction, as a result of thermal conductivity, along the field lines, flows downward and causes a weak, monochromatic glow at the heights of the F2 layer of the ionosphere [1]. Proton arcs are a consequence of the generation of EMIC waves that scatter protons at pitch

*Corresponding author: varlamovi@mail.ru

angles into a loss cone [2]. During precipitation, protons undergo a recharge, as a result of which excited hydrogen atoms are formed. In the visible part of the spectrum, the Hα and Hβ lines are emitted by which the precipitation of protons from the Earth is identified. EMIC waves manifest themselves in the form of variations in the components of the magnetic field in the region of generation and geomagnetic pulsations of the Pc1 range on the Earth.

In this work, the results of the analysis of data of synchronous registration of geomagnetic pulsations at Zhigansk and Yakutsk, as well as optical registration of proton auroras and SAR arc at the Maimaga subauroral station during the event on March 1, 2017, are presented.

2 Equipment and methods for analysis of observation data

Registration of geomagnetic pulsations was carried out at the Oybenkel radiophysical polygon (56°N, 199°E geom., L = 3.3) near Yakutsk and the complex geophysical station Zhigansk (62°N, 196°E geom., L = 4.5).

Measurements of magnetic field variations and natural electrical potentials at the Oybenkel polygon are carried out using a 24-bit geophysical electromagnetic measuring system METRONIX ADU-07e. The electric and magnetic field sensors are connected directly to the ADU-07 (Analog / Digital Signal Conditioning Unit), which is the main unit of the system. The sampling rate is 16 Hz.

Several different types of induction magnetometer sensors are used in the PWING project [3]. One is the preexisting sensors before the PWING project, i.e., those made by Tierra Technica (ULC-150) for Zhigansk. The sampling rate is 64 Hz. Geomagnetically northward (H-) component and eastward (D-) component are recorded in the first and second channels, respectively, while the riometer output is recorded at the third channel. The 1 pulse-per-second (PPS) signal generated by a GPS receiver is recorded in the fourth channel as a time standard.

Optical observations were carried out at the Maimaga station of the SHICRA SB RAS (58°N, 200°E geom., L = 4). The luminosity was recorded using a «Keo Sentry» all-sky imager. The camera has a PI Acton ProEM 1024B CCD device and is equipped with six cavity interference filters with a half-width transmission of 2 nm. The filters are centered at wavelengths 557.7 and 630.0 [OI], 470.9 (N2+), 486.1 (Hβ), 620.0 (continuum), 480.0 (continuum) nm.

The time-frequency spectral analysis of the geomagnetic pulsation data was carried out using the multistage Thomson method (MTM).

All-sky imager data are presented in this work in the form of all sky images and keograms projected onto the Earth's surface for luminosity heights of 110 km for emissions of 486.1 and 557.7 nm and 250 km for emissions of 630.0 nm. The signal intensity is given in kilorayleighs.

Due to the irregularity of the proton luminosity of arc along latitude, the keograms were plotted along a relatively narrow longitudinal sector, along the 198th geomagnetic meridian, where the luminosity of arc was most often observed in the 486.1 nm emission. The keograms are constructed without taking into account the Van Rijn effect.

Keograms for emissions 557.7 and 630.0 were obtained with the subtraction of the continuum in the vicinity of 620.0 nm, for the emission of 486.1, the subtraction of the intensity of the continuum in the region of 480.0 nm was made.

3 Results

The event under consideration occurs in the MLT pre-midnight sector and is not stationary, proceeding against the background of weak but regular magnetospheric activations (see Fig. 1). According to data from mid-latitude stations (not shown here, see Fig. 2 in [4]), the next substorm activation occurred at ~12:45 UT on the Magadan meridian.

The proton arc arising in the vicinity of the zenith of the optical observation station had an uneven glow along the latitude. Separate maxima moved along the arc, in an easterly direction.

The speed of the eastern motion of the maxima observed before 13:03 UT was ~ 600 m / s. After 13:03 UT, the irregularities stopped reaching the station meridian and were observed on the western horizon. Example of irregularity in the luminosity of an arc is shown on the left part of Figure 2.

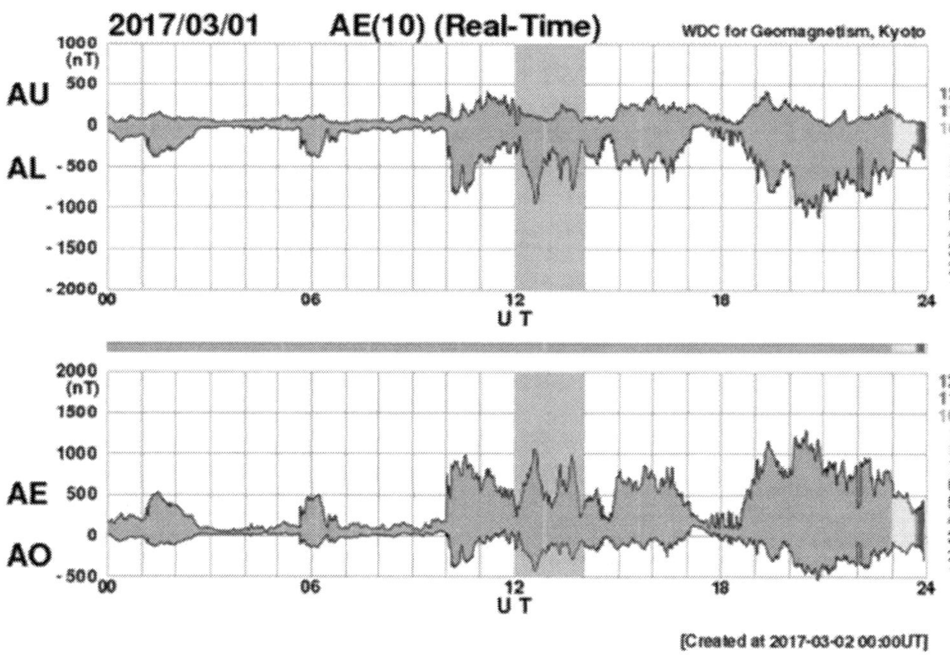

Fig. 1. Auroral indices on March 1, 2017 according to the Kyoto World Data Center for Geomagnetism.

The keograms shown in Figure 2 show the latitudinal dynamics of proton auroras and SAR arc. From the beginning of the considered period, a weak SAR arc was recorded ~ 200 km north of the station zenith. As the activity intensified, the red arc moved in the equatorial direction along the polar boundary of the diffuse aurora and crossed the zenith of the observation station at ~ 12:40 UT.

With the appearance of a proton arc at the same latitudes as the SAR-arc (SAR arc), the luminosity in the 630.0 nm emission intensified. Both structures moved synchronously along the latitude of the observation station. The total luminosity intensity in the 630.0 nm emission caused by the red and proton arcs reached ~ 150 Rayleigh. After the attenuation of the proton arc, the red arc continued to exist in the vicinity of the station zenith until 14:13 UT.

Also, among the features of the subauroral luminosity in the period under consideration, it can be noted that immediately after the proton arc decay, more polar than the SAR arc from

13: 23UT, STEVE began to form, accompanied by a picket fence (PF) (see the keograms in the 630.0 and 557.7 nm Fig. 2) [5].

In the Hβ (486.1 nm) emission from ~ 12: 20UT on the northern horizon, a wide band of proton aurora appeared (marked in the frames as a proton band), moving in the equatorial direction. At ~ 12: 55UT, for all emissions presented on the keograms, a sharp increase in the glow was recorded throughout the sky. After a brief flare, the equatorial boundary of the wide proton band began to recede.

From 12:35 UT, south of this band, at a distance of ~ 100 km from its near boundary, in the vicinity of the station zenith, along the existing SAR arc, a narrow proton arc appeared with an uneven distribution of the luminosity intensity along it. From the very beginning of the formation, the "non-uniform" arc began to move equatorially and by the time of the flare across the sky at ~ 12: 55UT, had already crossed the zenith of the station. During this initial period of time, the proton arc was most dynamic in the equatorial direction, its velocity reached values in ~150 m/s.

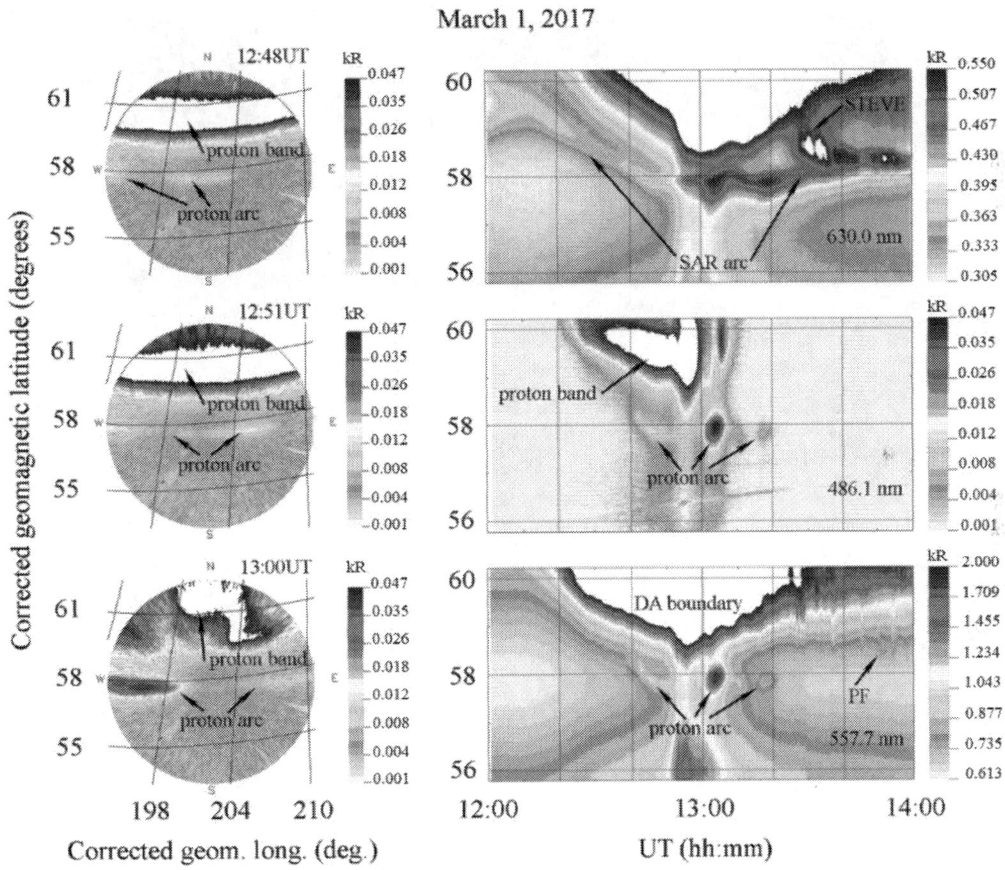

Fig. 2. Dynamics of subauroral luminosity at Maimaga station during observations on March 1, 2017. On the left side of the figure are images of all sky in emission of 486.1 nm for three points in time indicated in the images. The right side of the figure shows the latitudinal dynamics of the luminosity in the 630.0, 486.1, and 557.7 nm emissions in the form of keograms. All sky images and keograms projected onto the Earth's surface for luminosity heights of 110 km for emissions of 486.1 and 557.7 nm and 250 km for emissions of 630.0 nm. The signal intensity is given in kilorayleighs.

Further, the position of the arc has stabilized. The proton arc reached its culmination (maximum width and brightness) at 13: 03UT. At this moment, the arc width reached ~ 100 km, and the intensity in the Hβ emission was ~ 40 Rayleigh. By 13: 22UT, the proton arc, which existed at the latitudes of red, fades out.

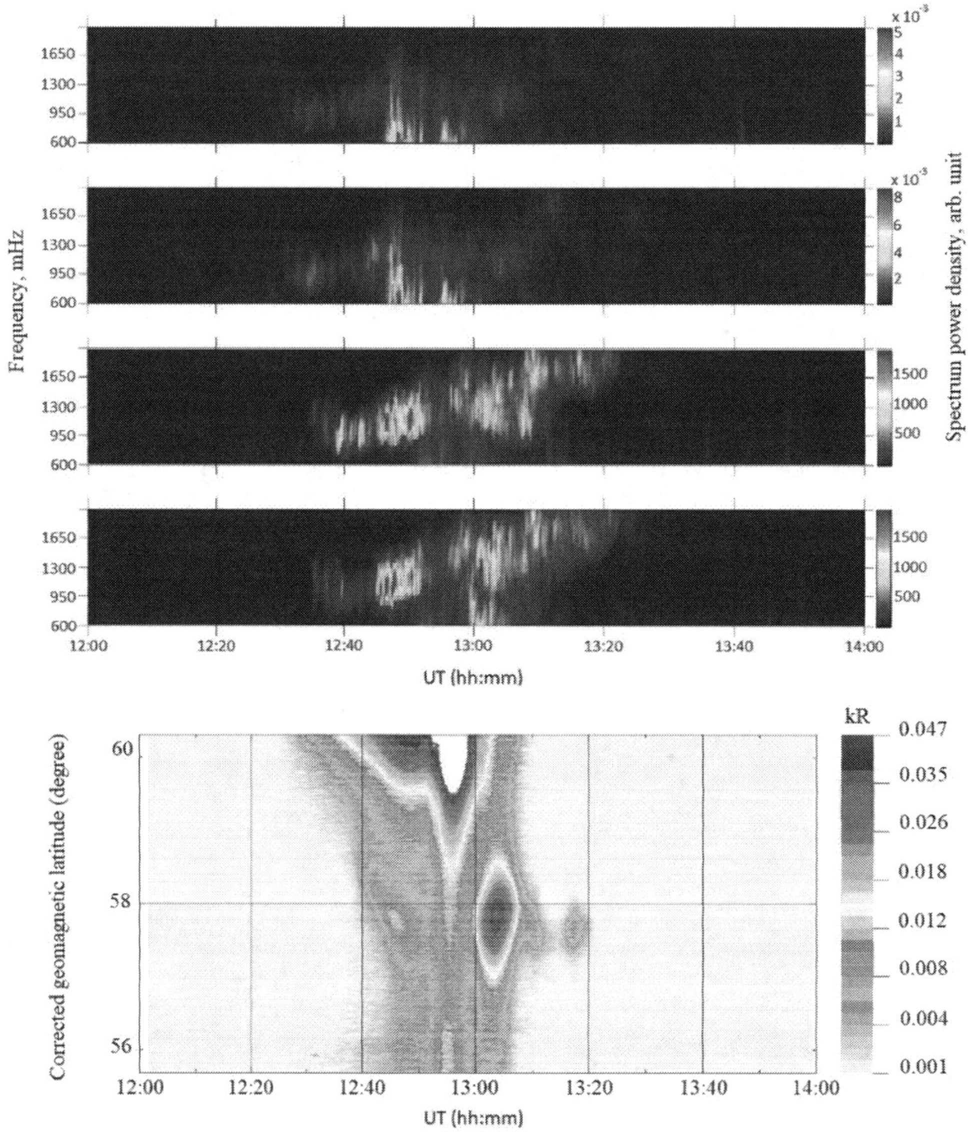

Fig. 3. From top to bottom, spectrograms of the H and D components at Zhigansk and Yakutsk stations, and the latitudinal keogram of the luminosity in the 486.1 nm emission at 110 km during observations on March 1, 2017 at 12:00-14:00 UT. The signal intensity is given in kilorayleighs.

In the 557.7 nm emission, the dynamics of the proton arc is expectedly similar to the dynamics in the 486.1 nm emission. The intensity of the proton arc in the green emission was much higher than in the Hβ emission and at the time of the culmination at 13:03 UT reached ~ 1000 Rayleighs. Also, in the 557.7 nm emission, it can be seen that by the time of the flare throughout the sky at ~ 12: 55 UT, the diffuse aurora boundary reached the zenith of the Maimaga station.

The spectrograms of the H and D components at Zhigansk and Yakutsk stations in Fig. 3 showed that in the interval 12:30-12:55 UT, geomagnetic pulsations Pc1 were recorded in the frequency range 0.9-1.3 Hz. Note that at Zhigansk station, Pc1 pulsations were recorded in time earlier than pulsations at Yakutsk station, the behavior and intensity of which coincided with the registration and dynamics of narrow proton arcs at the zenith of Maimaga station (starting from 12:35 UT). The fact of early registration of Pc1 pulsation at Zhigansk station indicates that the region of generation of narrow proton arcs was localized closer to the longitude of Zhigansk station than at the longitude of Yakutsk station, which is confirmed by direct optical observations of proton arcs at Maimaga station.

Figure 3 also clearly illustrates that the displacement of narrow proton arcs to low latitudes led to a noticeable increase in both the frequency and intensity of Pc1 pulsations at Yakutsk station, which, however, coincided in time with substorm activation, namely, the excitation of irregular geomagnetic pulsations at station Zhigansk.

A sharp increase in the luminosity over the entire sky at ~ 12:55 UT led to the attenuation of the Pc1 pulsations in the frequency range 0.9-1.3 Hz. Further, in the interval 12:58-13:10 UT at Zhigansk and Yakutsk stations, IPDP was recorded with an increase in frequency from 1 to 2 Hz in 12 minutes, possibly due to substorm activation at ~ 12:45 UT on the Magadan meridian. The IPDP intensity at Yakutsk station is more noticeable than at Zhigansk station, and it is associated with the proton arc recorded at the zenith of Maimaga station in the interval 12:58-13:10 UT, although there was no obvious radial drift in the arc dynamics.

In the interval 13:10-13:22 UT, only at Yakustk station, Pc1 pulsations were recorded in the frequency range 1.6-2 Hz, closely related to the observation of the proton arc at Maimaga station.

4 Discussion and summary

Data of synchronous geomagnetic pulsations at Zhigansk (L=4.5) and Yakutsk (L=3.3) and isolated proton arc registrations at Maimaga (L=4) showed one-to-one correspondence between magnetic and optical phenomena. Geomagnetic pulsations at Yakutsk in the interval of 12:35-12:55 UT coincided with isolated proton arc registrations at Maimaga and association of frequency growth of pulsations with equatorward motion of proton arc is clearly seen.

Irregular pulsations of the diminishing periods (IPDPs) in the range of 1-2 Hz in the interval of 12:58-13:10 UT were recorded simultaneously at Zhigansk and Yakutsk accompanying with intense isolated proton arc registrations at Maimaga. As shown by Baishev et al. [6] these magnetic and optical phenomena were associated with the injection of energetic protons after second substorm activation with onset at 12:45 UT.

New finding is optical observation of STEVE at Maimaga starting from13:23 UT after the decay of Pc1-associated proton arc registered in the interval 13:10-13:22 UT. Ground-based synchronous optical and geomagnetic observations showed complex physical phenomena in the course of a substorm.

Two-dimensional images of the isolated proton arc showed that its narrow (1 degree) latitudinal widths and limited longitudinal lengths are in agreement with measurements of isolated auroral arcs at Athabasca [7].

Thus it is shown that the proton arc and geomagnetic pulsations are a consequence of ion-cyclotron instability in the area of the outer plasmasphere overlapping by energetic protons.

A sharp increase in the luminosity over the entire sky for all emissions at ~12:55 UT and a short-term attenuation of geomagnetic pulsations Pc1, and STEVE occurrence starting from 13:23 UT require further studies using additional ground-based and satellite measurements.

This work was carried out with the financial support of the RFBR grant No. 21-55-500.

References

1. K. Cole, J. Geophys. Res., **70** (1965)
2. T. A. Yahnina, A. G. Yahnin, J. Kangas, J. Manninen, D. S. Evans, A. G. Demekhov, V. Y. Trakhtengerts, M. F. Thomsen, G. D. Reeves, and B. B. Gvozdevsky, Ann. Geophys., **21** (2003)
3. K. Shiokawa, Y. Katoh, Y. Hamaguchi et al, Earth Planets Space, **69**, 160 (2017)
4. S.G. Parnikov, I.B. Ievenko, D.G. Baishev, I.I. Koltovskoy *Proton aurora observation as a result of ion cyclotron instability*, in Proceedings of the SPIE 11560, 26th International Symposium on Atmospheric and Ocean Optics: Atmospheric Physics, 12 November 2020, Moscow, Russia (2020)
5. E.A. MacDonald, E. Donovan, Y. Nishimura, et al, Sci. Adv., **4** (2018)
6. D.G. Baishev, E.S. Barkova, S.I. Solovyev, K. Yumoto, A.G. Yahnin, IJGA, **2**, 2 (2000)
7. K. Sakaguchi, K. Shiokawa, Y. Miyoshi, Y. Otsuka, T. Ogawa, K. Asamura, M. Connors, J. Geophys. Res., **113** (2008)

Simultaneous lithospheric-atmospheric signals of acoustic emission at "Karymshina" site in Kamchatka

Igor Larionov[1,*], *Yuriy* Marapulets[1], and *Mikhail* Mishchenko[1]

[1]Institute of Cosmophysical Research and Radio Wave Propagation FEB RAS, Mirnaya, 7, Paratunka, Kamchatkskiy kray, 684034, Russia

Abstract. We present the results of complex lithospheric-atmospheric investigations of acoustic emission in a seismically active region (Kamchatka peninsula). A laser strainmeter-interferometer, a geophone, a wide-band acoustic system and a microbarometer, installed at Karymshina site (IKIR FEB RAS), are used in the monitoring. Rock deformation, acoustic emission in the near-surface rocks and in the atmosphere by the ground surface are under the consideration. Moreover, we suggest a method to detect acoustic signals recorded simultaneously in the near-surface rocks and in the atmosphere by the ground surface. The method consists in filtration of acoustic signals from the sensors at different frequency sub-ranges from fractions to the first hundreds of hertz followed by detection and accumulation of on 1-second interval. We analyze the data from September 2016 to December 2020. Examples of records of simultaneous acoustic signals in rocks and in the atmosphere are illustrated. The investigation is topical for the construction of a model of lithosphere-atmosphere interaction in a seismically active region.

1 Introduction

At the present time, one the most important directions in Geophysics is the construction of models for lithosphere-atmosphere-ionosphere-magnetosphere interaction. Acoustic waves, propagating in the earth crust-atmosphere system, play an important role in such an interaction. They can be excited during different processes of natural and anthropogenic character and are accompanied by intensive energy release [1]. Those may be different kinds of earth surface movements during a seismo-tectonic process [2–4], volcano eruptions [5, 6], explosions [7–9], and operation of powerful technical equipment and mechanisms, in particular, thumpers [10, 11]. At present there are experimental proves for the output of acoustic waves, excited by the near-surface sources, into the atmosphere and ionosphere (for example, [11–14]). Atmospheric-lithospheric interaction manifests the most evidently in the seismically active region at the final stage of earthquake preparation [15, 16]. High interest to the joint consideration of seismoacousic wave processes is determined by the fact that low-frequency acoustic waves are capable of propagating at long distances and reach ionospheric heights under favorable conditions (low attenuations and scattering, weak refraction in the atmosphere) [17]. Owing to that, they play a significant role in energy transfer between geospheres and

*e-mail: igor@ikir.ru

may be one of the reasons of earthquake precursor manifestation in the atmosphere and ionosphere.

Investigations of geoacoustic emission dynamics at the final stage of earthquake preparation have been carried out at IKIR FEB RAS since 1999. During the investigation the acoustic-emission effect was discovered. It consists in the increase of geoacoustic emission intensity in the frequency range from hundreds of hertz to the first tens of kilohertz during rock massive deformation rate increase [18–20]. It was shown that the effect manifests the most clearly in the earthquake preparation zone 1–3 days before a seismic event and is generated by geomechanics stress field change as a result of the transformation: stress – near-surface rock deformation.

One of the further directions of the investigation in this area is the consideration and confirmation of the possibility of output of the acoustic emission preceding earthquakes into the atmosphere. In view of this, complex lithospheric-atmospheric investigations of acoustic emission have been carried out at IKIR FEB RAS since 2015. To make such monitoring, we use a laser strainmeter-interferometer, a geophone, wide-band acoustic system and a microbarometer. Rock deformation, acoustic emission in the near-surface rocks and in the atmosphere by the ground surface are under the study.

Initially, in the course of the investigations it was shown that P and S waves from earthquakes enter the atmosphere and are recorded over the ground surface [21]. It was shown in the paper [22] that during deformation disturbances in rocks with relative deformations of $10^{-5} - 10^{-7}$, geoacoustic signals of pulse nature with the frequency of units-tens of Hz occur at the observation site. They may pass the earth-atmosphere boundary and be recorded in the air by the ground surface. In this paper we investigate occurrences of such signals over a long time period from September 2016 to December 2020.

2 Equipment and observation method

The investigations are carried out at Karymshina site, IKIR FEB RAS (52.83 N, 158.13 E), located in the zone of different-rank tectonic faults. A laser strainmeter-interferometer is used to record deformations. The measuring base length is 18 m and the sensitivity is 10^{-11}. When organizing deformation measurements, we chose the open-type scheme, i.e. without light guides and without weather effect compensation. Such a scheme imposed some restrictions on measurement accuracy. When recording relative deformations, it is not less than 10^{-8}. We should note that to investigate the local deformation process in sedimentary rocks, such accuracy turned to be enough [19, 23]. We consider rock relative deformation ε.

To record low-frequency geoacoustic signals at the measurement point of ground surface deformations, a three-component piezoceramic geophone is mounted on the streinmeter optical bench end surface [24]. It converts acoustic signals into electric voltage proportional to oscillatory acceleration in the frequency range of 0.2–400 Hz. The geophone was arranged in such a way that one of the orthogonal components of the recorded acoustic signal (Z component) coincides with the strainmeter measurement axis direction, and X component is vertical [25].

A system of four piezoceramic hydrophones is used for high-frequency geoacoustic measurements. They are installed in artificial reservoirs with the sizes of 1x1x1 m^3 and are oriented vertically downwards. We consider wide-band data in the frequency range of 0.1 Hz – 11 kHz and the signals filtered in seven sub-ranges of 0.1–10, 30–60, 70–200, 200–600, 600–2000, 2000–6500, 6500–11000 Hz followed by detection and accumulation on 1-second interval for each of 4 channels [19]. Three hydrophones (№2, 3, 4) form a triangle with the side of 5 m around the mount for the strainmeter with a reflector. The hydrophone №1 is installed about 50 m from this mount.

A microbarometer ISGM-03M, produced by the scientific-technical center "Geofizicheskie izmereniya", Novosibirsk, is used to measure atmospheric acoustic emission. It was installed 4 meters away from the strainmeter in a pit, which is 0.5 m deep and is covered from precipitation. The microbarometer ISGM-03M is aimed at converting air pressure variations into electric signal in the operating frequency band from 0.02 to 100 Hz in the amplitude range from 0.001 to 100 Pa. The mode of functioning is based on the measurement of elastic membrane shift relatively a fixed electrode from air pressure variation effect by displacement capacitance sensor [21].

The important peculiarity of the experiment is the fact that the sensors, installed not far from each other, are covered by a three-meter snow layer in winter-spring time that provides temperature stabilization. To analyze meteorological factor effect, we used the data from a weather station located 20 m from the strainmeter. We should note that wind velocity makes the greatest impact on the instrument reading. It manifests as signal amplitude oscillation amplification.

Ground surface deformations, geoacoustic emission of near-surface sedimentary rocks and atmospheric acoustic emission are recorded by a hardware-software complex described in detail in [26]. ADC sampling frequency is 1 kHz (48 kHz for high-frequency geoacoustic measurements). The important part of preliminary processing of the data from different recording systems is time synchronization. Owing to the installed standard time signal system including a high-accuracy GPS receiver and the same sampling frequency, we succeeded to synchronize the data series with the accuracy up to a millisecond.

At the stage of preliminary processing of the data from the microbarometer and the geophone, the recorded signals were filtered digitally applying Butterworth band filters with the passing frequency in the ranges of 0–0.13, 0.13–0.6, 0.6–5, 5–10, 10–30, 30–49, 52–99 Hz with further detection and accumulation on 1-second interval. When analyzing the data, in order to exclude the cases of the disturbances associated with earthquakes, all the detected simultaneous acoustic signals were compared with the data from regional earthquake catalogue of Kamchatka Branch of Geophysical Survey RAS. When such disturbances were detected, they were not taken into consideration.

The signals were analyzed simultaneously from four observation systems, thus it was necessary to chose a key time series relatively which all the rest of the data were considered. The microbarometer data were chosen as such a series. When these data were considered in detail and compared with meteorological parameters, it was discovered that wind velocity makes the greatest impact on the signal in the frequency range of 0–0.13 Hz, 0.13–0.6 and 0.6–5 (fig.1).

Preliminary analysis and comparison of the signals showed that sometimes sudden pulse jumps of signal amplitude are recorded in the microbarometer data in the frequency ranges of 0.6–5, 5–10 Hz and 10–30 Hz (fig.2). The disturbances in the ranges of 10–30, 30–49 Hz and 52–99 Hz in the geophone data correspond to them the most brightly (fig.3(a,b,c)). In this case, disturbances in geophone data manifest in X component the most. At that moment in high-frequency geoacoustic system data, the disturbances are recorded by the hydrophone №1 at the frequency ranges of 0.1–10, 30–60, 70–200 and 200–600 Hz (fig.4(a)). Deformations generating those signals are recorded both in wide-band data of high-frequency acoustic system and by the strainmeter-interferometer (fig.4(b)).

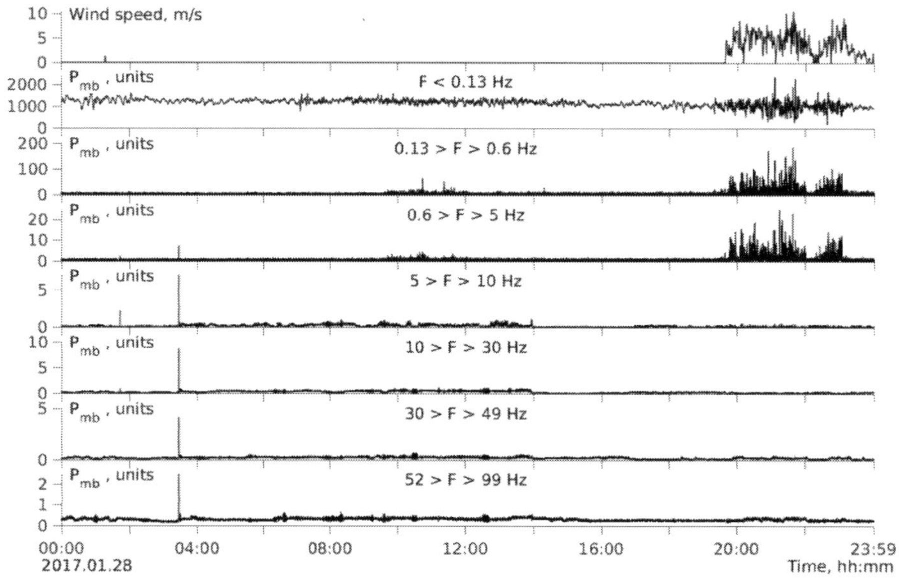

Figure 1: Example of wind effect on atmospheric acoustic signal variations P_{mb} recorded by the microbarograph in different frequency sub-ranges

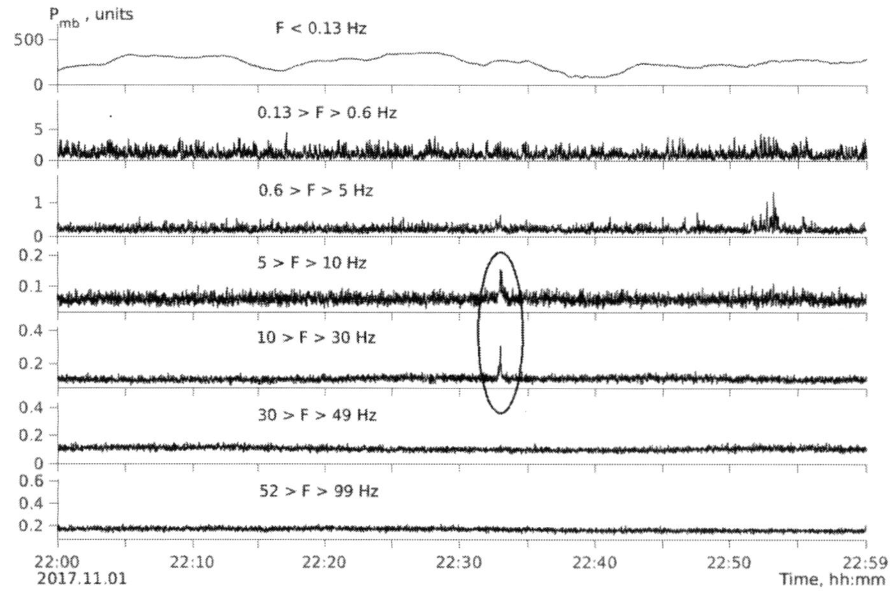

Figure 2: Atmospheric acoustic signal variations Pmb recorded by the microbarograph from 22:00 to 23:00 UT on November 1, 2017. Ellipse shows the amplitude pulse jump

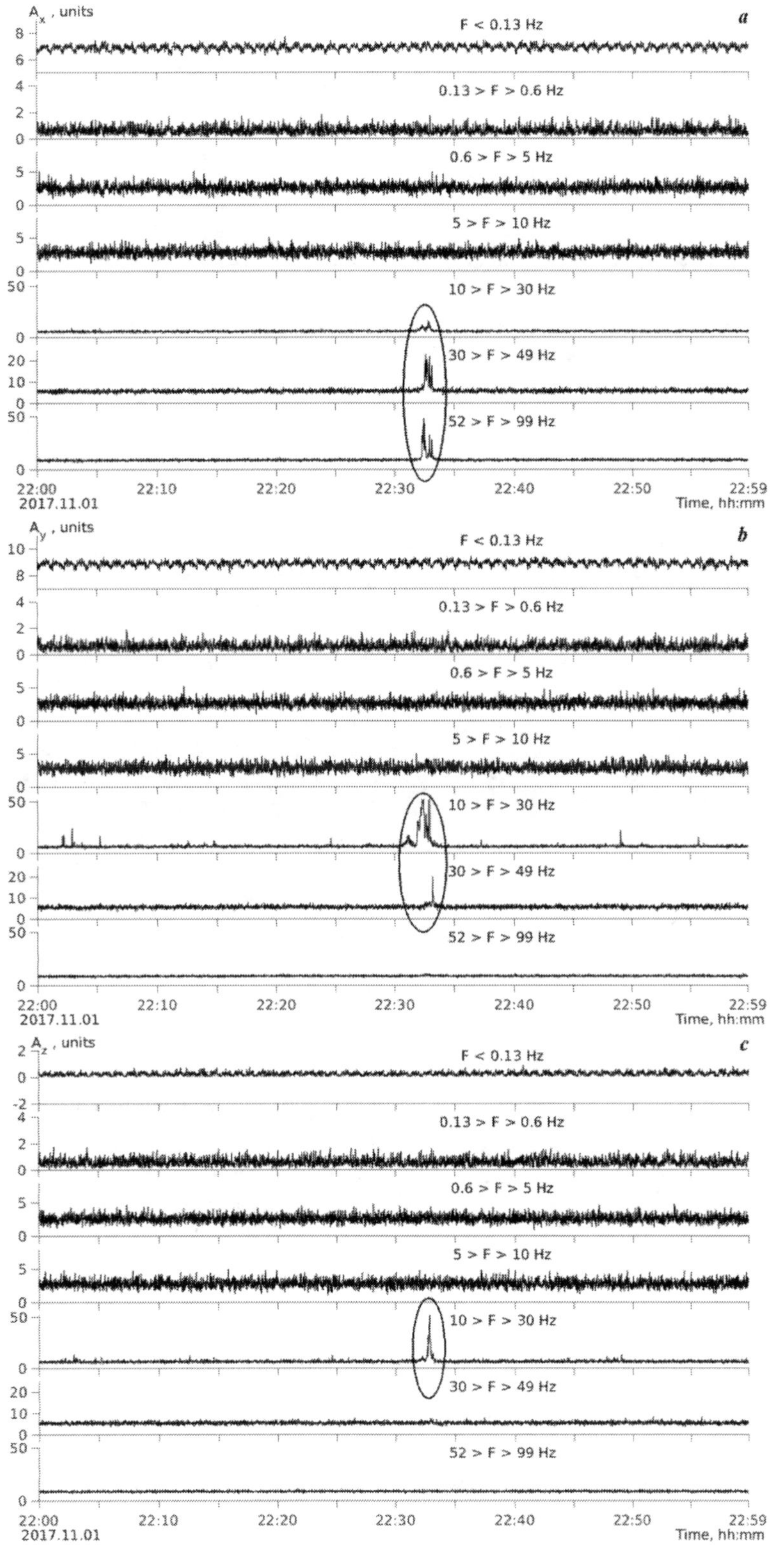

Figure 3: Geoacoustic signal variations A_x, A_y, A_z recorded by the three-component geophone from 22:00 to 23:00 UT on November 1, 2017. The components: X (a), Y (b), Z (c). Ellipses show amplitude pulse jump which was also recorded in microbarometer signal (Fig. 2)

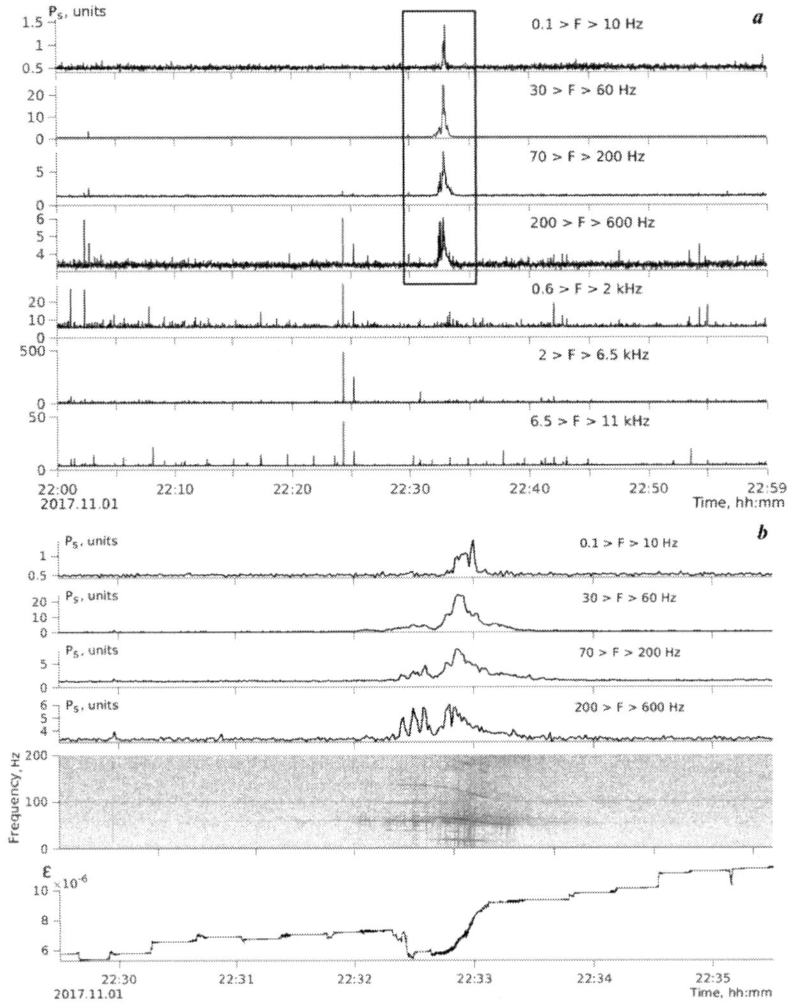

Figure 4: Geoacoustic signal variations P_s recorded by the hydrophone №1 of high-frequency geoacoustic system from 22:00 to 23:00 UT on November 1, 2017 (a). A record fragment of amplitude pulse jump of the acoustic signal which was also recorded in microbarometer and geophone signals (Fig. 2, 3) is marked by a rectangle. Development of the selected record fragment of amplitude pulse jump of acoustic signal P_s in four frequency sub-ranges, its spectrogram and rock relative deformation ε recorded by the laser strainmeter-interferometer (b)

Thus, in the further analysis we considered:

- microbarometer data in the frequency ranges of 5–10 Hz, 10–30 Hz (the range of 0.6–5 was not used due to strong wind effect);
- geophone data in the frequency ranges of 10–30, 30–49 Hz and 52–99 Hz;
- high-frequency geoacoustic system data (hydrophone №1) in the frequency ranges of 0.1–10, 30–60, 70–200 and 200–600 Hz;
- relative deformations recorded by the laser strainmeter-interferometer.

3 Observation results

We analyzed acoustic data for the period from September 2016 to December 2020. It was found out that during that period sudden jumps of signal amplitude were sometimes recorded simultaneously by all observation systems. Analysis of these signal time features did not discover any clear regularities. Such signals occurred during the whole period of observations in different seasons and at different time of a day. That fact almost excludes their man-caused nature as they would manifest stronger during day time. Moreover, since September 2016 until August 2019 the power for the equipment at Karymshina site was supplied by a diesel electric generator. In September 2019 a sun electric station was installed. However, that turn to noiseless power system did not affect the character of those acoustic signal occurrences. Fig.5,6,7 show the examples of simultaneous records of acoustic signal disturbances in the near-surface rocks and in the atmosphere by the ground surface during different time periods.

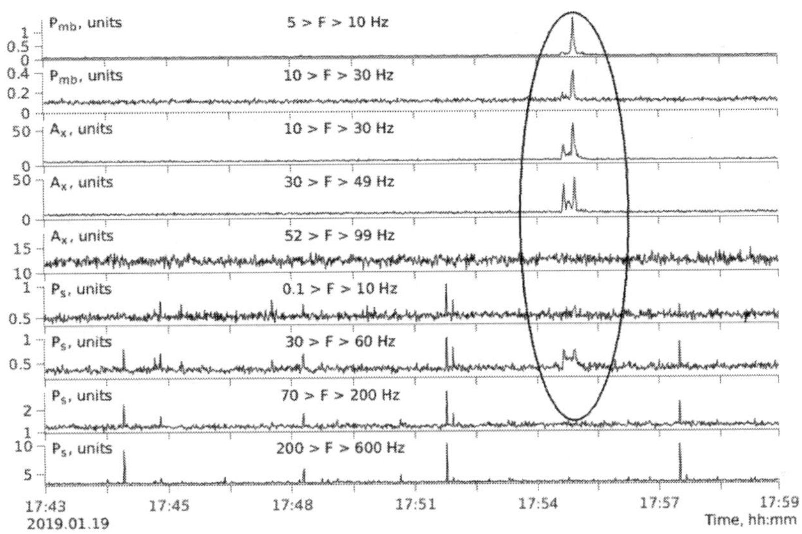

Figure 5: Atmospheric acoustic signal variations P_{mb} recorded by the microbarometer, geoacoustic signal A_x recorded by the geophone (X component), geoacoustic signal P_s recorded by the hydrophone №1 of high-frequency geoacoustic system from 17:43 to 17:59 UT on January 19, 2019. Ellipse shows the simultaneous amplitude jump.

Fig.5 illustrates a acoustic signal variations by the microbarometer, the geophone and by the high-frequency geoacoustic signal early in the morning (at about 5:00 LT) in winter 2019. In this case the pulse signal was detected in all analyzed ranges of the microbarometer and the geophone. The high-frequency geoacoustic system recorded the signal in the ranges of 0.1–10 Hz and 30–60 Hz (hydrophone №1).

Fig.6 shows a acoustic signal variations by the microbarometer, the geophone and by the high-frequency geoacoustic signal in spring 2019. In this case, the pulse signal was detected in the ranges of 5–10 Hz and 10–30 Hz of the microbarometer, in the ranges of 10–30 Hz and 30–49 Hz of geophone X channel and in all four analyzed ranges of hydrophone №1. The signal did not manifest in the range of 52–99 Hz of the geophone.

Fig.7 presents a acoustic signal variations by the microbarometer, the geophone and by the high-frequency geoacoustic signal in autumn 2019 when the source of power supply for all the equipment was changed to the solar electric station. In this case, pulse signals were detected in all the analyzed ranges of the microbarometer, the geophone and the high-frequency geoacoustic system.

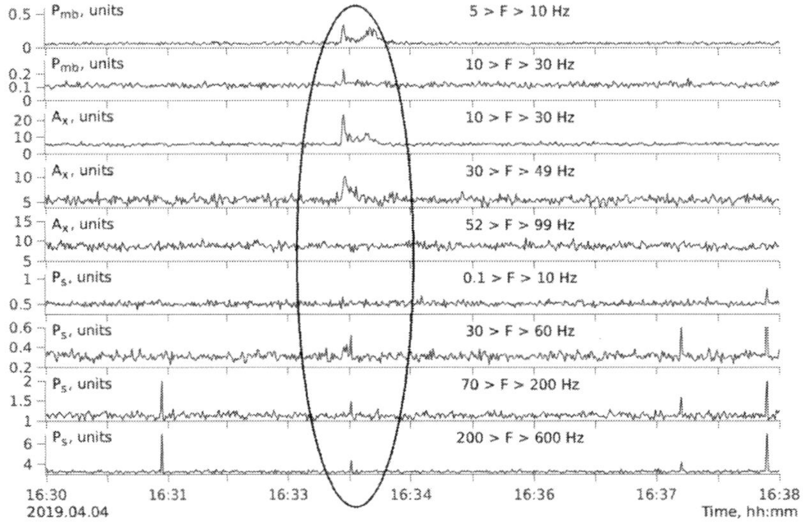

Figure 6: Atmospheric acoustic signal variations P_{mb} recorded by the microbarometer, geoacoustic signal A_x recorded by the geophone (X component), geoacoustic signal P_s recorded by the hydrophone №1 of high-frequency geoacoustic system from 16:30 to 16:38 UT on April 4, 2019. Ellipse shows the simultaneous amplitude jump

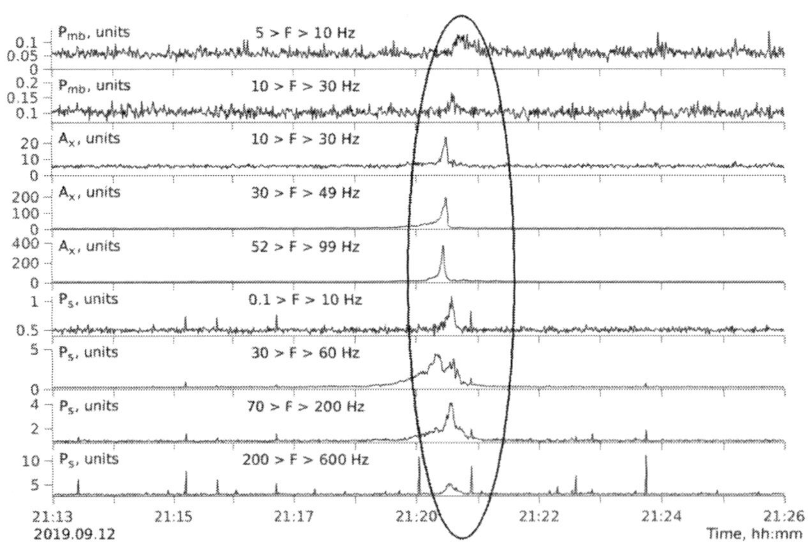

Figure 7: Atmospheric acoustic signal variations P_{mb} recorded by the microbarometer, geoacoustic signal A_x recorded by the geophone (X component), geoacoustic signal P_s recorded by the hydrophone №1 of high-frequency geoacoustic system from 21:13 to 21:26 UT on September 12, 2019. Ellipse shows the simultaneous amplitude jump

4 Conclusions

Thus, as the result of long-term lithospheric-atmospheric observations in the seismically active region, it was shown that acoustic emission in the range of units-tens of Hz occurs in the near-surface sedimentary rocks during deformation intensification. It penetrates into the atmosphere and is recorded by the ground surface. Such emission is not affected significantly by season or time of a day. Further direction of the investigation is the detection of possible relation between geoacoutic emission penetrating into the atmosphere and activation of seismotectonic process in the region.

The investigation is of interest for constructing a model of lithosphere-atmosphere relation in the seismically active region.

The work was carried out as part of the implementation of the state task AAAA-A21-121011290003-0

References

[1] Blanc E. Annals of Geophysics **3(5)**, 673–687 (1985)
[2] Gokhberg M.B., Pilipenko V.A., Pokhotelov O.A. Izvestiya of the Academy of Sciences of the USSR. Physics of the Solid Earth. **10**, 17–21 (1983)
[3] Garmash C.B., Lin'kov E.M., Petrova L.N., Shved G.M. Izvestiya of the Academy of Sciences of the USSR. Atmospheric and Oceanic Physics. **25(12)**, 1290–1299 (1989)
[4] Lin'kov E.M., Petrova L.N., Osipov K.S. Doklady of the USSR Academy of Sciences. **313(5)**, 1095–1098 (1990)
[5] Goerke V.H., Young J.M., Cook R.K. Jour. Geophys. Research. **70**, 6017–6022 (1965)
[6] Firstov P.P., Adushkin V.V., Storcheus A.B. Doklady of the USSR Academy of Sciences. **259(5)**, 1078–1081 (1978)
[7] Richie W.C., Chick D.R. Journ. Acoust Soc. Amer.**41(5)**, 1377–1378 (1967)
[8] Tseytlin Ya.I., Smoliy H.I. Seismic and shock waves from industrial bursts. – Moscow: Nedra, (1981)
[9] Shalimov S.L. Cosmic Research. **36(5)**, 542–545 (1998)
[10] Kuznetsov V.V., Plotkin V.V., Khomutov S.Y., Grekhov O.M., Pavlov A.F., Fedorov A.N. Physics and Chemistry of the Earth (A) **25(3)**, 325–328 (2000)
[11] Kovalevskiy V.V. Acoustical Physics. **51**, 92–102 (2005)
[12] Goshdzhanov M., Boltaev D., Karryev A., Kulieva R.N., Mukhanov M.B., Khismatullina N.R. Izvestiya of the Academy of Sciences of the USSR. Physics of the Solid Earth **2**, 90–95 (1996)
[13] Afraymovich E.L., Kosogorov E.A., Plotnikov A.B., Uralov A.M. Izvestiya. Physics of the Solid Earth **6**, 16–28 (2001)
[14] Al'perovich L.S., Ponomarev E.A., Fedorovich G.V. Izvestiya of the Academy of Sciences of the USSR. Physics of the Solid Earth **11**, 9–20 (1985)
[15] Adushkin V.V., Spivak A.A., Kishkina S.B., Loktev D.N., Solovev S.P. Izvestiya. Physics of the Solid Earth **42(7)**, 567–584 (2006)
[16] Gordienko V.A., Gordienko T.V., Krasnopistsev N.V., Kuptsov A.V., Larionov I.A., Marapulets Yu.V., Shevtsov B.M., Rutenko A.N. Acoustical physics **54(1)**, 82–93 (2008)
[17] Kuznecov V.V., Plotkin V.V., Homutov S.Yu., Dokl. Earth. Sci. **370(2)**, 243–248 (2000)
[18] Kuptsov A.V. Izvestiya. Physics of the Solid Earth **41(10)**, 825–831 (2005)
[19] Marapulets Yu.V., Shevtsov B.M., Larionov I.A., Mishchenko M.A., Shcherbina A.O., Solodchuk A.A. Russian Journal of Pacific Geology **6(6)**, 457–464 (2012)

[20] Marapulets Yu.V. Bulletin KRASEC. Phys. and Math. Sci. **10(1)**, 39–48 (2015)

[21] Larionov I.A., Marapulets Yu.V., Bulletin KRASEC. Phys. and Math. Sci, **26(1)**, 86–91 (2019)

[22] Larionov I.A., Marapulets Yu.V., Mishchenko M.A, E3S Web of Conferences, **127** (2019), 02023

[23] Larionov I.A., Marapulets Yu.V., Shevtsov B.M. Solid Earth **5**, 1293–1300 (2014)

[24] Seysmopriemniki p'ezoelektricheskie A16. Rukovodstvo po ekspluatatsii. 402152.004RE [Piezoelectrical seismometers A16. User manual. 402152.004RE], (ZAO Geoakustica, Moscow, 2006) 40

[25] Mishchenko M.A., Marapulets Yu.V., Larionov I.A., Bogomolov L.M., Sychev V.N. E3S Web of Conferences **62** (2018), 02009

[26] Larionov I.A., Malkin E.I., Marapulets Yu.V., Mishchenko M.A., Solodchuk A.A. Bulletin KRASEC. Phys. and Math. Sci. **24(4)**, 213–225 (2018)

Study of Chaotic and Regular Modes of the Fractional Dynamic System of Selkov

Roman Parivik[1,*], *Zafar* Rakhmonov[2], and *Rakhim* Zunnunov[3]

[1]Institute of Cosmophysical Research and Radio Wave Propagation FEB RAS
7, Mirnaya Str., Paratunka, Kamchata, 684043 Russia
[2]National University of Uzbekistan named after Mirzo Ulugbek, Tashkent, 100174 Uzbekistan
[3]Institute of Mechanics and Seismic Resistance of Structures named after M.T. Urazbayeva, Tashkent, 100125 Uzbekistan

Abstract. The paper investigates the dynamic modes of the Sel'kov fractional self-oscillating system in order to simulate the interaction of cracks. The spectra of the maximum Lyapunov exponents, constructed depending on the parameters of the dynamic system, are used as a research tool. The maximum Lyapunov exponents were constructed according to the Benettin-Wolf algorithm. It is shown that the existence of chaotic regimes is possible. In particular, the spectrum of the maximum Lyapunov exponents of the order of the fractional derivative contains positive values, which indicates the presence of a chaotic regime. Phase trajectories were also constructed to confirm these results. It was also confirmed that the orders of fractional derivatives are responsible for dissipation in the system under consideration.

1 Introduction

In [1], the authors proposed an interesting approach to describing the interaction of cracks in an elastic-brittle medium, which is based on the use of the Selkov's nonlinear dynamic system, which is investigated within the framework of the theory of biological systems [2].

The Selkov dynamic system [1], according to the authors, describes well the interaction of two types of cracks: the first type is seed cracks with a lower energy, which, when the critical level of their concentration is reached, transforms into the second type of cracks with a higher energy. Cracks of the second type are a source of microseismic phenomena (vibrations) and after the release of their energy, they partially pass into seed cracks. Further, this self-oscillatory process is repeated.

In the work of the authors [3], the generalized dynamic system of Selkov was considered, taking into account heredity (memory), which is studied within the framework of hereditary mechanics [4, 5]. This model takes into account the property of a dynamic system to remember the impact on it for some time. Accounting for the heredity of the system is determined using derived fractional orders in the sense of Gerasimov-Caputo. In [3], using the numerical method of Adams-Buschfort-Multon, the calculated curves were constructed and the rest points of the fractional Sel'kov system were investigated.

In this paper, we consider the dynamic modes of the Sel'kov fractional system by constructing the maximum Lyapunov exponents.

*e-mail: romanparovik@gmail.com

2 Basic concepts and definitions

Let's introduce some concepts and definitions.

Definition 1. A Sel'kov fractional dynamical system is a system of the form:

$$\begin{aligned} &\partial_{0t}^{\alpha_1} x(t) = -x(t) + ay(t) + bx^2(t)y(t), \\ &\partial_{0t}^{\alpha_2} y(t) = v - x(t) + ay(t) - bx^2(t)y(t), \\ &x(0) = x_0, \quad y(0) = y_0. \end{aligned} \tag{1}$$

where $x(t)$ is a function that determines the concentration of seed cracks of the first type; $y(t)$ is a function that determines the concentration of cracks of the second type that generate microseisms, $t \in [0; T]$ is a coordinate that is responsible for the current process time, $T > 0$ is a constant, modeling time; x_0, y_0, v, a, b – given positive constants; operators of fractional differentiation are understood in the sense of Gerasimov-Caputo of orders $0 < \alpha_i < 1$ and are defined:

$$\partial_{0t}^{\alpha_1} x(t) = \frac{1}{\Gamma(1-\alpha_1)} \int_0^t \frac{\dot{x}(\eta)\, d\eta}{(t-\eta)^{\alpha_1}}, \qquad \partial_{0t}^{\alpha_2} y(t) = \frac{1}{\Gamma(1-\alpha_2)} \int_0^t \frac{\dot{y}(\eta)\, d\eta}{(t-\eta)^{\alpha_2}}$$

properties that can be found in the monographs [6-8].

Consider a more general fractional non-autonomous dynamical system:

$$\begin{aligned} &\partial_{0t}^{\alpha_1} x(t) = f_1\left(x(t), y(t), t\right), \\ &\partial_{0t}^{\alpha_2} y(t) = f_2\left(x(t), y(t), t\right), \\ &x(0) = x_0, \quad y(0) = y_0. \end{aligned} \tag{2}$$

where the functions $f_1\left(x(t), y(t), t\right)$ and $f_2\left(x(t), y(t), t\right)$ possess the necessary smoothness properties for the existence of the first partial derivatives.

Definition 2. The system of equations in variations for the fractional dynamical system (2) is the following system of the form:

$$\begin{aligned} &\partial_{0t}^{\alpha_1} \Delta x(t) = \Delta x \frac{\partial f_1\left(x(t), y(t), t\right)}{\partial x} + \Delta y \frac{\partial f_1\left(x(t), y(t), t\right)}{\partial y}, \\ &\partial_{0t}^{\alpha_2} \Delta y(t) = \Delta x \frac{\partial f_2\left(x(t), y(t), t\right)}{\partial x} + \Delta y \frac{\partial f_2\left(x(t), y(t), t\right)}{\partial y}. \end{aligned} \tag{3}$$

Remark 1. For the Sel'kov fractional dynamical system (1), the variational equations, according to (3), will have the form:

$$\begin{aligned} &\partial_{0t}^{\alpha_1} \Delta x(t) = \left(2bx(t)y(t) - 1\right)\Delta x + \left(a + bx^2(t)\right)\Delta y, \\ &\partial_{0t}^{\alpha_2} \Delta y(t) = -2bx(t)y(t)\Delta x + \left(-a - bx^2(t)\right)\Delta y. \end{aligned} \tag{4}$$

The system of equations in variations (3) for the original dynamical system (1) is key in constructing the maximum Lyapunov exponents. These equations describe the evolution of an infinitesimal perturbation of the phase trajectory of the original dynamical system.

3 Algorithm for constructing the maximum Lyapunov exponents

Consider an algorithm for constructing the maximum Lyapunov exponents for the Selkov fractional dynamical system (1), which consists of the following stages:

1. We choose the starting point – the initial vector x_0 and together with it we will track the K perturbed trajectories. In our case $K = 2$.

2. We solve numerically, for example, by the Adams-Bashfort-Multon method [3,9], the original system (1) together with three sets of perturbed equations or equations in variations (4). The total number of equations to be solved can be determined by the formula $n(n+1)$, where n is the number of equations in the original system. As the initial vectors for the variational equations, it is necessary to choose a set of vectors $\tilde{x}_0^0, \tilde{y}_0^0$, which are orthogonal and normalized to unity.

3. After time T, the trajectory will go to the point – vector x_1, the perturbation vectors $\tilde{x}_1, \tilde{y}_1$ are renormalized using the Gram-Schmidt method according to the formulas: $\tilde{x}_1^0 = \dfrac{\tilde{x}_1}{\|\tilde{x}_1\|}$, $\tilde{y}_1^{'} = \tilde{y}_1 - \left\langle \tilde{y}_1, \tilde{x}_1^0 \right\rangle \tilde{x}_1^0$, $\tilde{y}_1^0 = \dfrac{\tilde{y}_1^{'}}{\left\|\tilde{y}_1^{'}\right\|}$. Here the notation $\langle \cdot, \cdot \rangle$ is the scalar product of vectors.

4. Then we continue counting from the point x_1 and the vectors of perturbations $\tilde{x}_1^0$, $\tilde{y}_1^0$. After the next time interval T, we obtain a new set of perturbation vectors $\tilde{x}_2, \tilde{y}_2$, which is subjected to orthogonalization and renormalization.

5. Stages 2 – 4 are repeated M times and in the course of calculations, the sums: $S_1 = \sum_{i=1}^{M} \ln \|\tilde{x}_i\|$, $S_2 = \sum_{i=1}^{M} \ln \|\tilde{y}_i\|$ are calculated, in which the disturbance vectors appear before renormalization, but after orthogonalization.

6. The maximum Lyapunov exponents are calculated by the formula:

$$\Lambda_{\max}^{i} = \frac{S_i}{MT}, \quad i = 1, \ldots, n.$$

Remark 2. The Gram-Schmidt orthogonalization procedure is necessary in order to exclude the dominance of the component of the maximum Lyapunov exponent at large times when calculating vectors along the phase trajectory. Otherwise, the task will be ill-conditioned.

Remark 3. It is important to study the spectrum of the maximum Lyapunov exponents, which is constructed depending on the values of the control parameter of interest to us. The following can be used as control parameters for system (1): $\upsilon, a, b, \alpha_1, \alpha_2$.

Therefore, we will investigate the following dependences of the maximum Lyapunov exponent: $\Lambda_{\max}(\upsilon)$, $\Lambda_{\max}(a)$, $\Lambda_{\max}(b)$, $\Lambda_{\max}(\alpha_1)$, $\Lambda_{\max}(\alpha_2)$. If the spectra of the maximum indicators have positive values, then there is a chaotic regime, and negative ones – a regular regime.

4 Research results

Of greatest interest is the spectrum of the maximum Lyapunov exponents $\Lambda_{\max}(\alpha_1)$, since there are positive values of the spectrum (figure 1).

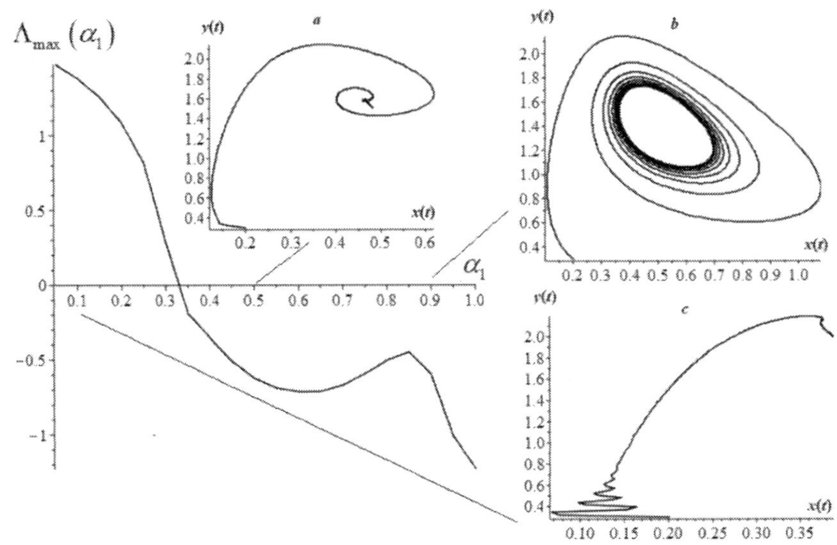

Figure 1. Spectrum of maximum Lyapunov exponents $\Lambda_{\max}(\alpha_1)$.

In figure 1 shows that there is a range of variation of the alpha parameter, at which it takes positive values. The spectrum was plotted for the following parameter values: $\upsilon = 0.5$, $a = 0.1$, $b = 1$, $t \in [0; 100]$, $x_0 = 0.2$, $y_0 = 0.3$ with a step $\tau = 0.05$. Also, figure 1 shows the phase trajectories for different values of α_1. At $\alpha_1 = 0.5$, the phase trajectory has the shape of a twisting spiral (stable focus) (figure 1a), for $\alpha_1 = 0.9$ the phase trajectory is a limit cycle (figure 1b), and for $\alpha_1 = 0.1$ the open phase trajectory has a complex (chaotic shape) (figure 1c).

Also, the phase trajectories in figure 1 show that the order of the fractional derivative α_1 is responsible for the energy dissipation in the system, i.e. as the value of alpha decreases, the limit cycle goes into a stable focus. Similarly, it can be shown that for order α_2 these conclusions remain valid.

The spectra of the remaining parameters are shown in figure 2.

We see that all the spectra of the maximum Lyapunov exponents take negative values, therefore, we have no chaotic regimes. The spectra were plotted according to the following parameter values ($t \in [0; 100]$, $x_0 = 0.2$, $y_0 = 0.3$ for all cases):

- $\Lambda_{\max}(a)$: $\upsilon = 0.5$, $\alpha_1 = \alpha_2 = 1$, $a \in (0; 10]$, with a step $\tau = 0.1$;
- $\Lambda_{\max}(\alpha_2)$: $\upsilon = 0.5$, $a = 0.1$, $\alpha_1 = b = 1$, $a \in (0; 3]$, with a step $\tau = 0.05$;
- $\Lambda_{\max}(b)$: $\upsilon = 0.5$, $a = 0.2$, $\alpha_1 = \alpha_2 = 1$, $b \in (0; 3]$, with a step $\tau = 0.1$;
- $\Lambda_{\max}(\upsilon)$: $a = 2$, $\alpha_1 = \alpha_2 = b = 1$, $\upsilon \in (0; 4]$, with a step $\tau = 0.1$.

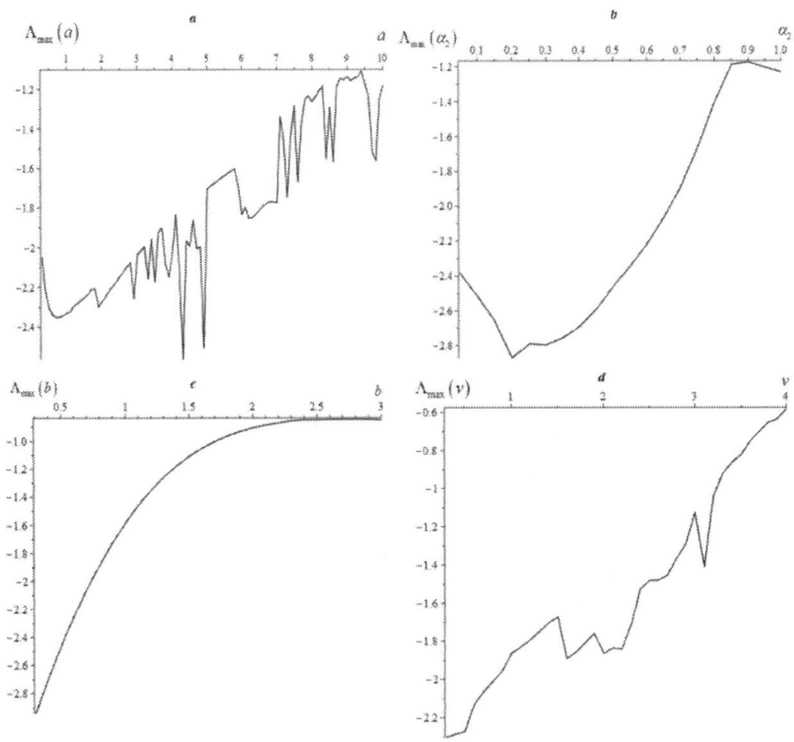

Figure 2. Spectra of maximum Lyapunov exponents: a – $\Lambda_{\max}(a)$, b – $\Lambda_{\max}(\alpha_2)$, c – $\Lambda_{\max}(b)$, d – $\Lambda_{\max}(v)$.

5 Conclusion

In this work, we investigated the spectra of the maximum Lyapunov exponents of the parameters of the Sel'kov fractional system (1). The $\Lambda_{\max}(\alpha_1)$ spectrum has positive values and, therefore, a chaotic regime is observed. The rest of the spectra of Lyapunov exponents have negative values, which indicates regular regimes. A more detailed picture of dynamic modes can be provided by building a map of dynamic modes, which will require large computational resources.

We can say that the orders of fractional derivatives introduce dissipation into the process under consideration (figure 1). In figure 1 it can be seen that with a decrease in the value of the parameter, the limit cycle is destroyed, and the phase trajectory becomes open. This confirms the results of studies of various fractional oscillatory systems [10-13].

Further continuation of the work can be based on a generalization of the orders of fractional derivatives in system (1) as functions of time: $\alpha_1(t)$, $\alpha_2(t)$.

Acknowledgement

The work was carried out within the framework of the subject research IKIR FEB RAS AAAA-A21-121011290003-0 «Physical processes in the system of near space and geospheres under solar and lithospheric influences».

References

[1] V.I. Makovetsky, I.P. Dudchenko, A.S. Zakupin, Geosistemy perehodnykh zon **4**, 1 (2017) (in Russian)
[2] P. Brechmann, A.D. Rendall, Mathematical Biosciences **306** (2018)
[3] R. Parovik, Z. Rakhmonov, R. Zunnunov, E3S Web of Conferences **196**, 02018 (2020) https://doi.org/10.1051/e3sconf/202019602018
[4] V. Volterra, Acta Mathematica **35**, 1 (1912).
[5] Yu.N. Rabotnov, Elements of hereditary solid mechanics (Mir, Moscow, 1980)
[6] Kilbas, H.M. Srivastava, J.J. Trujillo, Theory and Applications of Fractional Differential Equations (Elsevier, Amsterdam, 2006)
[7] K.B. Oldham, J. Spanier, The fractional calculus. Theory and applications of differentiation and integration to arbitrary order (Academic Press, London, 1974)
[8] K.S. Miller, B. Ross, An introduction to the fractional calculus and fractional differntial equations (A Wiley-Interscience publication, New York, 1993)
[9] R. Parovik, AIP Conference Proceedings 2365 (2021)
[10] O. Lipko, R. Parovik, E3S Web of Conferences **62**, 02017 (2018) https://doi.org/10.1051/e3sconf/20186202017
[11] R.I. Parovik, Vestnik Samarskogo Gosudarstvennogo Tekhnicheskogo Universiteta, Seriya Fiziko-Matematicheskie Nauki **23** (2019) (in Russian)
[12] R.I. Parovik, IOP Conference Series: Materials Science and Engineering **919**, 5 (2020).
[13] R.I. Parovik, Technical Physics Letters **45**, 7 (2019).

The modes of magnetic field generation in a low-mode model of $\alpha\Omega$-dynamo with α-generator varying intensity regulated by a function with an alternating kernel

Olga V. Sheremetyeva[1,*] and *Anna N.* Godomskaya[2,**]

[1]Institute of Cosmophysical Research and Radio Wave Propagation FEB RAS, RU-684034, Paratunka, Elizovskiy district, Kamchatka region, Russian Federation

[2]Municipal budgetary institution of supplementary education «Center«Luch»», RU-684000, Yelizovo, Elizovskiy district, Kamchatka region, Russian Federation

Abstract. The low-mode model $\alpha\Omega$-dynamo is used in this paper to simulate the modes of magnetic field generation with insignificant changes in the velocity field of a viscous fluid. In the framework of those model the α-effect intensity is regulated by the process that is included in the magnetohydrodynamic system (MHD-system) as an additive correction as a functional $Z(t)$ depended on the magnetic field energy. Function that determines damped oscillations with variable damping frequency and constant damping coefficient, taken equal to one, is selected as kernel $J(t)$ of functional $Z(t)$. The research of the behavior of the magnetic field is carried out on large time scales, therefore, a rescaled and dimensionless MHD-system with the unit of time iquel the time of the magnetic field dissipation (10^4 years) for numerical calculations is used. The control parameters of the system are the Reynolds number and the amplitude of the α-effect, that include information about the large-scale and turbulent generators, respectively. Numerical simulation of the magnetic field generation modes was carried out for the values of the damping coefficient $b = 1$ and frequency $a = 0.1, 0.5, 1, 5, 10$. According to the results of numerical simulation, an increase in the values of the damping frequency, when the damping coefficient is equal to one, is characterized by a decrease in the inhibitory effect of the process $Z(t)$ on the α-effect and an increase in the region of divergence of the magnetic field on the phase plane of the control parameters. In a comparative analysis with the results of the authors' work, where the change of the α-effect intensity was determined by the function $Z(t)$ with an exponential kernel and the same value of the damping coefficient, the following differences were noted: an increase in oscillations in both a magnetic and a velocity fields, the appearance of a chaotic regime of magnetic field generation at the value of the damping frequency equal to one, and also insignificant narrowing of the region of α-effect suppression at values of the damping frequency increasing to one.

Keywords: $\alpha\Omega$-dynamo, magnetohydrodynamic system, α-effect additive correction, α-effect intensity variation, modes of magnetic field generation, reversal.

*e-mail: sheremeteva@ikir.ru
**e-mail: anna_antonenko@mail.ru

1 Introduction

The magnetic field affects all spheres of human activity, and a change in the sign of the dipole component of the field (a change of polarity or reversal) can lead to global changes on the planet, therefore, constantly measurements of the magnetic field are being carried out and models of convection in the liquid core that can generating magnetic fields similar to real ones are being built [1–3]. Such models are used to study the nature of the occurrence of the magnetic field, its structure and to make predictions about the possible changes in the field in the near future. Despite of achievements in the theory of geomagnetism, the problem of the Earth's magnetic field origin (and other space objects) cannot be considered as completely solved, because the exact solution of the geodynamo equations is unknown. In this regard, the task of studying various approximate solutions for the generation of a geomagnetic field close in morphology to the observed one remains urgent.

The magnetic fields of space objects undergo changes over time both of a chaotic (for example, the Earth) and of a periodic nature (for example, the Sun), including those with reversals. The duration of the time intervals between reversals (polarity intervals) may differ by several orders of magnitude [4]. Model studies of the behavior of MHD-systems are limited by computational capabilities. Direct numerical modeling allows us to either solve the problem of constructing MHD-flows on small time scales and reproduce numerous inversion modes, but does not allow us to understand their causes [5, 6], or reproduce the evolution of large-scale spatial structures and explain the physical cause, signs, and most important properties of the phenomenon of polarity (reversal) [1, 7–9]. In the models of the second type, simple dynamical systems of small dimension are used, which reduces the number of variable parameters, some of which carry information about the magnetic field and allow us to understand the nature of the changes occurring in it.

The main mechanisms describing the dynamo systems of space objects are $\alpha\Omega$, α^2 and the most common $\alpha^2\Omega$ [8, 10, 11]. The model $\alpha\Omega$-dynamo is used to describe an Earth dynamo, whose characteristic properties are a strong differential rotation of the object and the turbulent nature of the conducting medium motion [7, 8, 10–14].

This work is a continuation of the cycle of long-term research on the dynamo theory conducted by the laboratory of the IKIR FEB RAS [15]. Within the framework of the low-mode model $\alpha\Omega$-dynamo [14, 16] with the introduced additive correction of the α-effect intensity, the modes of magnetic field generation are simulated. The investigation is carried out on time scales comparable to the time of magnetic field attenuation, therefore, a rescaled and dimensionless MHD-system is used for the study, which reduces the possibility of computational errors during calculations over long time intervals, overloading numerical solutions with large values and shifting to «machine» infinity. The control parameters of the system are the Reynolds number and the α-effect amplitude, which contain information about large-scale and turbulent generators, respectively. The α-effect intensity is regulated by a process that is determined by a functional depends on time and on two parameters – the coefficient b and the frequency a of damping.

The variation of the two control parameters responsible for the effect of each of the generators and the parameter a of the process that inhibition the α-effect intensity makes it possible to simulate various modes of magnetic field generation, including those with a change in the polarity of the magnetic field that is not associated with the rearrangement of convection. The analysis of the simulation results and their comparison with the phase portraits obtained for the exponential kernel in the paper [17] is carried out.

2 Formulation of the problem

We'll give a summery description of the model, a complete investigation of which and also the main provisions and limitations are presented in the work [17]. The investigation is conducted within the framework of the model $\alpha\Omega$-dynamo. We assume that the liquid core of the Earth is bounded by a spherical shell (r_1 – inner and r_2 – outer radii of the shell) rotating with a constant angular velocity $\mathbf{\Omega}$ (values varying within $10^{-1} \div 10$ 1/s).

The velocity field $\mathbf{v}$ of a viscous incompressible fluid of the outer core and the generated magnetic field are axially symmetric. At the boundaries of the shell, the velocity field of a viscous liquid $\mathbf{v}$ is zero (boundary conditions of adhesion), the magnetic permeability of the inner core is equal to the permeability of the outer core, the medium beyond the boundary of the outer core is not conductive. In accordance with the accepted dynamo model, the average flow of $\overline{\mathbf{v}}$ has the character of differential rotation in the field of external forces [7, 8, 11, 12].

The physical parameters of the shell are assumed to be unchanged and the turbulence in the core is isotropic. The turbulent α-effect is antisymmetric with respect to the equator and a scalar parametrization in the form of a function $\alpha(r,\theta) = \alpha_0 a(r)\cos\theta = \alpha_0\alpha'(r,\theta)$ is used to represent it, where the positive coefficient α_0 defines a constant intensity α-effect, $\alpha'(r,\theta)$ – dimensionless quantity, $a(r)$ – radial component, that satisfies the following condition $r_1 \leq |a(r)| \leq r_2$. Based on the accepted constraints, the low-mode model is described by MHD-equations including the Navier-Stokes equation, the magnetic field induction equation $\mathbf{B}$ taking into account the turbulent α-effect, the conditions of continuity of the velocity field $\mathbf{v}$ and the solenoidality of the magnetic field, as well as boundary conditions, and in the Boussinesq approximation takes the form:

$$
\begin{aligned}
&\frac{\partial \mathbf{v}}{\partial t} + (\mathbf{v}\nabla)\mathbf{v} + \mathbf{f}_c = \nu\Delta\mathbf{v} - \frac{1}{\rho_0}\nabla P - \mathbf{f}_K + \mathbf{f}_{out} + \mathbf{f}_L,\\
&\frac{\partial \mathbf{B}}{\partial t} = \nabla\times(\mathbf{v}\times\mathbf{B}) + \nabla\times(\alpha(r,\theta)\,\mathbf{B}) + \nu_m\Delta\mathbf{B},\\
&\nabla\mathbf{v} = 0,\\
&\nabla\mathbf{B} = 0,\\
&\mathbf{v}(\mathbf{r}_1) = \mathbf{v}(\mathbf{r}_2) = \mathbf{0},
\end{aligned}
\tag{1}
$$

where $\mathbf{f}_c$ is centrifugal acceleration forces, $\mathbf{f}_K$ is mass density of the Coriolis force, $\mathbf{f}_{out}$ is mass density of the external force field (source of poloidal velocity), $\mathbf{f}_L$ is mass density of the Lorentz forces, ν is kinematic viscosity (limits of change $10^{-6} \div 10^2$ m^2/s), $\rho_0 = 7 \cdot 10^3$ kg/m^3 is density, P is pressure, ν_m is magnetic viscosity (varies within $1 \div 20$ m^2/s), $\mathbf{r}_1$ and $\mathbf{r}_2$ are radii-vectors of the inner and outer boundaries of the spherical shell of the liquid core, respectively.

We introduce the characteristic values of the velocity of u_0 [m/s] and the linear size of the region of L [m] (the radius of the outer core r_2), through which we express the units of magnetic induction B_0 [Tl], pressure P_0 [Pa], the mass density of external forces f_0 [N/kg] and angular velocity Ω_0 [1/s]. Let is move on to a new system of units, where as a unit of time we use the time of dissipation of the magnetic field L^2/ν_m [s], the order of which is 10^4 years, and as the control parameters we take the magnetic Reynolds number $Re_m = \dfrac{u_0 L}{\nu_m}$, responsible for the effect of a large-scale generator, and the dimensionless amplitude of the α-effect $R_\alpha = \dfrac{\alpha_0 L}{\nu_m}$, which contains information about the turbulent generator [17]. Then, in the rescaled form, the system (1) will be written as follows

$$
\begin{aligned}
&\frac{\partial \mathbf{v}}{\partial t} = P_m \Delta \mathbf{v} - \nabla P - E^{-1} P_m (\mathbf{e}_z \times \mathbf{v}) + \mathbf{f}_{out} + (\nabla \times \mathbf{B}) \times \mathbf{B}, \\
&\frac{\partial \mathbf{B}}{\partial t} = Re_m [\nabla \times (\mathbf{v} \times \mathbf{B})] + R_\alpha [\nabla \times \alpha(r,\theta)\mathbf{B})] + \Delta \mathbf{B}, \\
&\nabla \mathbf{v} = 0, \\
&\nabla \mathbf{B} = 0, \\
&\mathbf{v}\Big(\frac{\mathbf{r}_1}{L}\Big) = \mathbf{v}(\mathbf{e}_2) = \mathbf{0}.
\end{aligned} \tag{2}
$$

The intensity of the α-effect changes over time under the influence of external forces $\mathbf{f}_{out}$. The information about the α-effect intensity in the rescaled system is contained in the control parameter R_α. In our model, the process that regulates the degree of influence of the turbulent effect is defined as a functional $Z(t)$, the effect of which is determined by the quadratic form $Q(\mathbf{B}(\tau), \mathbf{B}(\tau))$, and the nature of the those effect is determined by the dimensionless kernel $J(t-\tau)$:

$$
Z(t) = \int_0^t J(t-\tau) Q(\mathbf{B}(\tau), \mathbf{B}(\tau)) d\tau. \tag{3}
$$

Considering the additive correction $Z(t)$ of the α-effect intensity the MHD-system can be written in the following form

$$
\begin{aligned}
&\frac{\partial \mathbf{v}}{\partial t} = P_m \Delta \mathbf{v} - \nabla P - E^{-1} P_m (\mathbf{e}_z \times \mathbf{v}) + \mathbf{f}_{out} + (\nabla \times \mathbf{B}) \times \mathbf{B}, \\
&\frac{\partial \mathbf{B}}{\partial t} = Re_m [\nabla \times (\mathbf{v} \times \mathbf{B})] + (R_\alpha - Z(t)) [\nabla \times \alpha(r,\theta)\mathbf{B})] + \Delta \mathbf{B}, \\
&\nabla \mathbf{v} = 0, \\
&\nabla \mathbf{B} = 0, \\
&\mathbf{v}\Big(\frac{\mathbf{r}_1}{L}\Big) = \mathbf{v}(\mathbf{e}_2) = \mathbf{0}.
\end{aligned} \tag{4}
$$

3 The numerical simulation

One of the approaches to the investigation of MHD-systems is the construction of low-mode approximations of the Galerkin type [1]. Note, that different versions of the dynamo mechanism have a unified approach in that the source of the toroidal (poloidal) component of the field **B** is the nonlinear interaction of the poloidal (toroidal) component of the same field with the fluid flow [8, 11]. In this regard, one of the possible truncations is the selection of four components (modes): one hydrodynamic and three magnetic – one toroidal and two poloidal [14, 18], which we use for the numerical solution of the system (4),

$$
\begin{aligned}
\mathbf{v} = u(t)\mathbf{v_0} = u(t)(\alpha_1 \mathbf{v}^T_{0,1,0} + \alpha_2 \mathbf{v}^P_{0,2,0} + \alpha_3 \mathbf{v}^T_{0,3,0} + \alpha_{11} \mathbf{v}^T_{1,1,0} + \alpha_{13} \mathbf{v}^T_{1,3,0}), \\
\mathbf{B} = B^T_2(t)\mathbf{B}^T_{0,2,0}(\mathbf{r}) + B^P_1(t)\mathbf{B}^P_{0,1,0}(\mathbf{r}) + B^P_3(t)\mathbf{B}^P_{0,3,0}(\mathbf{r}),
\end{aligned} \tag{5}
$$

where $\mathbf{v_0}$ is the Poincare mode [19], $u(t)$ is the velocity amplitude, $\mathbf{B}^P_{0,1,0}(\mathbf{r})$ is the dipole component of the magnetic field, which under the influence of differential rotation generates toroidal $\mathbf{B}^T_{0,2,0}(\mathbf{r})$ and poloidal $\mathbf{B}^P_{0,3,0}(\mathbf{r})$ components. We consider the components of the velocity and of the magnetic fields to be independent.

The functional $Z(t)$ (3) is given as the integral of the product of the magnetic field energy and the alternating function $J(t)$, which determines the damped oscillations with a damping coefficient b and the frequency of damped oscillations a,

$$Z(t) = \int_0^t e^{-b(t-\tau)} \cos(a(t-\tau))\mathbf{B}^2(\tau)d\tau. \tag{6}$$

The choice of an alternating damping function as the kernel $J(t)$ defines a process with a finite waiting time (memory), the overwhelming influence of which on the change in the intensity of the α-effect is included at the initial moment of time and gradually accumulates during the waiting time t_0. Variation of the parameters changes the influence nature of the inhibition functional. In this paper, we will investigate the model with a constant damping coefficient of b and a variation of the parameter a, which is responsible for the frequency of changing the sign of the functional kernel. The waiting time t_0 was calculated based on the condition that 95% of the inhibition functional is concentrated on the segment $[0,\ t_0]$,

$$\int_0^{t_0} e^{-bt}|\cos(at)|dt = 0.95 \int_0^{\infty} e^{-bt}|\cos(at)|dt. \tag{7}$$

We substitute the decompositions of (5) and the function (6) into the system (4), after which we apply the Galerkin method. As a result of the transformations, we obtain a system that is then used for numerical implementation by the backward Euler method,

$$\begin{aligned}
\frac{\partial u}{\partial t} &= -P_m u(t) \sum_k \alpha_k^2 \lambda_k + f_{out} + \sum_{i,j,k} \alpha_i L_{ijk} B_j B_k,\\
\frac{\partial B_i}{\partial t} &= Re_m u(t) \sum_{j,k} \alpha_j W_{ijk} B_k - \mu_i B_i + (R_\alpha - Z) \sum_k W_{ik}^\alpha B_k,\\
\frac{\partial Z}{\partial t} &= \sum_k B_k^2 - bZ - aZ_s,\ \frac{\partial Z_s}{\partial t} = aZ - bZ_s,\ \ i,\ j,\ k = 1,\ 2,\ 3.
\end{aligned} \tag{8}$$

where λ_i is eigenvalues of the Poincare mode [20], coefficients L_{ijk}, W_{ijk}, W_{ij}^α are the volume integrals of the fields under consideration [21], μ_i is the viscous dissipation parameter [20], the values of the coefficients α_i were calculated from the Poincare problem on free small oscillation of a viscous rotating fluid [19].

The accepted value of the sampling step h is two orders of magnitude less than the smallest of the damping times of the magnetic field and the velocity field, which are determined by the values of the coefficients λ_i and μ_i. The choice of initial conditions ensures the output of the system (8) from the rest point at the initial moment of time $t = 0$ and further mutual generation of the components of the magnetic and the velocity fields

$$u(0) = 1,\ B_2^T(0) = 0,\ B_1^P(0) = 1,\ B_3^P(0) = 0,\ Z(0) = 0,\ Z_s(0) = 0. \tag{9}$$

Table 1. The parameters depending on the function J(t)

a	t_0	$\int_0^\infty \lvert J(t)\rvert dt$	$\int_0^\infty J(t)dt$
0.1	2.922	0.99	0.99
0.5	1.871	0.817	0.8
1	1.108	0.604	0.5

The damping effect in the velocity field is compensated by the influence of external forces, therefore, in the numerical implementation, the value of the mass density of external forces f_{out} is assumed to be equal to the coefficient at $u(t)$ [17]. The numerical model is considered for control parameters, that varying in the following ranges $Re_m \in (0,\ 1000]$ and $R_\alpha \in (0,\ 100]$, in increments of five. Damping frequency values a and the characteristics of the inhibition process $Z(t)$, depened on it, are presented in table 1, the damping coefficient b is constant and assumed to be equal to one. The results of numerical simulation for the given model parameters are presented on the phase plane of the control parameters Re_m, R_α (fig. 2).

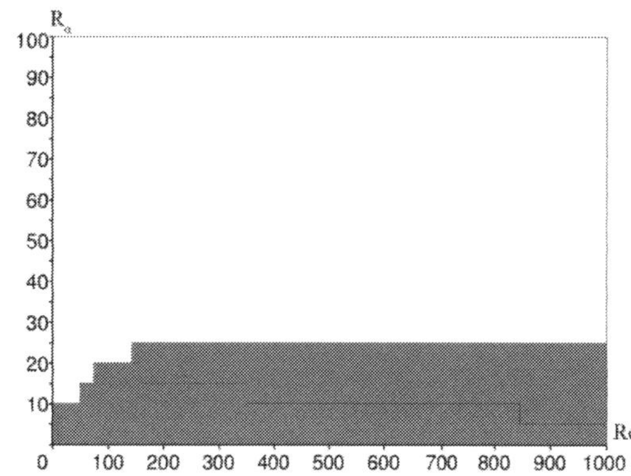

Figure 1. The nature of magnetic field generation depending on the parameters R_α and Re_m. The α-effect intensity constant – α_0. The white region is the generation of the increases infinitely magnetic field, the red one – is the generation of a damped magnetic field, the green is the steady mode of magnetic field generation [17].

4 Simulation results and discussion

The simulation results showed that the inclusion in the MHD-system of influence that regulates the α-effect intensity in the form of a process determined by the functional $Z(t)$ with an alternating kernel, lead to a narrowing of the divergence region of the magnetic field and to an increase in the regions with modes of magnetic field generation without unlimited growth, including those with reversals, under the condition of a weakly varying velocity field. And also it's caused the emergence of new modes of magnetic field generation, in addition to damped, steady and unlimited growth of the magnetic field, which are formed in the case of a constant intensity of the α-effect, i. e. in the absence of a supression effect (fig. 1) [17].

We compare the simulation results (fig. 2 b, c, d) with those obtained in the case of a constant intensity of the α-effect. With a constant damping coefficient $b = 1$ and an increase in the damping frequency a of the alternating kernel $J(t)$ of the functional $Z(t)$ on the phase plane of the control parameters Re_m, R_α the region of the damped generation mode (with or without oscillations) is preserved within the same boundaries, a slight increase in which occurs, when $a = 0.1$ and $Re_m \in [100,\ 250]$, $R_\alpha \in [15,\ 20]$, as a result of inhibition of the steady mode of magnetic field generation. For the values of the damping frequency $a \leq 1$, the divergence region of the magnetic field decreases becouse the regions of steady, steady-state, chaotic, vacillation and dynamo-burst modes appearance on the interval $R_\alpha \in [10,\ 80]$. The largest part of the phase plane (in the $Re_m \in [80,\ 1000]$ interval) is occupied by the region of the steady mode of magnetic field generation. At the values of $Re_m < 100$, small regions of the steady-state and vacillation modes are formed on the phase plane, and for $a = 1$ are formed a dynamo-burst regions. If the values of $Re_m > 100$ and $a = 1$, then a small region of chaotic mode generation is formed. With a further increase in the damping frequency a to a value of 10, the divergence region of the magnetic field increases again due to a significant decrease in the region of the steady mode, although in this case the regions of the steady-state, chaotic and vacillation modes are increase. Note, that at values of $a \in (1;\ 10]$, changes

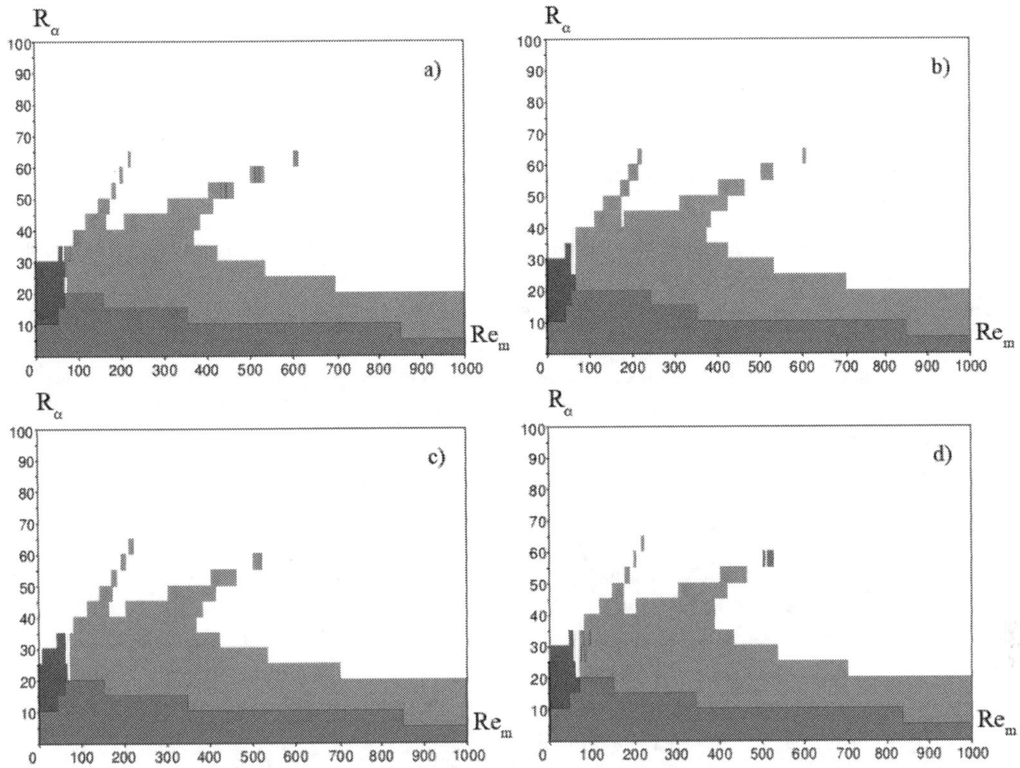

Figure 2. The nature of magnetic field generation depending on the parameters R_α and Re_m. The α-effect intensity is defined by the function $Z(t)$ with the kernel: a) $J(t) = e^{-t}$ [17], b) $J(t) = e^{-t}\cos(0.1t)$, c) $J(t) = e^{-t}\cos(0.5t)$, d) $J(t) = e^{-t}\cos(t)$. The white region is the generation of the increases infinitely magnetic field, the red one – is the generation of a damped magnetic field, the green is the steady mode of magnetic field generation, the blue is the steady-state mode, the yellow is the vacillation, the lilac is the dynamo-burst, the grey is the chaotic magnetic field generation mode.

of the magnetic field components amplitudes occur with a high frequency of the oscillation, which causes significant changes in the amplitude of the velocity field ($0.25 \leq \Delta u \leq 1$) up to a sign reversal for $a = 10$. Thus, the applicability of the consideration model is constraints by the values of the parameter $a \leq 1$, therefore, the further discussion and comparison of the simulation results will be carried out for these values of the damping frequency.

Comparison of the simulation results for the values of the parameter a of the kernel $J(t)$ of the inhibition functional $Z(t)$ vary from 0.1 to 1, revealed a tendency of decrease of the regions of limited oscillations and of the insignificant change their boundaries (fig. 2 b, c, d). For all of the parameter a values the steady-state mode on the interval $Re_m \in (0,\ 100)$ is replaced by the vacillation mode, which forms a small transition region, and then – by the steady mode, which forms the main region of the limited oscillations within the values of the control parameter $Re_m \in [100,\ 1000]$. At the value of the damping frequency $a = 1$ in the transition region (the region of the vacillation mode), the small inclusions of the dynamo-burst regions appear in front of the region of the steady mode of magnetic field generation and the region of the steady mode, located on the phase plane in the square $Re_m \in [510,\ 530]$, $R_\alpha \in [55,\ 60]$ at $a < 1$, is replaced by the region of the chaotic mode of magnetic field generation.

Changes on the phase plane are related to the characteristics of the inhibition functional $Z(t)$ (6), depended on the choice of the kernel $J(t)$ and the values of its parameters (table 1). An increase in the values of the damping frequency a at a constant damping coefficient $b = 1$ leads to a decrease in the values of the waiting time t_0 and integrals of the function $J(t)$ and its modulus. However, in the case of $a = 0.1$, the values of the integrals practically coincide (up to the eighth decimal place), that indicates a very weak effect of damped oscillations in the tail on the total value of the integral. In other cases, this influence is more significant and this does not allow suppressing the α-effect in a larger region than at $a = 0.1$, which is displayed on the narrowing of the region of magnetic field generation limited modes in the interval $R_\alpha \in [50, 70]$ on the phase plane. A more significant influence of the tail for $a = 1$ also causes the appearance of new modes of generation: dynamo-burst and chaotic. Thus, the short-term intense integral action of the process $Z(t)$ during the waiting time ($t_0 \approx 3$), which changes insignificantly during the rest of the time, leads to suppression of the field divergence in the most of the considered region on the phase plane control parameters. Nevertheless, in all cases, the main contribution to the suppression of the α-effect is made at the initial timing steps, where the function $J(t)$ is positive and the integral of it takes the greatest value, which in a large extent affects the similarity of the patterns on phase plane for all considered values of the damping frequency a.

If we compare the obtained results with the simulation one's for the inhibition functional $Z(t)$ with an exponential kernel and damping coefficient $b = 1$ (fig. 2 a) [17], then the obtained patterns on the phase plane are similar. On the whole, the contours of the main regions (steady, damping, steady-state modes) are preserved. But this patterns have also differences. For values $a = 0.1 - 0.5$ (fig. 2 b, c), there are no dynamo-burst regions. Instead of dynamo-burst regions, in one case, there is a vacillation region in the transition from steady-state regime region to steady one, in the other, there is a the region of a steady regime in the transition to the region of divergence of the magnetic field. For the case $a = 1$ in the transition to the divergence region of the field, the dynamo-burst is replaced by the chaotic regime region (fig. 2 d). Thus, the use of an oscillating damping function instead of an exponential one as the kernel $J(t)$ of the functional $Z(t)$ caused to an increase in oscillations in both the magnetic field and the velocity field, that, on the one side, reduces or exclud the possibility of appearance of a dynamo-burst regions, on the other side, leads to irregular behavior and to the emergence of a chaotic regime at the boundary with the divergence region of the magnetic field.

5 Conclusion

The consideration of the MHD-system with an additive correction of the α-effect intensity in the form of a functional $Z(t)$ with an alternating kernel $J(t)$ induced an increase in the magnetic field oscillations frequency, which led to an increase in the amplitude and/or the frequency of oscillations in the velocity field, including those with a change in sign at values of the damping frequency $1 < a \leq 10$. Thus, in this case the model on consideration is applicable only at values of the damping frequency $a \leq 1$.

Using the process $Z(t)$ for regulate α-effect intensity led to a decrease in the magnetic field divergence region compared with the case of constant intensity of α-effect and to emergence of new modes of magnetic field generation: vacillation, dynamo-burst, chaotic, steady-state.

The nature of the influence of the process $Z(t)$ on the α-effect intensity is determined by the tow parameters of the kernel $J(t)$ – the damping coefficient and frequency. The waiting time t_0 decreases with increasing frequency values a, when the damping coefficient is constant $b = 1$. As a consequence, a greater contribution is made by alternating tails, the influence on the α-effect intensity decreases, which leads to an increase in the magnetic field divergence region, to change in the configuration of the regions of the obtained modes of magnetic field generation and to emergence of new modes of magnetic field generation under conditions of a weakly varying velosity field.

The characteristics of the alternating kernel of functional $Z(t)$ at the permissible values of the damping frequency differ insignificantly from the those of the exponential kernel. Therefore, the patterns on the phase planes in both cases are similar in the configuration and in the regions location of the obtained modes of magnetic field generation. However, in our case, a decrease in the damping frequency a leads to the disappearance of the dynamo-burst regions obtained due to choosing the exponential kernel and to an increase in the vacillation region, while an increase in frequency leads to a decrease in the vacillation region and to the appearance of the chaotic and dynamo-burst regions.

The paper was carried out within the framework on the subject «Physical processes in the system of near space and geospheres under solar and lithospheric influences» (AAAA-A21-121011290003-0).

References

[1] E. B. Gledzer, F. V. Dolzhanskiy, A. M. Obukhov, *Sistemy gidrodinamicheskogo tipa i ikh primenenie* [Hydrodynamic Type Systems and Their Application] (Nauka, Moscow, 1981) 368 [in Russian]

[2] M. Kono, P.H. Roberts, Reviews of Geophysics **40**, B1–B41 (2002)

[3] D. D. Sokoloff, R. A. Stepanov, P. G. Frick, UFN **184**, 313–335 (2014) https://doi.org/10.3367/UFNr.0184.201403g.0313

[4] R. T. Merril, M. W. McElhinny, P. L. McFadden, *The Magnetic Field of the Earth: Paleomagnetism, the Core, and the Deep Mantle* (Academic Press, London, 1996) 531

[5] G. Glatzmaier, P. Roberts, Nature **377**, 203–209 (1995)

[6] W. Kuang, J. Bloxman, Nature **389**, 371–374 (1997)

[7] M. Steenbek, F. Krause, Astron. Nachr. **291**, 49–84 (1969)

[8] Ya. B. Zeldovich, A. A.Rusmaikin, D.D. Sokoloff, *Magnetic fields in astrophysics. The Fluid Mechanics of Astrophysics and Geophysics* (Gordon and Breach, New York, 1983) 382

[9] L. K. Feschenko, G. M. Vodinchar, Nonlinear Processes in Geophysics **22**, 361-369 (2015) https://doi.org/10.5194/npg-22-361-2015

[10] E.N. Parker, Astrophys. J. **122**, 293–314 (1955)

[11] F. Krause, K.H. Rädler, *Mean-filed magnetohydrodynamics and dynamo theory* (Pergamon Press, Oxford, 1980) 271

[12] M. Steenbek, F. Krause, K.H. Rädler, Z. Naturforsch. **21**, 369–376 (1996)

[13] D. D. Sokoloff, S. N. Nefedov, Num. Meth. Prog. **8**, 195–204 (2007)

[14] G. M. Vodinchar, L. K. Feshchenko, Magnetohydrodynamics **52**, 287-300 (2016)

[15] G. M. Vodinchar, R. I. Parovik, A. S. Perezhogin, O. V. Sheremetyeva, Istoriya nauki i tehniki [History of Science and Engineering] **8**, 100–112 (2017) [in Russian]

[16] G. M. Vodinchar, Vestnik KRAUNTS. Fiz.-Mat. Nauki [Bulletin KRASEC. Phys. & Math. Sci.] **2(7)**, 33–42 (2017) [in Russian] https://doi.org/10.18454/2079-6641-2013-7-2-33-42

[17] O. V. Sheremetyeva, A. N. Godomskaya, Bulletin of the South Ural State University, Series «Mathematical Modelling, Programming & Computer Software» **14**, 27–38 (2021) https://doi.org/10.14529/mmp210203

[18] G. M. Vodinchar, A. N. Godomskaya, O. V. Sheremetyeva, E3S Web of Conferences, **11**, 00024 (2016) https://doi.org/10.1051/e3sconf/ 20161100024

[19] G. M. Vodinchar, L. K. Feshenko, Bulletin KRASEC. Phys. & Math. Sci., **2(11)**, 41–50 (2015) https://doi.org/10.18454/2313-0156-2015-11-2-41-50

[20] G. M. Vodinchar, *Baza dannyh «Parametry sobstvennyh mod svobodnyh kolebanij MGD polej v yadre Zemli»* [Database «Parameters of natural modes of normal oscillations of MHD fields in the Earth's core»], RU 2019620054 (2019) [in Russian]

[21] G. M. Vodinchar, L. K. Feshenko, *Biblioteka programm dlya issledovaniya «Malomodovoj modeli geodinamo»* [Library of programs for researching «Low-mode geodynamo model» «LowModedGeodinamoModel»], RU 50201100092 (2011) [in Russian]

The source parameters of earthquakes of Bishkek geodynamic proving ground (Northern Tien Shan)

Naylya Sycheva *

Science Research Station of the RAS in Bishkek city, Kyrgyzstan

Abstract. Based on the method of polarity of signs of *P*-waves, the focal mechanisms of 1674 earthquakes with $M \geq 1.6$, which occurred on the territory of the Bishkek geodynamic proving ground (BGPG) from 1994 to 2020, were determined. Some characteristics of the complete catalogue are presented. Quantitative distributions by the type of mechanisms and diagrams of azimuths of the main stress axes are constructed. A variety of focal mechanism of earthquake is observed, most of them are reverse fault, oblique reverse fault, and horizontal strike-slip fault. The compression axis for most of the events has a north-northwest direction and a sub-horizontal position. For 183, dynamic parameters (DP, source parameters) were obtained: spectral density Ω_0, corner frequency f_0, scalar seismic moment M_0, source radius (Brune radius) r, and stress drop $\Delta\sigma$. The correlations between DP and energy characteristic (magnitude) and scalar seismic moment are investigated. The smallest correlation coefficient was obtained for stress drop.

Introduction

The Kyrgyz seismological network, installed in 1991 on the territory of the Northern Tien Shan, consists of 10 digital, broadband stations, each of which is equipped with an STS-2 sensor (Fig. 1).

It was found that the stations of the KNET network are low-noise [1] and site under stations does not generate resonance effects [2]. The catalogue of earthquakes created on the basis of the KNET network database is representative [3], which makes it possible to carry out various seismological studies on its basis.

The propagation of seismic waves in rocks is a complex process characterized by a number of parameters. It is conventionally accepted to divide the parameters into kinematic and dynamic. On the basis of seismic waves, various source parameters of earthquakes are calculated, which are also usually divided into kinematic and dynamic ones.

The focal mechanism is a geomechanical model of an earthquake source based on instrumental data from seismic records. The source model, presented in three-dimensional form, characterizes the orientation of the axes of the main relieved stresses and the position of the nodal planes of displacement along the rupture in the seismic source and is indirectly

* Corresponding author: ivtran@mail.ru

related to tectonic stresses. The focal mechanism refers to the kinematic characteristic of the source.

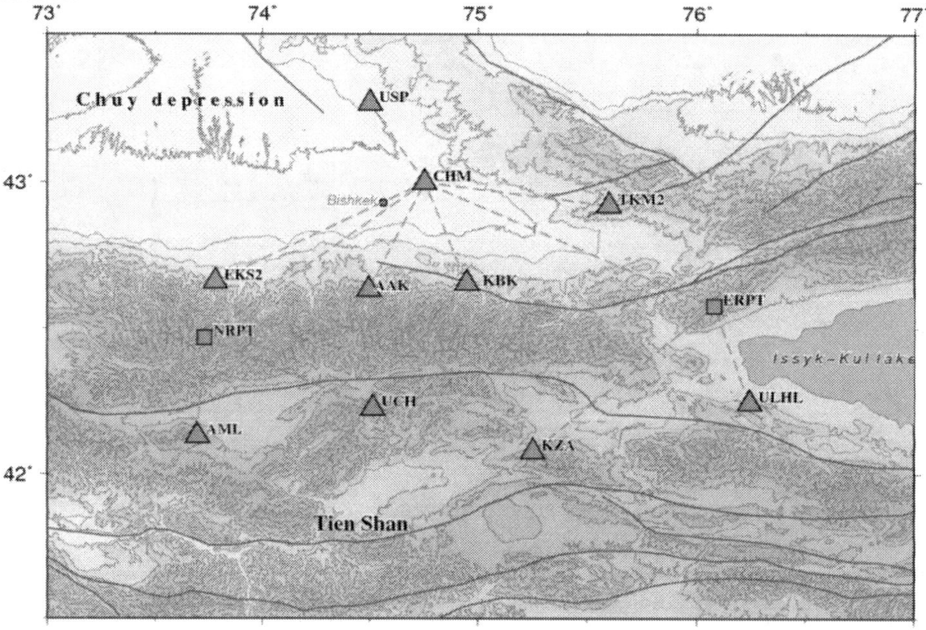

Fig. 1. Position of KNET network stations (triangles) and repeaters (squares).

The dynamic characteristics of seismic waves make it possible to determine such source parameters of earthquakes as the source radius r, scalar seismic moment M_0, stress drop $\Delta\sigma$ etc., which are related to dynamic parameters.

The purpose of this work is to calculate the source parameters of earthquakes for their further use in determining the stressed-deformed state of the earth's crust of Northern Tien Shan.

Methodology

Focal mechanisms (FM). The overwhelming amount of experimental information accumulated in modern seismology on the mechanisms of earthquake sources was obtained on the basis of data on the polarity of arrivals of seismic waves. At the present time, in the conditions of equipment of an increasing number of seismic stations with digital equipment, this source continues to be the most important [4-6]. When interpreting the picture of the first arrivals, a two-dipole seismic source is considered according to [7]:

$$m_{ij} = (t_i t_j - p_i p_j) / \sqrt{2}, \qquad (1)$$

where t, p are unit direction vectors for the tension and compression axes. The nodal surface for P-waves are two mutually perpendicular planes in the considered case of an isotropic elastic medium. As a rule, the focal mechanism is characterized by the orientation in space of its main axes.

One of the conditions for the reliable determination of the focal mechanism from the signs of the arrival of P-waves is the completeness of the surrounding of the epicenter by observation points. In the conditions of the KNET network, for the calculation of the focal mechanism, only those earthquakes that occurred in the territory bounded by the coordinates of the network edge stations can be considered: 42.0–43.0° N. and 73.75–76.0° E, which is a small area of 100 × 300 km^2 located on the territory of the Northern Tien

Shan (green rectangle in Fig. 2). Potentially, the focal mechanism can be determined for earthquakes marked in black. The earthquakes for which the focal mechanisms were obtained are marked in green. They account for more than 55% of the number of events that occurred in the territory bounded by the green rectangle.

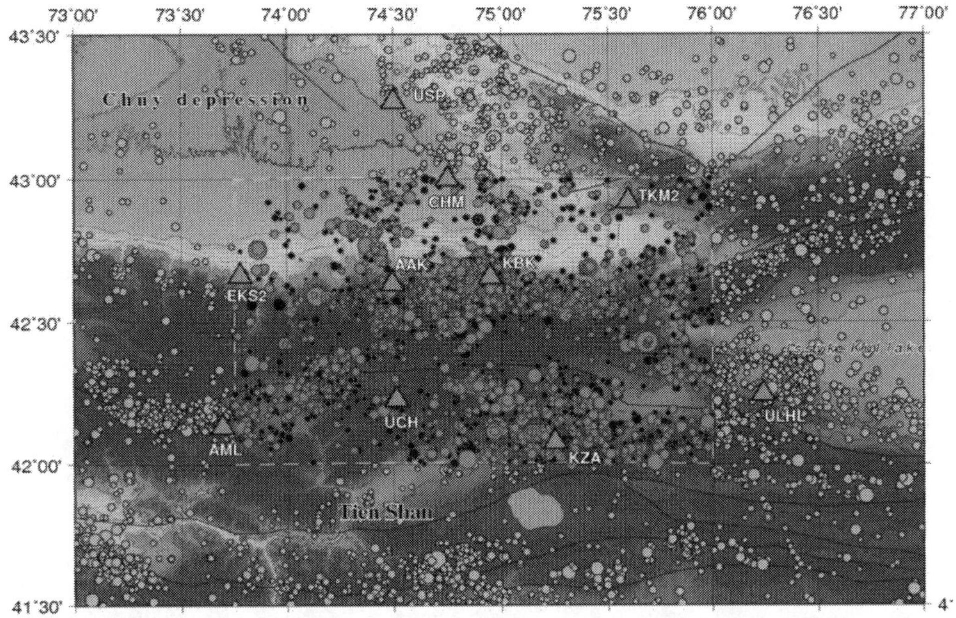

Fig. 2. The earthquakes location of according to the data of the KNET network. The green rectangle (dashed line) is the territory bounded by the coordinates of the edge stations of the KNET network. Earthquakes for which the focal mechanisms are determined by the polarity of arrivals of seismic waves (1674 events) are marked in green.

To calculate the focal mechanisms, the FPFIT program was used [8-12], which finds solutions for a double pair of rupture planes (initial model) that best suit the given set of recorded values of the first arrivals of an earthquake.

Dynamic parameters. To calculate the dynamic parameters, it is necessary to construct a source spectrum and determine its characteristics - spectral density Ω_0 and corner frequency f_0 (time of rupture in the source). The method for calculating the source spectrum in a KNET network is presented in detail in [13-14] and is not considered here. The spectral density Ω_0 is used to determine the scalar seismic moment M_0. To calculate M_0 the expression [15] is used:

$$M_0 = 4\pi\rho R V_S^3 \Omega_0 / \Psi_0 \quad (2)$$

where R is the distance from the source to the reference point, where the spectrum is determined, ρ is the density of rocks in the source area, Ψ_0 is the directional factor of radiation from the source (for a source described by a displacement jump on the rupture plane), the average value of this factor is taken as 0.64 [16].
The source radius is determined by the formula [17-19]

$$r = kV_S/f_0, \quad (3)$$

where k is a numerical coefficient depending on the source rupture model. The work uses the Brune model [14], for which this coefficient has a value of k = 0.37. To calculate the source radius, it is necessary to know the corner frequency, which is determined from the source spectrum and corresponds to the rupture time in the source point.

The value of the stress drop $\Delta\sigma$ characterizes the average difference between the shear stress in the fault before the earthquake, σ_0, and σ_1 after it, [20]. To calculate the stress drop, the expression [21] is used:

$$\Delta\sigma = 7M_0/16r^3. \tag{4}$$

Seismic energy can be estimated from the spectral parameters Ω_0 and f_0 of seismic waves at a reference point at a distance from the source [22], with the contribution of *S*-waves being the most significant. Compact expressions for E_S can be obtained if the directivity of radiation from the source is not taken into account [22-23]. A convenient expression relating seismic energy, seismic moment and corner frequency [23]:

$$E_S = \frac{1}{2}\pi^2\Psi_0^2 M_0^2 f_0^3 / \rho V_S^5 \approx 2M_0^2 f_0^3 / \rho V_S^5. \tag{5}$$

Since the velocity of *S*-waves is nothing more than $(G/\rho)^{1/2}$, it follows from expression (5) that the given seismic energy is proportional to factor $(f_0/V_S)^3$

$$e_{PR} = 2\cdot(M_0/G)f_0^3/V_S^3. \tag{6}$$

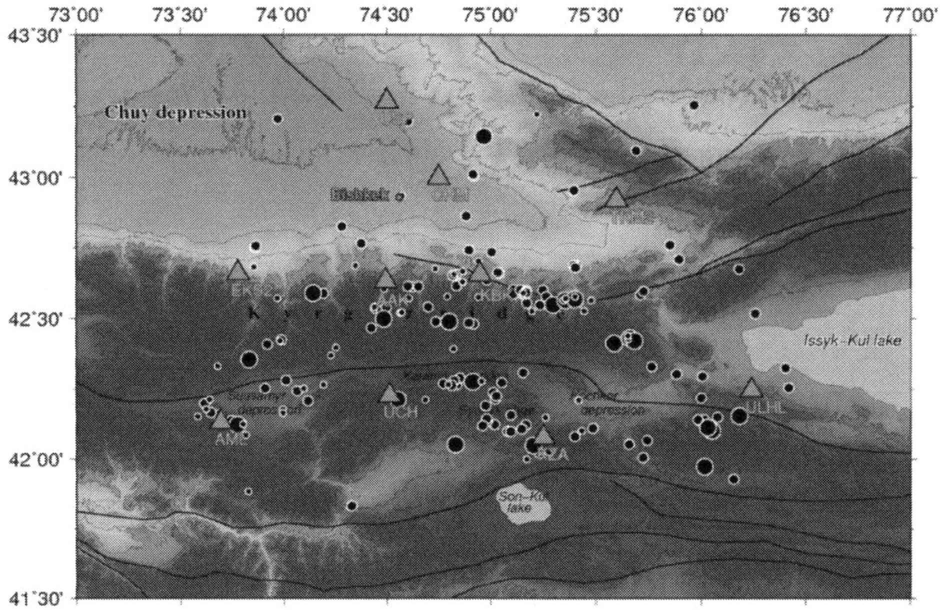

Fig. 3. Earthquakes epicentral location of (183 events).The minimum circle corresponds to the event with M = 2.7, the maximum - with M = 6.0 Triangles - KNET stations location. Black lines are regional faults.

As the initial data, we used the waveforms of earthquakes that occurred in the territory bounded by the coordinates of the edge stations of the KNET network for 1998–2017 with $K = 8.7 - 14.8$. Geological faults, geographical names of some ridges in the study area and the epicentral location of the seismic events under consideration are shown in Fig. 3. The main part of earthquake epicenters is located on the northern slopes of the Kyrgyz ridge, ridge Karamoynok, mount Sandyk and in the area of the Suusamyr and Kochkor depressions.

https://doi.org/10.1051/epjconf/202125402016

Results

Focal mechanisms. As a result of the application of the method for determining the focal mechanisms based on the polarity of the arrival of *P*-waves at the station, the focal mechanisms of 1674 earthquakes were determined and a catalogue was compiled (Fig. 4). The main part of the catalogue of focal mechanisms is made up of weak events (M = 1.6–2.5) that occurred at a depth of 5–15 km. Most of the earthquakes included in the catalogue occurred from 1999 to 2020. Solutions were received for 7 or more components, 43% of solutions were received for 9-10 components.

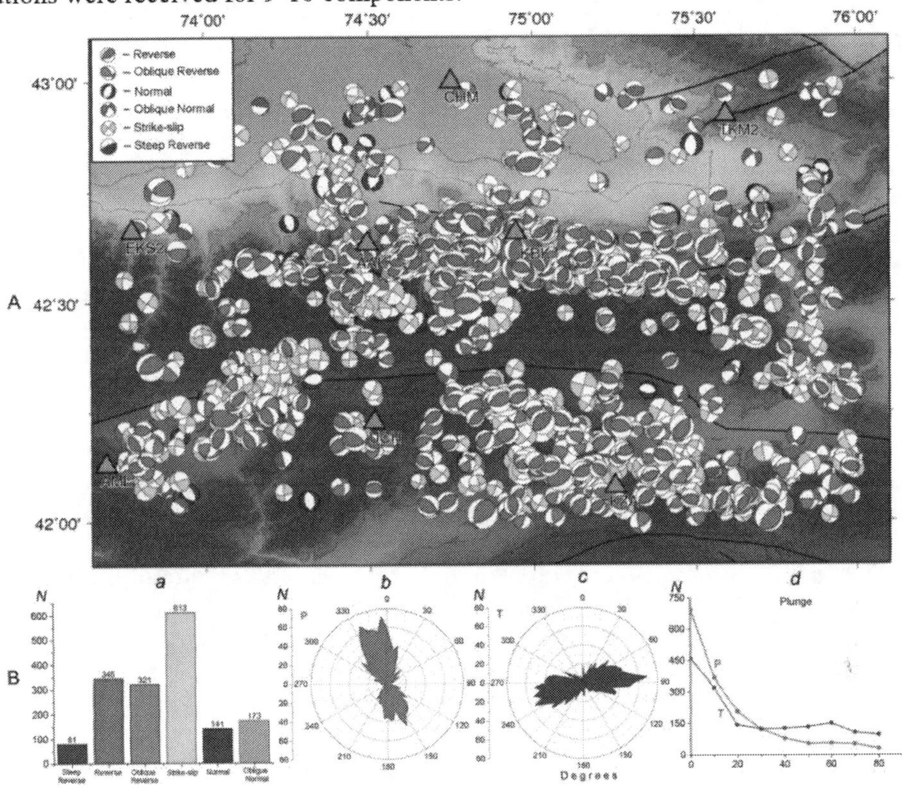

Fig. 4. A – epicentral position and focal mechanisms of earthquake centers (1674). Δ - Stations of the KNET network. B: *a* - ratio of the number of earthquakes by type of movement; *b*- diagrams of distribution of compression axes; *c* - axis of tension; *d* - the angles of immersion of these axes.

To build a map of focal mechanisms with a designation of the type of movement by color, samples were formed based on the analysis of the plunge angles of the main stress axes (plunge) *P*pl and *T*pl. The sampling criterion and quantitative results are presented in table 1.

Table 1. Criteria for choosing the type of mechanism, the number of earthquakes according to the criterion for choosing the type of mechanism for the entire catalogue and for individual depths

Options	Mechanism types											
	A		B		C		D		E		F	
Selection criterion	*P*pl < 30° *T*pl > 60°		*P*pl < 30° *T*pl > 60°		*P*pl < 30° *T*pl < 30°		*P*pl < 30° 30 ≤ *T*pl ≤ 60		30 ≤ *P*pl ≤ 60 *T*pl < 30°		30 ≤ *P*pl ≤ 60 30 ≤ *T*pl ≤ 60	
Quantity	*N*	%	*N*	%	*N*	%	*N*	%	*N*	%	*N*	%
All events	345	21	141	8.4	613	36.6	321	19.1	173	10.3	81	4.8
H = 0-5 km	8	9.4	12	14.1	24	28.2	15	17.6	16	18.8	10	11.7

H = 5-10 km	95	16.6	84	14.7	203	35.6	89	15.6	70	12.2	29	5.1
H = 10-15 km	129	23.6	25	4.5	204	37.3	120	21.9	46	8.4	22	4
H > 15 km	113	23.8	20	4.2	182	38.4	97	20.5	41	8.6	20	4.2

Note. A – Reverse fault; B – Normal fault; C – horizontal Strike-slip fault; D – Oblique reverse fault; E – oblique normal fault; F – Steep reverse fault.

In fig. 4A shows the position of the stations of the KNET network and earthquakes with an indication of the type of mechanism (see the inset in Fig. 3A). A variety of focal mechanisms is noted – reverse fault, normal fault, horizontal strike-slip faults, oblique reverse fault and oblique normal fault. A separate part of the events was attributed to steep reverse fault. 40% of events are reverse fault and oblique reverse fault, 41% are horizontal strike-slip faults and steep reverse fault. Normal fault and oblique normal fault are few in number and total 18% (Fig. 4Ba, Table 1.).

The statistical characteristics of the parameters of the axes of the principal stresses make it possible to determine some patterns of deformation processes. Diagrams of the distribution of the azimuths of the compression and extension axes, as well as the diagrams of the dependence of the number of earthquakes on the angle of immersion of these axes, have been constructed. When constructing, the values of the strike angles were averaged with a step of 5 degrees, and the angles of immersion were averaged with a step of 10 degrees.

According to the results obtained, the direction of the compression axes changes from north-north-west to north-north-east (the scatter is limited to the 330° –10° sector) (Fig. 4B*b*). For extension axes, the direction of the axes changes from northeast to southeast (the scatter is limited to a sector of 60° –110°) (Fig. 4B*c*). This is in good agreement with previously published results [24-26]. The sub-horizontal position (up to 30°) has 75% of the compression axes and 53% of the tension axes (Fig. 4B*d*).

Reverse fault and normal fault. In fig. 5A separately presents the distribution of reverse fault earthquakes (345 events, red) and normal fault type (141 events, blue). A joint map of reverse fault and normal fault events makes it possible to assess the ratio of their positions in the study area. The reverse fault seismic events are concentrated in three sublatitudinal zones: in the southern part of the Chuya depression up to its junction with the Chu-Ili mountains, along the northern slope of the central and eastern part of the Kyrgyz ridge and south of the Kyrgyz ridge from the Suusamyr to Kochkor depression through the Sandyk, Karamoinok and Kyzart mountains. The normal fault events are mostly isolated, and their concentration is observed only in the central part of the Kyrgyz ridge at the meridian level of 74.5° E; several events are located in the area of the Suusamyr and Chuya depressions and along the northern slope of the eastern part of the Kyrgyz ridge to the Chu-Ili Mountains. Numerically, normal fault events predominate in the western part of the territory under consideration. The area of normal fault events concentration south of the AAK station and in the south-eastern part of the Chui Depression. In the first case, the zone of reverse fault and normal fault events overlap, in the second they are allocated to a separate area.

In fig. 5B shows diagrams of the *P*- and *T*-axes directions for the reverse fault and normal fault mechanisms. *P*-axes for reverse fault events have three directions: north-north-west, north and north-north-east, *T*-axes are mostly oriented sublatitudinally. The *P*-axes for normal fault mechanisms have many directions, the most representative is the northeast direction, and for the *T*-axes the western direction is the most representative.

It is noted in [27] that, having different rheological properties, the layers of the earth's crust of the Tien Shan react differently to the tectonic compression they experience. According to the study of the azimuths of the compression axes at all considered depths (see Table 1) changes in the 330° – 10° sector, however, the maximum peak value of the number of events with depth shifts from 330° (northwest direction) to 360° (north). The direction of the extension axes for all depths varies from northeast to southeast.

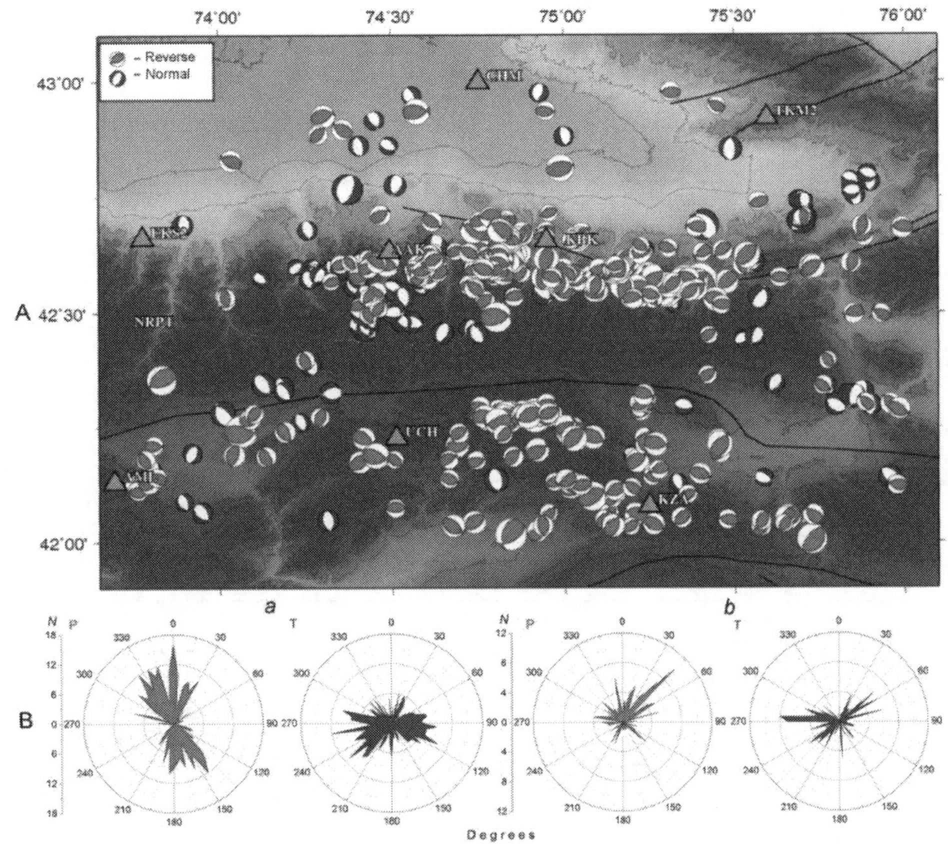

Fig. 5. A - the earthquakes position of the reverse fault and normal fault types. Δ - KNET stations. B - diagrams of the distribution of the azimuths of the compression (*P*) and tension (*T*) axes for the mechanisms of reverse fault events (*a*, 345 events) and normal fault events (*b*, 141 events).

Dynamic parameters. The dynamic parameters for 183 earthquakes of the Northern Tien Shan have been determined. When calculating the scalar seismic moment M_0, source radius r, stress drop $\Delta\sigma$, emitted seismic energy E_S and reduced seismic energy e_{PR} according to formulas (2) - (6), the values of the rock density in the source and the shear wave velocities were taken to be equal, respectively, $\rho = 2.6$ g / cm^3 [28], $V_S = 3.5$ km/s [29]. The boundary values of the obtained DP values are presented in table. 2. In the further, the DP of 182 earthquakes will be considered. The earthquake of December 25, 2006 with $K = 14.7$, $M_0 \sim 2\cdot10^{17}$ N·m and $\Delta\sigma \sim 214$ MPa (which radically differs in value from other events) is excluded from the analysis.

Table 2. Boundary values of dynamic parameters (182 earthquakes).

Parameters	K	M	f_0, Hz	M_0, N·m	r, m	$\Delta\sigma$, MPa	E_S, J	e_{PR}
Min	8.7	2.6	1.7	$6.2\cdot10^{12}$	210	0.2	$1.08\cdot10^{7}$	$1.75\cdot10^{-6}$
Max	13.7	5.4	6.0	$2.7\cdot10^{16}$	690	42	$8.03\cdot10^{12}$	$0.33\cdot10^{-3}$

The dependences of the dynamic parameters on the scalar seismic moment, the magnitude M, recalculated from the class K, were constructed, and the correlation coefficients between the DP and the magnitude, between the DP and the scalar seismic moment were calculated. Table 3 the obtained correlation coefficients are presented, as well as some regression dependences. The largest correlation coefficient between the seismic moment M_0 and the seismic energy E_S ($k = 0.98$), the smallest between the seismic moment M_0 and the stress drop $\Delta\sigma$ ($k = 0.74$). Let us consider the features of the distributions of the parameters of the spectral characteristics Ω_0, f_0 and the dynamic parameters for 182 events.

Table. 3. Regression dependences and correlation coefficients of dynamic parameters of earthquake sources and their magnitude and seismic moment.

№	Parameter	Parameter	Correlation coefficient, *k*	Model
1	$\lg \Omega_0$	M	0.88	$\lg \Omega_0 = 1.12M - 8.27$
2	$\lg e_{PR}$	M	0.78	$\lg e_{PR} = 0.63 \cdot M_0 - 7.1$
3	$\lg M_0$	M	0.88	$\lg M_0 = 1.10M + 10.1$
4	e_{PR}	M_0	0.74	$e_{PR} = 7 \cdot 10^{-14} M_0^{0.59}$
5	$\Delta\sigma$	M_0	0.74	$\Delta\sigma = 6.1 \cdot 10^{-7} \cdot M_0^{0.47}$
5	$\Delta\sigma$	$\lg M_0$	0.78	$\Delta\sigma = 2.3 \cdot 10^{-19} \cdot \lg M_0^{16}$
6	f_0	$\lg M_0$	-0.85	$f_0 = 21 - 1.16 \cdot \lg M_0$
7	$\lg E_S$	$\lg M_0$	0.98	$\lg E_S = 1.59 \cdot \lg M_0 - 13.18$
8	$\lg e_{PR}$	$\lg M_0$	0.92	$\lg e_{PR} = 0.59 \cdot \lg M_0 - 13.15$
9	r	$\lg M_0$	0.84	$r = 116 \cdot \lg M_0 - 1327$

Spectral density and scalar seismic moment. The distribution of the parameter Ω_0 (on a logarithmic scale) depending on the magnitude is shown in Fig. 6a, and the distribution of the scalar seismic moment for the same events ($M_0 \sim \Omega_0$) is shown in Fig. 6b. These logarithmic distributions are well described by the line model. For a scalar seismic moment, the regression dependence on magnitude (best model) can be written in the form:

$$\lg M_0 = 1.1M + 10.1 \quad (7)$$

The correlation coefficient between the values of $\lg M_0$ and the magnitudes M is 0.88. The deviation of the values relative to the linear model (regression lines in Fig. 6b) is uniform over the entire range of the considered values.

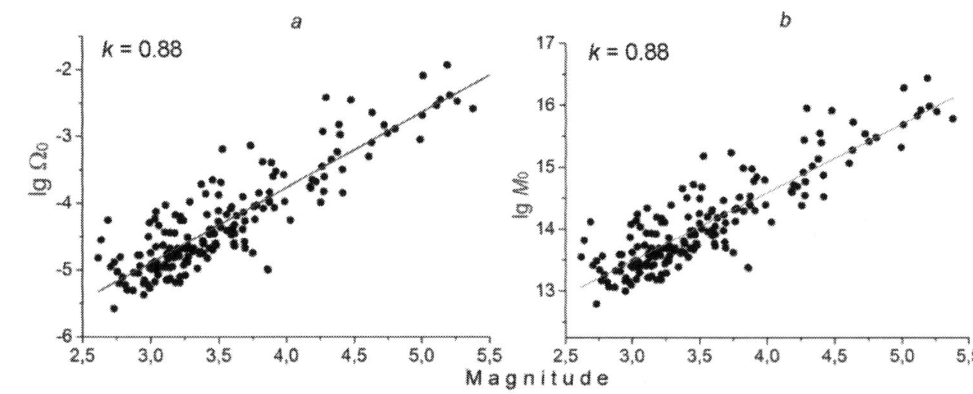

Fig. 6. Logarithmic distributions of spectral density (*a*) and scalar seismic moment (*b*) versus magnitude. *k* is the correlation coefficient.

For the events under consideration, the range of magnitude variation is $2.2 \leq M \leq 6.0$, and the values of M_0 lie in the range from $6.2 \cdot 10^{12}$ to $2.2 \cdot 10^{17}$ N·m. It should be noted that, due to the presence of a scatter and a relatively narrow range of magnitudes, the distribution of $\lg M_0$ also agrees (with an accuracy of 10%) with the well-known expression $\lg M_0 = 1.6M + 8.4$ [16].

Taking into account the high value of the correlation coefficient between the magnitudes and the values of the logarithm of the seismic moment, in the future, in a number of cases, we will consider the distribution of DP depending on lg M_0

Corner frequency and source radius. In fig. 7a shows the distribution of f_0 as a function of lg M_0. The corner frequency is characterized by a decrease with increasing lg M_0 or magnitude. The source radius was determined from the values of the corner frequency using formula (2). The distribution of r values depending on the logarithm of the seismic moment is shown in Fig. 7b. The Brune radius r various from approximately 210 to 640 m.

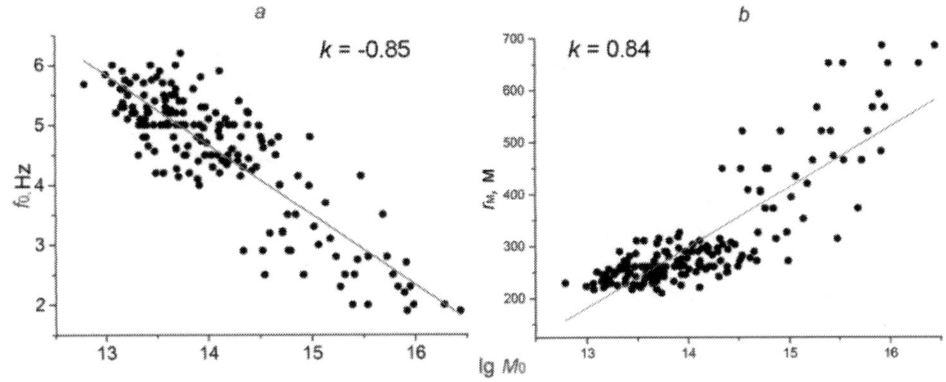

Fig. 7. The values of the corner frequency (*a*) and the source radius (*b*) in comparison with lg M_0.

Distribution of sources radius in Fig. 7b allows linear approximation in terms of lg M_0: lg M_0: $r = 116 \cdot \lg M_0 - 1327$. The deviations of r values from the linear model for events with lg M_0 <14.5 turned out to be smaller than for events with lg M_0 > 14.5. The correlation coefficient between the r and lg M_0 values is about 0.84.

Seismic energy and reduced seismic energy. The logarithmic distribution of the seismic energy E_S and the reduced seismic energy e_{PR} in comparison with the values of lg M_0 is shown in Fig. 8. Seismic energy varies in the range from $\sim 1.0 \cdot 10^7$ to $\sim 8.0 \cdot 10^{12}$J (see Table 2).

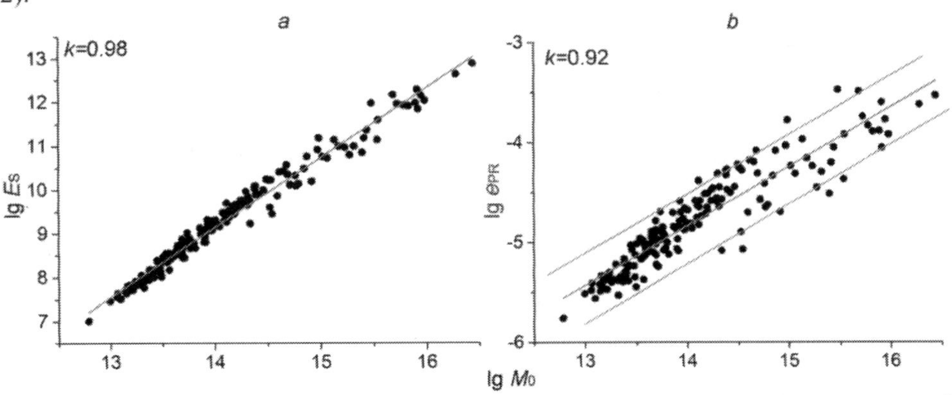

Fig. 8. Distribution of seismic (a) and reduced seismic energy (b) versus seismic moment on a double logarithmic scale.

The linear model describing the E_S distribution has the form lg $E_S = 1.6 \cdot \lg M_0 - 13.2$, and the correlation coefficient between seismic energy and scalar moment is $k = 0.98$. For the events under consideration, the e_{PR} parameter is on the order of $1.8 \cdot 10^{-6}$ to $\sim 0.3 \cdot 10^{-3}$ (see Table 2). The distribution of lg e_{PR} is also described by a linear model, although for it the scatter of values relative to the regression line is greater in comparison with the case in Fig. 8a. The expression for the linear model (regression dependence) has the form

$$\lg e_{PR} = 0.6 \cdot \lg M_0 - 13, \ M_0 \,(\text{N}\cdot\text{m}), \quad (8)$$

moreover, the correlation coefficient between the values of lg e_{PR} and the logarithm of the scalar seismic moment is 0.92.

Fig. 8*b* shows that the e_{PR} values are localized in a strip, the boundaries of which are parallel to the approximating straight line: about 95% of the imaging points fall inside such strips. The boundaries of the localization band of the lg e_{PR} values in Fig. 8b are determined by the ratio: $0.59 \cdot \lg M_0 - 13.6 \leq \lg e_{PR} \leq 0.59 \cdot \lg M_0 - 12.9$.

Stress drop. The values of the stress drop are in the range from 0.2 to ~ 40 MPa (Table 2). There is only one earthquake on 25.12.2006, $M = 6$ with $\Delta\sigma \approx 214$, which is excluded from the analysis. The distribution of the stress drop together with the values of lg M_0 is shown in Fig. 9a. The best approximation of the $\Delta\sigma$ distribution is a power-law dependence on lg M_0. Let us explain this circumstance. According to expressions (6), (4), (2), the values of the dropped stresses and reduced seismic energy are proportional: $\Delta\sigma \sim e_{PR}$. Consequently, the distribution lg $\Delta\sigma$ will differ from that shown in Fig. 8b of the logarithmic distribution of e_{PR} only by a vertical shift by some constant. For the calculation results for $\Delta\sigma$, the following approximation is obtained: $\Delta\sigma = 3.0 \cdot 10^{-8} \cdot M_0^{0.59}$. The correlation coefficient between drop stress and seismic moment is $k = 0.74$ (Table 3).

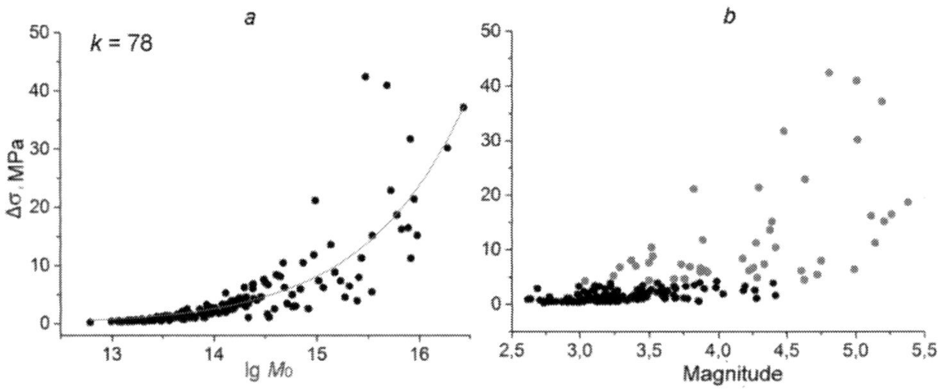

Fig. 9. Distribution of dropped stresses in comparison with the values of lg M_0 (*a*) and the result of clustering (*b*) of the representing points on the graph $\Delta\sigma(M)$.

Fig. 9*a*, it can be seen that the scatter of $\Delta\sigma$ values relative to the regression line increases for events with lg $M_0 > 14.5$ (magnitudes M> 4.0). Taking this into account, the issue of correlations between $\Delta\sigma$ and lg M_0 was analyzed in more detail, for which several samples of events were, formed (Table 4), and for each of them the correlation coefficients were determined. The first two samples were made according to the values of lg M_0, which characterizes the scale of earthquakes: events with lg $M_0 < 14.5$ constituted the first sample, and events with lg $M_0 \geq 14.5$ - the second. Two more samples were formed on the basis of stress drop: the third sample included events with $\Delta\sigma < 10$ MPa, and the fourth - $\Delta\sigma \geq 10$ MPa. The distinction based on the $\Delta\sigma$ feature corresponds with good accuracy to the selection of two clusters in the distribution of significant points in Fig. 9*b* using the clustering algorithm [30-31]. The distribution of stress drop for the first two samples is shown in Fig. 10*a*, the latter in Fig. 10*b*. The color of the icons indicates that the events belong to one of the selections.

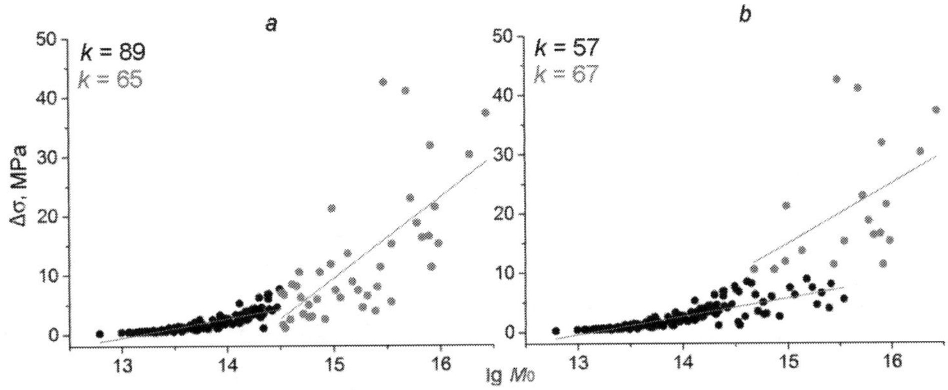

Fig. 10. Distribution of stress drop depending on the logarithm of the scalar moment: a - for events with lg M_0 < 14.5 - black color, lg $M_0 \geq$ 14.5 - crimson color; *b* - for events with Δσ < 10 - black color, for Δσ ≥ 10 MPa - crimson.

Table 4. Regression dependences and correlation coefficients of dynamic parameters of earthquake sources and their magnitude and seismic moment.

№	Sample for Δσ	Parameter	Correlation coefficient, *k*	Model
1	Все Δσ	M_0	0.78	$\Delta\sigma = 3\cdot10^{-8}\cdot M_0^{0.59}$
2	lg M_0 < 14.5	M_0	0.89	$\Delta\sigma = 4\cdot10^{-11}\cdot M_0^{0.8}$
3	lg $M_0 \geq$ 14.5	M_0	0.65	$\Delta\sigma = 6\cdot10^{-7}\cdot M_0^{0.50}$
4	Δσ < 10 MPa	M_0	0.57	$\Delta\sigma = 4\cdot10^{-8}\cdot M_0^{0.58}$
5	Δσ ≥ 10 MPa	M_0	0.67	$\Delta\sigma = 8\cdot10^{-5}\cdot M_0^{0.37}$

Table 4 shows the values of the correlation coefficients between the stress drop and the logarithm of the seismic moment for the above-described samples, as well as for the general sample of 182 events. For samples with lg $M_0 \geq$ 14.5 and with Δσ ≥ 10, the correlation coefficients are lower than for the general sample and there is a significant scatter of values relative to the regression line. The high value of the correlation coefficient for events with lg M_0 < 14.5 is reflected in the graph by the fact that the points of the values of Δσ in Fig. 10*a* are more localized near the regression line than points for another sample of events (crimson). For Δσ < 10 MPa, the correlation coefficient is low, which is explained by a small change in the values of the stress drop for the considered sample of events (black dots in Fig.10*b*). Such formal correlations convey physical meaning to a lesser extent than the correlations between the stress drop and M_0, or between the logarithms of these quantities. The obtained regression dependences, in which the exponent of M_0 is not less than 0.37, as well as Fig. 10 indicate the absence of self-similarity of earthquake sources on the territory of the BGPG in terms of stress drop and reduced seismic energy (Δσ ~ e_{PR}).

Conclusion

For weak earthquakes that occurred on the territory bounded by the coordinates of the edge stations of the KNET network (completely surrounded by stations), the focal mechanisms of the sources were determined by the signs of the arrival of *P*-waves and a catalogue of focal mechanisms was formed, which includes 1674 earthquakes that occurred from 1994 to 2020 biennium. In the region under consideration, mainly three groups of sources are formed, differing in the nature of the movements occurring in them: reverse fault, oblique reverse fault, and horizontal strike-slip faults. The Normal fault and oblique

normal fault events occur less frequently and are localized to a greater extent. A significant part of the events are reverse fault and oblique reverse fault events, most of which have a near-horizontal compression axis. For weak events, the prevailing direction of the azimuths of the compression axes is north-north-west (with depth this direction changes from north-north-west to north), and for extension axes - east-north-east.
A data bank has been formed on the dynamic parameters of earthquakes that occurred on the territory of the Northern Tien Shan (BGPG) in 1998–2017 (183 events). Distributions of dynamic parameters from magnitude M and scalar seismic moment M_0 are constructed, regression equations and correlation coefficients of dynamic parameters and scalar seismic moment are obtained. For a scalar seismic moment, the regression dependence on magnitude has the form: $\lg M_0 = 1.1M + 10.1$ for BGPG, which is in good agreement with the well-known relationship given by Yu.V. Riznichenko. For earthquake sources on the territory of BGPG, regression was established - a power-law dependence of stress drop on the scalar seismic moment with a correlation coefficient of 0.74. The main contribution to the correlation is made by seismic events with $M_0 < 3\cdot10^{14}$ N·m ($\lg M_0 < 14.5$). For stronger earthquakes with $\lg M_0 \geq 14.5$, the correlation coefficient between the values of $\Delta\sigma$ and M_0 is no more than 2/3, and this makes it possible to understand the results of works in which no significant correlations were found between the stress drop and the seismic moment (they did not consider relatively weak developments). The presence of an increasing dependence of the stress drop and reduced seismic energy on the seismic moment, established for the BGPG, is consistent with the results on the power-law dependence of $e_{PR}(M_0)$ in a number of other regions. The get data on focal mechanisms and dynamic parameters can be used to assess the stress-strain state.

Acknowledgements
The research was partially carried out within the framework of the state assignment of Research Station of the Russian Academy of Sciences in Bishkek sity (thems № AAAA-A19-119020190064-9 and № AAAA-A19-119020190066-3).

References

1. N.A. Sycheva, A.N. Mansurov, I.V. Sychev, S.I Kuzikov, Vulkanologiya I Sesmologiya, **3** (2020) (in Russian)
2. N.A. Sycheva, A.N. Mansurov, Geophysical Research, **19**, (2018)
3. N.A. Sycheva Dis. ... Cand. physical-mat. Sciences (Moscow) (2005)
4. V.G. Trifonov, O.V. Soboleva, R.V. Trifonov, G.A. Vostrikov Modern geodynamics of the Alpine-Himalayan collision belt (GEOS, Moscow, 2002)
5. A.A. Lukk, V.I. Shevchenko, V.G. Leonova, Physics of the Solid Earth, **51**, 6 (2015).
6. A.A. Lukk, V.I. Shevchenko Seismicity Izvestiya Physics of the Solid Earth, **55,** 4 (2019)
7. A.V. Vvedenskaya, Study of stresses and ruptures in earthquake sources using the theory of dislocations (Nauka, Moscow, 1969) (in Russian)
8. J.A Snoke, J.W. Munsey, A.C. Teague, G.A Bollinger, Earthquake Notes, **55**, 3 (1984)
9. J.A. Snoke, Earthquake Mechanisms, Encyclopedia of Geophysics (Van Nostrand Reinhold Company, 1989)
10. J.A Snoke, Seismological Research Letters, **61** (1990)
11. J.A. Snoke, FOCMEC FOCal MEChanism determinations. A manual (2000)

12. P.A. Reasenberg, D. Oppenheimer, FPFIT, FPPLOT and FPPAGE: FORTRAN Computer Programs for Calculating and Displaying Earthquake Fault-Plane Solutions (US Geological Survey Open-File Report, 1985)
13. N.A. Sycheva, L.M. Bogomolov, Russian Geology and Geophysics, **57**, 11 (2016)
14. N.A. Sycheva, L.M. Bogomolov, S.I. Kuzikov, Computational technologies in seismological research (on the example of KNET, Northern Tian Shan). (Yuzhno-Sakhalinsk: IMGG FEB RAS, 2020)
15. K. Aki, P. Richards, Quantitative seismology. Theory and methods (Mir, Moscow, 1983)
16. Yu.V. Riznichenko, Problems of seismology (Nauka, Moscow, 1985)
17. C.H. Scholz The mechanics of earthquakes and faulting (Cambridge Univ. Press, Cambridge, 2019)
18. R.E. Abercrombie, J.R. Rice, Geophysical Journal International, **162** (2005)
19. M.M. Scuderi, C. Marone, E. Tinti, G. Di Stefano, C. Collettini, Nature Geoscience, **9**, 9 (2016)
20. S.J. Gibowicz, A. Kijko, An introduction to mining seismology. (Academic Press, San Diego, 1994)
21. J.D. Eshelby, Proceedings of the Royal Society of London **A241**, 1226, (1957)
22. J. Boatwright, Bulletin of the Seismological Society of America, **70** (1980)
23. R. Madariaga, Earthquake scaling laws In: R.A. Meyers, ed. Extreme environmental events: Complexity in forecasting and early warning (Springer, 2011)
24. V.N. Krestnikov, E.I. Shishkin, D.V. Shtange, S.L Yunga, Izvestiya Academy of Sciences of the USSR Physics of the Earth, **3** (1987) (in Russian)
25. S.L. Yunga, Methods and results of studies of seismotectonic deformations (Nauka, Moscow, 1990)
26. A.K. Kurskeev, Earthquakes and seismic safety of Kazakhstan (Evero, Almaty, 2004)
27. F.N. Yudakhin, T.Ya. Belenovich, Izvestiya Academy of Sciences Kirg. SSR Fiz Tech. and matem. Science, **1** (1989)
28. D.K Chedia, Morphostructures and the latest tectogenesis of Tian Shan (Ilim, Frunze, 1986)
29. S.W. Roecker, T.M. Sabitova, L.P. Vinnik., Y.A. Burmakov, V.I. Golvanov, R. Mamatkanova., L. Munirova, Journal Geophysical Research, **98** B9 (1993)
30. R.F. Ling, The Computer Journal, **15**, 4 (1972)
31. J. Sander, M. Ester, H. Kriegel, X. Xu Data, Mining and Knowledge Discovery, **2**, 2 (1998)

Universal dynamo paradigm for solar activity, Higgs fields and disasters

Boris Shevtsov[1,*]

[1]Institute of Cosmophysical Research and Radio Wave Propagation, FEB RAS, 7, Mirnaya st., Paratunka, Kamchatka, Russia, 684034

Abstract. There is still a problem of a correct and accurate description of the dynamo and its uses in various fields of physics. To solve this problem, a special and universal representation of dynamo is proposed. The magnetic induction equation of dynamo is presented in the form of a Lienard relaxation oscillator with cubic nonlinear restoring force corresponding to the Mexican hat or champagne bottle potential which is used to determine the Higgs fields which are considered here in its general sense. Universal dynamo paradigm in field theory which can be used to describe disasters is proposed. Using solar activity as an example, it is shown how a dynamo induces a magnetic analogue of the Higgs fields with a broken symmetry of the magnetic field. Various dynamo modes are considered and different dynamo numbers are estimated. The dynamo effect can be used in field theory as an alternative to spontaneous symmetry breaking. Opportunities for the promotion of the new dynamo paradigm are discussed.

1 Introduction

Many physical systems are represented by relaxation generators. An example of this is Dynamo, the development of the theory of which largely determines the state of modern physics. As another confirmation of the universality of relaxation mechanisms, let us consider the analogies between solar activity, Higgs fields and catastrophes that follow from the representation of a magnetic dynamo in the form of a multidimensional Lienard relaxation oscillator [1] with a cubic nonlinearity of the returning force corresponding to the Mexican hat or champagne bottle potential which is used to determine the Higgs fields and to discribe the disasters.

Along with the field generation, the dynamo effect gives a forced symmetry breaking, which can be considered as an alternative to the spontaneous breaking used in field theory. How this works at different scales will be shown by the example of solar activity, for which previously unknown features of nonlinear oscillations of the Lienard generator, Higgs fields and catastrophes will be obtained. It will be shown that the nonlinearity of oscillations increases at small scales, which justifies the use of the dynamo effect to explain the forced symmetry breaking in field theory. It is noteworthy that the nonlinearities of the dynamo are determined by the gravitational field, according to which it is possible to compare the mass of the object responsible for the dynamo effect.

*e-mail: bshev@ikir.ru

The considered analogies give a dynamo paradigm of elementary particles, according to which particles are formed as fluctuations on different scales as a result of the dynamo effect. Elementary particles can be defined as the inner edge of the scaling, and the universe and galaxies as its outer edge. The dynamo effect consolidates naturally the physical movements on all scales that corresponds to the Mach principle. While independent motions on different scales correspond to the usual understanding of particles in the first approximation.

The disasters theory in the light of the dynamo paradigm can become an alternative to the Big Bang theory and a fruitful concept for understanding the sudden, rapid and spontaneous emergence of life at the molecular and cellular levels. The theory of catastrophes describes explosive processes and combines well the evolutionary and revolutionary phases of development. Inflation and collapse on the scales depend on the ratio of different types of energy, and the main blow of disasters is the embodiment of the singularity. If we know what galaxies and cyclones look like, then we can imagine what atoms and elementary particles look like in the theory of catastrophes with the dynamo paradigm according to the similarity principle.

The analogue of the Higgs boson can be represented as displaced weak oscillations, known in dynamo as excursions of magnetic poles. In this view, the displacement is explained by the dynamo effect, and not by a spontaneous breaking of symmetry. Highly nonlinear reversible oscillations can be compared to the generalized Higgs boson.

The development of the dynamo concept largely reflects the success of modern physics. A special representation of the dynamo by a Lienard relaxation oscillator with a cubic nonlinearity of the returning force opens up new opportunities in the development of field theory and allows us to revise numerous dynamo models in order to determine the effectiveness of various approximations.

The relaxation oscillator model was proposed in [2–4], and its development is in [5, 6]. A generalization to the multidimensional case, which can be used in nonlinear field theory, is considered in [1].

The idea of a relaxation generator in one form or another was used to create the theory of a magnetic dynamo [7–22] and see additionally the reviews [23–25], but there is no convection in the early models. The low-mode models with convection have a particular place in this theory [26–29], but only the representation of the dynamo in the form of the multidimensional Lienard oscillator [1] defines correctly the symmetry of the system and opens up new ways in the development of nonlinear physics thanks to the universality and productivity of the new dynamo paradigm and the specificity of relaxation oscillations.

The dynamo representation which will be discussed below can be called canonical because it is obtained without any approximations, reveals the mechanism of dynamo and takes into account all the nuances of field generation. It is not difficult to generalize the dynamo effect to any fields.

2 Magnetic dynamo

We write the equation of magnetic induction **B** in the form of an oscillator equation [?]:

$$\partial_t \mathbf{B} = \mathbf{Y}, \tag{1}$$

$$\partial_t \mathbf{Y} = rot(\partial_t \mathbf{v} \times \mathbf{B} + \mathbf{v} \times \mathbf{Y}) + \nu_m \Delta \mathbf{Y},$$

where $\mathbf{v}$ is the velocity of a medium, $\nu_m = 1/\mu\mu_0\sigma$ is the magnetic viscosity, μ and μ_0 are the relative and absolute magnetic permeabilities, σ is the electrical conductivity of a medium, $\mathbf{Y} = rot(\mathbf{v} \times \mathbf{B}) + \nu_m \Delta \mathbf{B}$.

We write the Navier-Stokes equation in the Boussinesq approximation as the diffusion equation:

$$\partial_t \mathbf{v} = \hat{D}\mathbf{v} + \mathbf{f} \tag{2}$$

$$\hat{D}\mathbf{v} = -\rho^{-1}\nabla P - 2\Omega \mathbf{e}_z \times \mathbf{v} - (\mathbf{v}, \nabla \mathbf{v}) + \nu \Delta \mathbf{v},$$

$$\mathbf{f} = g\beta T \mathbf{e}_r - (\rho\mu\mu_0)^{-1}(\mathbf{B} \times rot\mathbf{B}),$$

where P is the pressure, ν is the kinematic viscosity, Ω is the angular velocity of the convective layer rotation, ρ is the density of a medium, g is the acceleration of gravity, β is the coefficient of thermal volumetric expansion, $\mathbf{e}_z$ and $\mathbf{e}_r$ are the unit vectors of a spherical coordinate system, T is the temperature deviation from the stationary value T_0 in the convection layer at $\mathbf{v} = 0$.

The solution of equations (2) in the form of the Stokes' law

$$\mathbf{v}(\mathbf{r}, t) = \hat{G}\mathbf{f}(\mathbf{r}, t) \tag{3}$$

allows us to rewrite first equation (2) as

$$\partial_t \mathbf{v} = (\hat{D}\hat{G} + \hat{1})\mathbf{f}, \tag{4}$$

where $\hat{1}$ is the unit operator.

Substituting (3) and (4) in (1), we obtain the equation of magnetic induction in the form of the multidimensional Lienard equation [1, 5]:

$$\partial_t \mathbf{B} = \mathbf{Y}, \tag{5}$$

$$\partial_t \mathbf{Y} = rot[\hat{G}\mathbf{f} \times \mathbf{Y} + (\hat{D}\hat{G} + \hat{1})\mathbf{f} \times \mathbf{B}] + \nu_m \Delta \mathbf{Y}.$$

Equations (5) are a symbolic notation of the magnetic induction equation with integral operator kinematic coefficients $\hat{D}$ and $\hat{G}$. It generalises the concept of dynamo numbers, the physical meaning of which is revealed depending on problem conditions.

The $\hat{D}\hat{G}$ composition in (4) and (5) defines kinematic waves correlations and specifics of their oscillations. $\hat{D}$ represents diffusion and scales consolidation. $\hat{G}$ represents turbulence and scales fragmentation. $\hat{D}\hat{G}$ combines two processes.

The magnetic feedback $\mathbf{f}$, which is a balans between Archimedes and Lorentz forces, affects damping and frequencies of oscillations. The attenuation and amplification of the magnetic field is determined by $\hat{G}g\beta T\mathbf{e}_r$. This kinematic temperature tandem is a convection wave.

The equations (5) are a multidimensional genearalisation of Van der Pol [2–4] and Lienard [5] systems, and are an example of a dynamic system of fast-slow type [6], if its nonlinearity parameter is large.

In accordance with the magnetic nonlinearity and symmetry of the equation (5), the dynamo can be considered as a relaxation generator of the magnetic analogue of Higgs fields. The deflecting force in the equation (5) is cubic, and the potential will be a fourth-degree polynomial as in the Higgs equation. The nonlinear effect provides the magnetic field generation and the symmetry breaking. If the equation of magnetic induction does not have the magnetic symmetry of equation (5), then it cannot represent a dynamo. Primary dynamo models did not possess this symmetry.

Equations (5) are closed by the thermal diffusion equation

$$\partial_t T + (\hat{G}\mathbf{f}, \nabla)(T + T_0) = k\Delta(T + T_0), \tag{6}$$

where k is the thermal conductivity coefficient, T_0 is the solution of this equation at $\mathbf{v} = 0$, which corresponds to the stationary state of the system.

Thus, we obtained the magnetic dynamo equations (5) and (6) with magnetic feedback $\mathbf{f}$, which affects on the convection velocity through kinematic coefficient $\hat{G}$, which is mobility in (3) and feedback gain in (5) and (6).

In addition, the core of the integral operator $\hat{G}$ is convection waves, which are responsible for memory and nonlocality effects. Local approximation for $\hat{G}$ allows us to reduce the integral operator kinematic coefficients $\hat{D}$ and $\hat{G}$ to dynamo numbers that reduces the variables number and opens the way to small-dimention models with hidden kinematic dimensions. In the low-mode case with weak convection, this reducing leads to a simple dynamical system.

The equations (5) and (6) represent the magnetic-temperature dynemo waves. Kinematics is included implicitly in these equations through the operator kinematic coefficients $\hat{D}$ and $\hat{G}$ which are determened by complex friction (acoustic, Coriolis, turbulent, and viscous). Let us explain that acoustic or pressure waves carry away the impulse and act as effective friction.

In an incompressible medium, there are stationary waves of pressure that transfer momentum to the boundaries of the dynamo region with infinite speed.

We don't stop specially on boundary and other conditions for equations (1), (2), and (6), assuming that they provide a variety of solutions to these equations and actions of corresponding operators.

In order to obtain the magnetic dynamo equations (5) and (6), we used symbolic mathematics methods, which allow us to estimate the system parameters. We will do this for solar activity to show the effectiveness of the relaxation approach to the dynamo problem.

3 Magnetic fields

The condition $\mathbf{f} = 0$ allows us to determine the configuration of the magnetic field, the Lorentz force of which is in balance with the Archimedes force. Depending on the speed of convection, nonlinear reversible oscillations of the magnetic field will be excited around this configuration.

In the case when the Archimedes and Lorentz forces are collinear and form a mutually orthogonal triple of vectors with $\mathbf{B}$ and $rot\mathbf{B} = \mathbf{j}$, where $\mathbf{j}$ is the electric current, the condition $\mathbf{f} = 0$ gives $B_\pm \simeq \pm\sqrt{\rho\mu\mu_0 g\beta T L/2\pi}$, where L is the wave motion scale.

Taking into account $|T| \leq T_0$ (the saturation condition of the alternating variable T, which in amplitude cannot be greater than the stationary temperature T_0, otherwise the sum $T + T_0$ will take negative value) and the ratio $\beta \simeq 1/T_0$ (as for ideal gas), we obtain the upper estimate for the Archimedes force $g\beta T \simeq g$ and the magnetic field $\hat{B}_\pm \simeq \pm\sqrt{\rho\mu\mu_0 g L/2\pi}$. Here gL is the moment of gravitational forces per mass unit. It is responsible for magnetic field generation in a convective cell and it coincides in dimension with work per mass unit.

Substituting for the density of the solar atmosphere in the convection layer $\rho = 10^{-3} kg/m^3$ (10^{-3} from the Earth's atmosphere density), $\mu = 1$, $\mu_0 = 4\pi \cdot 10^{-7} H/m$, $g = 3 \cdot 10^2 m/s^2$ (acceleration of free fall on the Sun surface), and $L = D_S = 1.4 \cdot 10^9 m$ (the Sun diameter), we obtain $\hat{B}_\pm \simeq \pm 10T$.

This is four orders of magnitude greater than the observed fields on the sun's surface. If such intense fields are understood as internal, then the emergence of their flux tubes with high magnetic energy can explain sunspots, flares and coronal ejections on the Sun.

The condition $\mathbf{f} = 0$ at $|B| \simeq \sqrt{\rho\mu\mu_0 g L/2\pi}$ defines dynamical equilibrium for a free convection at the scale L. The kinematic approximation is applicable under the condition $|B| \ll \sqrt{\rho\mu\mu_0 g L/2\pi}$ which gives the unlimitedly growing solution, so it can't be used to describe solar activity.

4 Oscillations

On the period T_A of solar activity, the magnetic field oscillates between the values $\hat{B}_\pm \simeq \pm\sqrt{(\rho\mu\mu_0)gL/2\pi}$. The oscillation factor λ (cyclic gain and damping) is determined as $\lambda T_A = 2\pi$. Amplification and damping λ, according to the induction equation (1), are determined by the distortion of the convection velocity $\lambda \simeq 2\pi v/L$ (differential effect).

On the scale $L = D_S = 1.4 \cdot 10^9 m$, we have $\lambda \simeq 2\pi v/D_S = 2\pi/T_A$. If $T_A = 3.42 \cdot 10^8 s$ (11 years), then $v = D_S/T_A \simeq 4m/s$. It is the speed of meridian cerculation, $2T_A$ is a full cycle with double pass $2D_S$.

Taking into account the local, linear and isotropic approximation for (3) $\mathbf{v} = \hat{G}\mathbf{f} \to \mathbf{v} \simeq G_0\mathbf{f}$ and the upper estimate $\hat{f} \simeq f|_{B=0} = g\beta T \simeq g$ (see above $\beta T \simeq 1$), we find the mobility $G_0 \simeq v/g \simeq 10^{-2}s$. This time (acceleration or free run) determines the damping and non-linearity of the return force (or oscillation frequency) of the equation (5). The free run length at $v \simeq 4m/s$ will be about $10^{-2}m$.

Let us estimate the square of the oscillation frequency $\omega^2 = \left|rot(\hat{D}\hat{G} + \hat{1})\mathbf{f}\right|$ in (5). If $\hat{f} \simeq f|_{B=0} = g\beta T \simeq g$, $v \simeq G_0 f$, and $\hat{D} \simeq 2\Omega$ (predominance of the Coriolis effect on large convection scales is assumed), where $2\Omega = 4\pi/T_R$ (T_R is the Sun rotation period which is one month or $2.6 \cdot 10^6 s$), the estimate of the first term in ω^2 is $\omega_{DG}^2 = rot\hat{D}\hat{G}f \simeq 4\pi\Omega v/L \simeq 4\pi\Omega G_0 g/L$.

We should note, $\omega_{DG} \simeq \sqrt{4\pi\Omega v/L}$ is the geometric mean of zonal Ω and meridian $4\pi v/L$ rotation velocities which presents the reletive magnetic oscillation frequency through the weak inclination of the rotation $\omega_{DG}/\Omega \simeq \sqrt{4\pi v/L\Omega}$.

For $L = D_S$, $\omega_{DG}^2 = (2\pi/T_{DG})^2 \simeq 4\pi\Omega G_0 g/D_S = 2\Omega G_0\omega_g^2$, were $\omega_g^2 = 2\pi g/D_S$ is the estimation of the second term in ω^2. We see that $\omega_g^2 = 2\pi g/D_S$ is the gravitational constant normalized on the scale D_S and the square of the pendulum oscillations frequency in gravitational field. This is the result of the upper estimation of the Archimedes force.

Damping can be presented in these units as $\lambda = 2\pi v/D_S = 2\pi G_0 g/D_S = G_0\omega_g^2$.

For $D_S = 1.4 \cdot 10^9 m$ and $T_R = 2.6 \cdot 10^6 s$, $T_g = 2\pi/\omega_g = 2\pi 10^3 s$ (about 2 hours) and $T_{DG} \simeq 2 \cdot 10^7 s$ (2/3 year, this is the period of solar activity carrier oscillations).

Thus, we have three types of oscillations with the periods: T_A, T_{DG} and T_g. They are a slow dynamo wave, a fast dynamo wave, and an internal gravity wave. We should note, there is very little time G_0. This is mobility in Stokes' law (3). So the scaling of solar activity is more than 10^{10}.

The ratio $T_{DG}/T_A = 1/16.5$ is a pitch of winding, vice versa $T_A/T_{DG} = 16.5$ is the relative filling oscillation frequency in the relaxation pulse. These oscillations are observed in the Wolf numbers, figure 1 and the magnetic field, figure 2. About 16.5 oscillations per solar cycle is the fundamental result of the relaxation theory in the light of the universal dynamo paradigm.

In the composition of numbers $\omega_{DG}^2 \simeq 4\pi\Omega G_0 g/D_S = 2\Omega G_0\omega_g^2$, the small dimensionless parameter $2\Omega G_0$ of the evolutionary operator $\hat{D}\hat{G}$ is a phasa of Sun rotation on $2G_0$ period. For two scales of the dynamo, we have $\omega_{DG}^2/\omega_g^2 \simeq 2\Omega G_0 \ll 1$ that justifies the diffusion approximation for kinematic waves.

The composition of parameters $\omega_{DG}^2 \simeq 4\pi\Omega v/D_S$ can be also understood as the product of convection speed $v = G_0 g$ and differential rotation Ω/D_S. The latter value can be understood as a parameter of the omega effect. However, it should be noted that the representation of the magnetic induction equation (1) differs significantly from what was considered in [**?**]. There is no full analogy with the dynamic numbers of this work. Here, we selected two main dynamo numbers G_0 and Ω/D_S, which determine damping and nonlinear oscillation frequencies in (5).

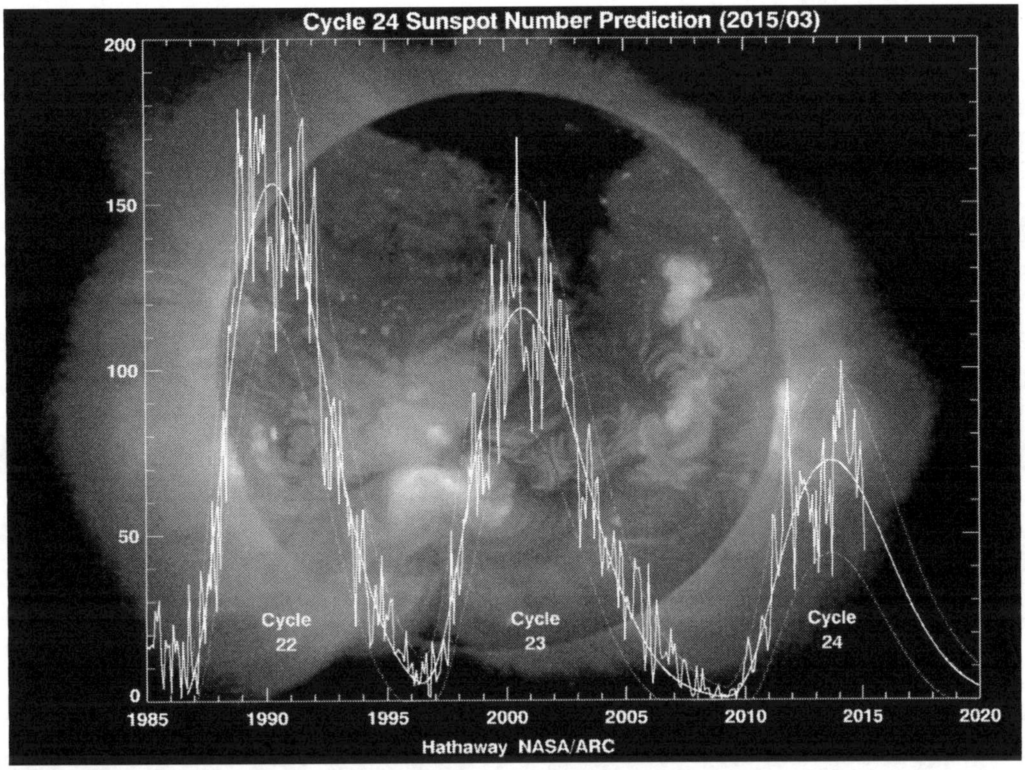

Figure 1. Oscillations in the solar cycle with a period of about 2/3 years in the Wolf numbers.

5 Nonlinearity parameter

The nonlinearity parameter $\mu = \lambda/\omega_{DG} = T_{DG}/T_A \simeq \sqrt{\pi G_0 g/\Omega D_S}$ is the winding pitch which must be more than a unit for the strong nonlinear mode. The relaxation impulse without filling is the sign of strong nonlinearity. This is a video impulse.

We should note that $\mu = \sqrt{2\pi Ro}$, were Ro is Rossby number.

For the Sun, $\mu \simeq 0.06$, therefore, the solar reversal oscillations are quasi-harmonic.

For more massive stars at the same diameter, the oscillations may be strong nonlinear.

If we turn to the scale $L < D_S$, for example to the sunspots scale $10^{-2} D_S$, where $\mu \simeq 0.6$, the oscillations there will be quasi-harmonic too.

But on the scale of $10^3 km$ ($10^{-3} D_S$), $\mu \simeq 2$. At a convection velocity of about 1 m/s, the Coriolis effects are comparable to kinematic turbulent ones on this scale.

Thus the non-linearity of oscillations increases when the scale decreases, and this is determined by kinematic turbulence which is the nonlinear component of complex friction in (3). As a result, the solar dynamo has a hybrid oscillation mode with inhomogeneous scaling, a quasi-harmonic oscillation mode at large scales, and a nonlinear mode at small scales. $10^{-3} D_S$ is the critical scale.

If we compare the magnetic dynamo with a violin then the magnetic field and the convective temperature wave are like the strings and a bow, and then a kinematic turbulence viscosity is like a rosin which affects the feedback gain. The kinematic variables play the role of the rosin.

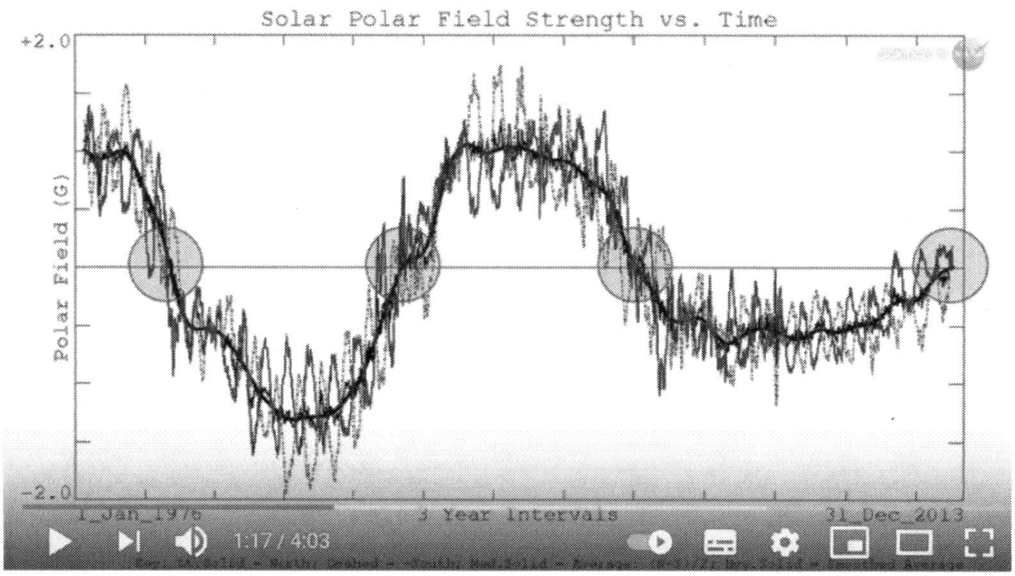

Figure 2. Oscillations in the solar cycle with a period of about 2/3 years in the magnetic field.

6 Three regimes of convection and oscillations

A nonlinear system will be of a fast-slow type if the nonlinearity parameter (the ratio of damping to oscillation frequency) is large. This is the case of the high convection speed $\mu^2 = \pi G_0 g/\Omega D_S = \pi v/\Omega D_S > 1$ and strong nonlinear oscillations. Such a system represents fast changes which may be compared to the electric discharge.

The nonlinearity parameter $\mu^2 = \pi v/\Omega D_S < 1$ is small for weak convection. In this case, quasi-harmonic oscillations are realized. That can be compared with a low-mode laser. The slow amplitudes of such magnetic-temperature oscillations will obey the Lorentz-like dynamical system if we exclude fast changes by averaging over high frequency oscillations. Another way to obtian the equations for a low-mode dynamo is the Galerkin's method [26–29]. But this method has a harmonic number selection problem that cannot be solved without a general relaxation approach.

Under very weak convection, the magnetic field is small and is determined by magnetic diffusion, and the reversible oscillations regress into excursus which are the oscillations about a stationary magnetic point without a reverse. Only in this case, the kinematic approximation for dynamo problem may be used. This is not applicable for solar activity but may be used for planetary dynamo [26–29].

The activation of very weak convection will determine the statistics of the waiting time of random reversals, between which there will be fading of reversible oscillations - this is the planetary dynamo mode. The minimum of Maunder solar activity - the fading of the reversible oscillations too.

Thus, three dynamo modes depending on convection are excursus, quasi-tonal and stron nonlinear ones. The two latter modes have reverses. The latter mode has fast reverse (magnetic switching).

Three dynamo modes may be compared with the same modes of brake pads, a violin, a spark gap, an earthquake, or a quantum generator. These comparisons help us to understand better how the dynamo works.

7 Reverse

Two reversive dynamo modes differ by the sets of axial and nonaxial dipole components. For example, the axial reverse of the magnetic dipole is observed in solar activity [30], figure 3. While nonaxial magnetic dipole component is suppressed.

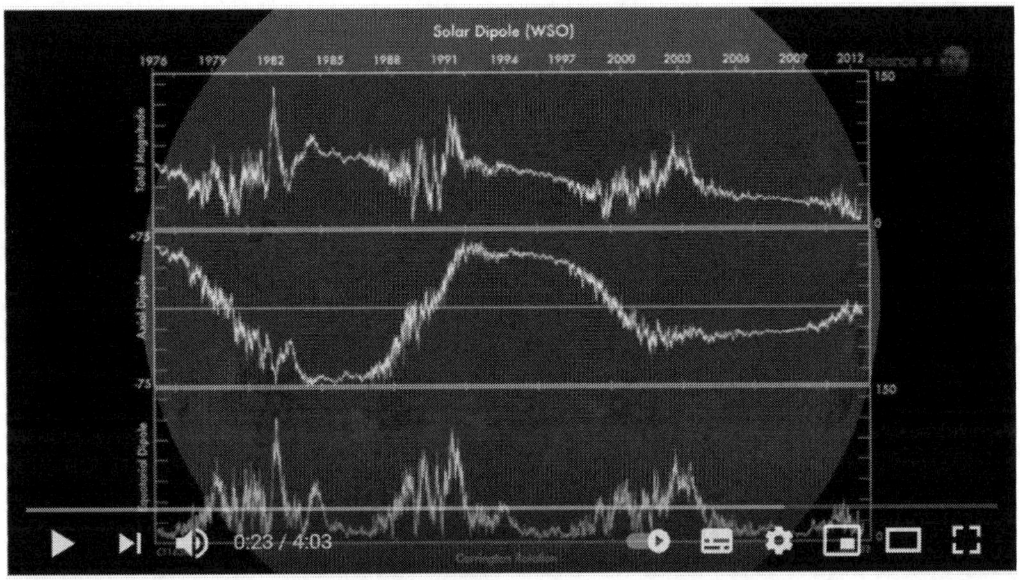

Figure 3. The ratio of the axial and non-axial dipole components of the magnetic field during reverse.

The synchronization of axial spheric harmonics in the defferent orders is responsible for the latitudinal localization of sunspots. It takes at least a hundred of spherical harmonics.

Assuming that only quasi axial modes are synchronized, Eq. (5) can be reduced to a one-dimensional Lienard equation. However, this assumption is not correct for strong convection.

The magnetic dynamo in the strong convection may be compared with a quantum generator passive Q-switching mode or seismic unloading [31, 32]. Such fast reverse is a magnetic switching or fall. This is the generating mechanism of a superstar, quasar or Black Hole.

All relaxation oscillations have a specific shape of Berlage impulse which is also in the solar activity, see figure 1.

8 Conclusion

The new dynamo paradigm is a special representation of the dynamo by a magnetic analogue of Higgs fields.

The magnetic dynamo is presented by the Lienard relaxation oscillator with the cubic nonlinearity of restoring force corresponding to the potential of the Mexican hat or champagne bottle which is used to determine the Higgs fields.

Strongly nonlinear oscillations in the magnetic analogue of Higgs fields represent catastrophe events.

The dynamo effect can be used in field theory as an alternative to spontaneous symmetry breaking.

The relaxation dynamo mechanism is the balance between Archimedes and Lorentz forces. This is the key to understanding the dynamo paradigm.

The dynamo has three oscillation modes depending on convection speed, they are excursus, quasi-harmonic and strong nonlinear ones.

The dynamo numbers are estimated for the solar activity and it was shown that Sun magnetic oscillations are quasi-harmonic for large scales and critical for small scales.

Oscillations in the solar cycle with a period of about 2/3 years are a fundamental result of the new dynamo paradigm.

The relaxation approach to the dynamo problem gives the explanations of numerous small-dimension and low-mode models, the variety of which is generated by the properties of kinematic coefficients and their approximations which should be reviewed from a general point of view.

Sudden changes in the properties of the kinematic coefficients can give stable structures or cause local disasters. It's like a loss of strength in a solid. This is the key to understanding the catastrophes of living, weather and climatic systems.

The relaxation approach provides a comprehensive understanding of the magnetic dynamo and it may be useful for many areas of Physics, especially where the fast-slow nonlinear dynamical systems are considered to describe catastrophic events.

The representation of the dynamo by magnetic-temperature waves reduces the initial dimension and gives the effective approximation method in the study of nonlinear systems.

The dynamo paradigm can be generalized to any fields.

Then the dynamo effect can explain the scaling and baryon asymmetry of the Universe.

The elementary particles can be considered as the space and time dislocations on the internal scale, as the catastrophes of the fold type.

A better understanding of subhadronic structures can be achieved through the examples of solar activity and many other relaxation systems.

The collapse and inflation of dark and light matter at different scales of the Universe can be presented as turbulent pulsations with disasters.

The Universe in the light of the catastrophes theory can become an alternative to the concept of the Big Bang.

Acknowledgments

The paper was carried out on the subject Physical Processes in the System of Near Space and Geospheres under Solar and Lithospheric Impacts: AAAA-A 21-121011290003-0.

References

[1] B. M. Shevtsov, E3S Web of Conferences **196**, 02026 (2020)

[2] B. van der Pol, The London, Edinburgh, and Dublin Philosophical Magazine and Journal of Science Ser.7, **2** 978 (1926)

[3] B. van der Pol, The London, Edinburgh, and Dublin Philosophical Magazine and Journal of Science Ser.7, **3** 65 (1927)

[4] B. van der Pol, J. van der Mark, Nature **120** 363 (1927)

[5] A. Liénard, Etude des oscillations entretenues. Revue générale de l'électricité **23** 901 and 946 (1928)

[6] M. Diener, The Mathematical Intelligencer **6** 38 (1984)

[7] E. C. Bullard, H. Gellman, Philosophical Transactions of the Royal Society of London A **247** 213 (1954)

[8] E. C. Bullard, Proc. Cambridge Philos. Soc. **51** 744 (1955)

[9] T. Rikitake, Proceedings of the Cambridge Philosophical Society **54** 89 (1958)

[10] E. N. Parker, Astrophysical Journal **122** 293 (1955)
[11] H. W. Babcock, Astrophysical Journal **133** 572Â (1961)
[12] Max Steenbeck, Fritz Krause, Karl-Heinz Rädler, Z. Naturforsch **21a** 369 (1966)
[13] R. B. Leighton, Astrophysical Journal **156** 1 (1969)
[14] N. I. Kliorin, A. A. Ruzmaikin, Magn. hydrodyn. **18** 116 (1982)
[15] K. M. Kuzanyan, D. D. Sokoloff, Astron. Rep. **40** 249 (1996)
[16] N. Meunier, M. R. E. Proctor, D. D. Sokoloff, A. M. Soward, S. M. Tobias, Geophys. Astrophys. Fluid Dyn. **86** 249 (1997)
[17] A. P. Bassom, K. M. Kuzanyan, A. M. Soward, Proc. R. Soc. Lond. A **455** 1443 (1999)
[18] Ya. B. Zel'dovich, A. A. Ruzmaikin, Sov. Phys. JETP **51** 493 (1980)
[19] A. A. Ruzmaikin, Comments on Astrophysics **9** 85 (1981)
[20] Ya. B. Zeldovich, A. A. Ruzmaikin, Astrophys. and Space Phys. Rev. **2** 333 (1983)
[21] N. O. Weiss, F. Cattaneo, C. A. Jones, Geophys. Astrophys. Fluid Dyn. **30** 305 (1984)
[22] C. A. Jones, N. O. Weiss, F. Cattaneo, Physica. Ser. D **14** 161 (1985)
[23] Ya. B. Zeldovich, A. A. Ruzmaikin, UFN **152** 263 (1987)
[24] P. Charbonneau, Annual Review of Astronomy and Astrophysics **52** 251 (2014)
[25] H. P. Popova, UFN **186** 577 (2016)
[26] G. M.Vodinchar, B. M. Shevtsov, Vychisl. Tekhnol. **14** 3 (2009)
[27] G.M. Vodinchar, L.K. Kruteva, Computational Technologies **16** 35 (2011)
[28] L. K. Feschenko, G. M. Vodinchar, Nonlin. Processes Geophys. **22** 361 (2015)
[29] L. K. Feschenko, G. M. Vodinchar, Magnetohydrodynemics **52** 287 (2016)
[30] NASA, The Sun's Magnetic Field is About to Flip (2013) http://science.nasa.gov/science-news/...
[31] R. Burridge, L. Knopoff, Bull. Seismol. Soc. Am. **57** 341 (1967)
[32] J. M. Carlson, J. S. Langer, Phys. Rev. Lett. **62** 2632 (1989)

Vector Overhauser magnetometer POS-4: experience and prospects of application

Sergey Y. Khomutov[1,*], *Vladimir A.* Sapunov[2,**], *Alexey Y.* Denisov[2], *Pavel B.* Borodin[3], *Dmitry V.* Kudin[4,5], *Roman V.* Sidorov[4], *Alexey S.* Bebnev[3], *Tatyana A.* Cherepanova[3], *Andrey L.* Kotikov[4,6], *Konstantin K.* Kanonidi[7], *Yury G.* Turbin[8], and *Pavel E.* Baryshev[8]

[1]Institute of Cosmophysical Research and Radio Wave Propagation FEB RAS, Paratunka, Kamchatka, Russia

[2]Ural Federal University, Ekaterinburg, Russia

[3]Institute of Geophysics UB RAS, Ekaterinburg, Russia

[4]Geophysical Center RAS, Moscow, Russia

[5]Gorno-Altaisk State University, Gorno-Altaysk, Altai Republic, Russia

[6]St.-Petersburg branch, Pushkov Institute of Terrestrial Magnetism, Ionosphere and Radio Wave Propagation of RAS, Saint-Petersburg, Russia

[7]Pushkov Institute of Terrestrial Magnetism, Ionosphere and Radio Wave Propagation RAS, Moscow, Russia

[8]The Arctic and Antarctic Research Institute (AARI) of the Russian Federal Service for Hydrometeorology and Environmental Monitoring, Roshydromet, Saint-Petersburg, Russia

Abstract. The results of practical use of a POS-4 vector magnetometer, developed by the Research Laboratory of Quantum Magnetometry, UrFU (Yekaterinburg) and based on POS Overhauser sensors, are presented. Continuous measurements by POS-4 have been carried out at the Paratunka observatory (IKIR FEB RAS, Kamchatka) since 2015, were done at the Saint Petersburg observatory (GC RAS / IZMIRAN SPb Branch, Leningrad Region) in 2017-2018 and have been performed at the Arti observatory (Institute of Geophysics, UB RAS, Sverdlovsk Region) since 2020. On the new high-latitude observatory White Sea (IAGA code WSE, GC RAS / MSU, Nikolai Pertsov White Sea Biological Station , Karelia), POS-4 is used as a main variometer for magnetic measurements. In April 2019, the magnetometer was successfully used for field measurements on ice during the TRANSARCTIC expedition in the Barents Sea (AARI, Roshydromet). At the beginning of 2021 IZMIRAN started testing two POS-4 magnetometers at the Moskow observatory. According to the results of field and observatory measurements it was possible to identify the advantages and disadvantages of the magnetometer and provide the information for its developers for further modernization in order to improve its efficiency and reliability. Many years of experience in POS-4 application determine the areas where its scientific and applied usage will provide important results, for example, for magnetic measurements in the Arctic regions or for monitoring of active zones around volcanoes.

*e-mail: khomutov@ikir.ru

**e-mail: vasapunov@gmail.com

1 Introduction

Ground-based magnetic measurements at observatories, Repeat stations and during surveys of various ranks are an important part of the study of the Earth's magnetic field. The need to obtain results with ever higher accuracy and reliability leads to new requirements for measurement technology, processing and analysis. Technical tools such as magnetometers, data collection and transmission systems are very important.

In recent decades, the development of magnetometers has made significant progress, especially with regard to devices for measuring variations and the total field intensity. However, with respect to magnetometers for measuring the full vector of magnetic induction (absolute measurements), progress is not so significant. Almost everywhere, a set of two magnetometers is used for this kind of measurement — a scalar, usually based on an Overhauser or quantum sensor, and a DI-magnetometer (DIflux) using a fluxgate sensor mounted on a non-magnetic theodolite tube. Such a set is the standard for the observatories of the INTERMAGNET network [1]. However, with the relative stability of the results, DIflux requires manual work of a highly qualified magnetologist, has low productivity and some limitations due to the observation technology.

One of the approaches for creating an automatic magnetometer for absolute observations is the automation of DIflux. AutoDIF and GyroDIF magnetometers have been developed and are already being used in observational practice, completely new devices with an automatic laser binding system to a remote reference point, a sensor rotation system and high-precision positioning [2, 3]. In addition, there are DIflux based on standard non-magnetic theodolites, in which manual manipulations and visual reading of the scale readings are performed automatically [4]. Together with a scalar magnetometer, an automatic or semi-automatic system for measuring the full field vector is obtained.

Another approach is based on the use of scalar sensors that measure the total field intensity F, placed in a system of coils that create compensating or additional fields in specified directions. Such systems were developed quite a long time ago and were used at observatories for absolute measurements of the horizontal and vertical components of the vector F, but they required careful tuning, calibration and a significant amount of manual procedures and time [5].

With the advent of sensors using the Overhauser effect and new technological approaches, it became possible to significantly automate measurements and calculations, minimize methodological errors and perform measurements with a frequency comparable to ordinary variometers. In practice, this approach was implemented on the basis of the developments of L. Hegimegy (Hungary) at GEM Systems (Canada) as the dIdD GSM-19FD magnetometer [6] and in the Laboratory of Quantum Magnetometry of UrFU (QMLab, Yekaterinburg, Russia) in POS-3 and POS-4 magnetometers [7].

2 Description of the magnetometer POS-4

The POS-4 absolute vector magnetometer was developed in the Laboratory of Quantum Magnetometry of the Ural Federal University on the basis of a high-precision scalar Overhauser magnetometer POS-1 (registered in the state Register of the SI of the Russian Federation, No. 79451-20). The device combines absolute measurements of the total field intensity with the accuracy of a scalar magnetometer [8, 9] and measurements of the vertical and horizontal intensity of Earth's magnetic field [10].

The magnetometer consists of three key components: a magnetizing system, a primary probe (an Overhauser sensor) placed in it, and an electronics unit separate from the magnetic system. The electronics unit is connected to a computer that controls the magnetometer, processes and stores data (see figure 1).

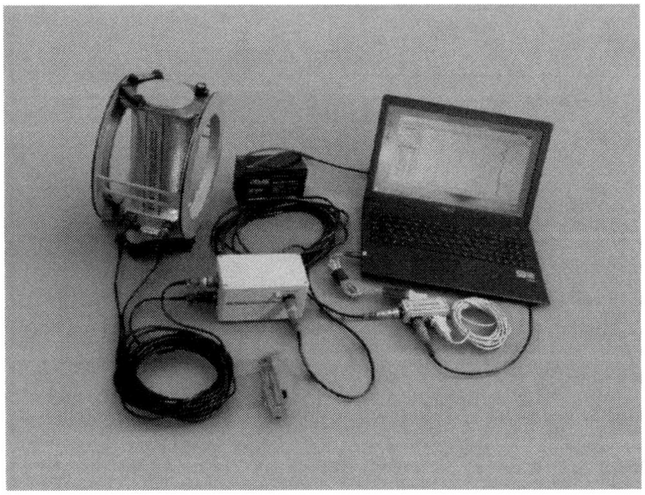

Figure 1. Assembly of the POS-4 magnetometer (from left to right: a solenoid with a coil system and a sensor, an electronics unit with a battery, an interface switching unit and a recording laptop).

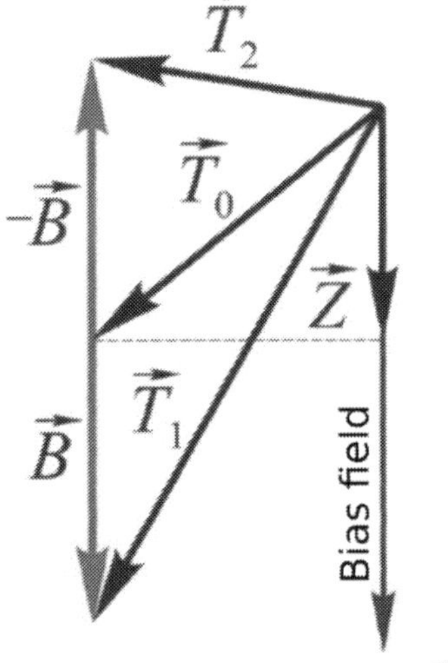

Figure 2. Field switching scheme in a vertical magnetizing system.

The magnetic field is measured by a primary probe that generates a signal of free precession of the nuclear magnetization of the working substance, followed by the calculation of the total field intensity in the electronics unit.

The magnetic system creates a magnetizing field (uniform in the sensor area) in the direction of the measured field components. The vertical magnetizing field is formed by a vertical Garrett solenoid. The verticality of the solenoid installation is ensured with high accuracy by two mutually perpendicular liquid levels at the top of the solenoid. Additionally, POS-4 contains a magnetic coil system that forms a horizontal magnetizing field. The recommended orientation (according to the criterion of the maximum sensitivity of the sensor) of this magnetization system corresponds to the choice of the direction of the axis of the coils perpendicular to the magnetic meridian plane, for example, using the reference compass. The concept of operation of POS-4 is based on measurements of the total field intensity by a scalar magnetometer placed in a magnetizing system that creates switched highly homogeneous magnetic fields. The magnetic system includes a vertical Garrett solenoid and a system of four coils that creates a horizontal field.

In each of the magnetic subsystems, a connected controlled source of stable switched current creates a magnetizing field by setting the current: $i = +I, -I, 0$. The upgraded software for POS magnetometers provides current switching and recording of the obtained scalar values of the total field intensity with subsequent calculation and visualization of the vertical Z and horizontal H components. In particular, the measurement of the vertical component (see figure 2) is carried out according to the following formulas:

$$Z = \frac{T_1^2 - T_2^2}{4B}; \qquad B = \sqrt{\frac{T_1^2 + T_2^2}{2} - T_0^2}.$$

Similarly, the horizontal component H is measured. It should be noted that the "raw" data $T_0, T_1, T_2, T_1', T_2'$ (T' are fields measured by secondary system) are stored in the data file for the possibility of further more detailed processing.

The high absoluteness of the POS-4 vector magnetometer results are due, firstly, to the absoluteness of the POS-1 Overhauser sensor as a probe and, secondly, to the technology of the magnetic system produce based on a titanium frame.

3 Geophysical Observatory Paratunka, IKIR FEB RAS, Kamchatka

The POS-4 magnetometer was purchased by IKIR FEB RAS as part of the implementation of the RSF project, including to assess the possibility to upgrade the observatory's magnetic monitoring to Russian-made equipment. The tasks of testing of magnetometer with the prospect of installation at the observatory Cape Schmidt (CPS) at Chukotka were also solved.

To install the magnetometer, a small 2x2 m wooden pavilion was built with an inner layer of expanded polystyrene for passive thermal stabilization. Such installation of the POS-4 provided full-fledged maintenance and carrying out various works, without interfering with other magnetometers of the observatory used for regular monitoring of the magnetic field. The noise influence during the generation of additional fields in the coil system and solenoid of POS-4 was also excluded. Inside the pavilion, a pillar was installed, in the lower part cast from a special low-magnetic mixture, and in the upper part made of glass blocks. There is no 220V power supply to the pavilion, there is no lighting and heating. To control the thermal conditions, three digital temperature sensors DS18B20 with a sensitivity of 0.1^oC are installed: (a) in a thermally insulated air gap around the sunken part of the pillar (to control freezing, the gap is filled with water in the spring), (b) at the height of the sensor and (c) on the housing of the electronics unit (to control overheating).

The POS-4 solenoid with a standard non-magnetic stand from theodolite is levelled using two high-precision bubble levels on its upper part. The horizontal axis of the magnetometer is oriented to the geographical East, so the values of the horizontal component are close to the current value of the component Y of the magnetic field. A recording laptop with a standard software POS Manager and a power supply system are installed in the technical pavilion at a distance of about 20 m. The magnetometer timer is synchronized when the measurements are started or manually according to the laptop's system time, which is tied to UTC every hour by the institute's time server. The measurement of single total field intensity value is 2 seconds, so the full cycle (measurement of the full magnetic field vector) takes 10 seconds. The photo and diagram of the pavilion, the photo of the sensor on the pillar are shown in figure 3, a detailed description is given in [7, 11].

Regular measurements using POS-4 were started in the spring of 2015 and continue continuously to the present. There was a short break in November 2016, when the magnetometer was unmounted for the period of work on strengthening the lower part of the pillar (strengthening was required due to the low stability of the pillar, especially during the freezing and defrosting of the surrounding soil). No special work was carried out with the magnetometer. But occasionally, with a noticeable departure of the levels, a corrective leveling of solenoid was performed. Also every year in the autumn the air gap between the ground and the pillar was insulated (in early summer, the insulation was removed).

The magnetometer POS-4 was mainly considered as a variometer and complete processing was performed for it, as for the other magnetometers of the observatory: check of raw records, cleaning from noise, calculation of baseline values and minute values. In some cases, POS-4 data was used to fill the gaps in the records of the main magnetometers, if the data of the backup devices were also missing. In figure 4 the comparison of the raw POS-4 data with the records of the FGE-DTU fluxgate variometer, which is the main one at the observatory,

 https://doi.org/10.1051/epjconf/202125402018

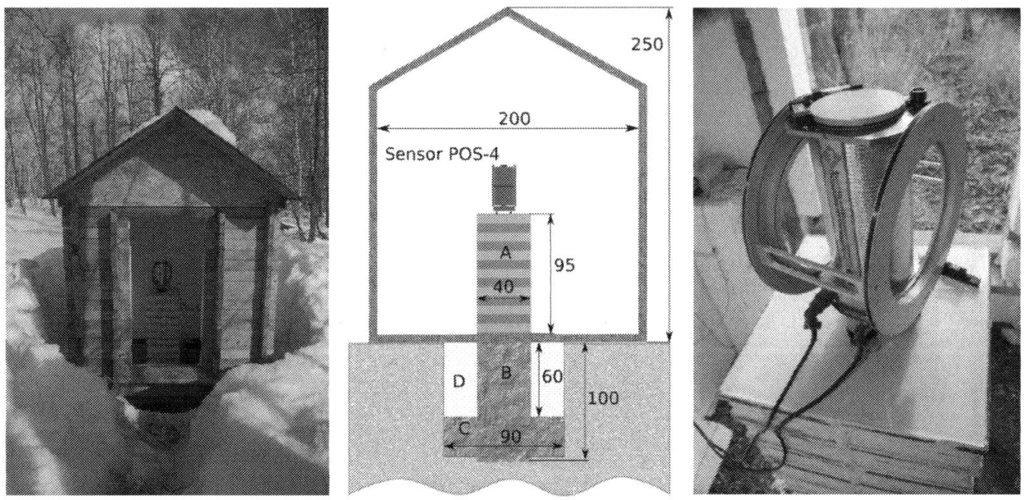

(a) (b) (c)

Figure 3. Location of the magnetometer POS-4 at the Observatory Paratunka: (a) the external appearance of the pavilion; (b) a design of the pavilion and the pillar (dimensions in cm); (c) a sensor on the pillar, the level for rough control of the pillar slopes is visible on the right.

is presented for 12.05.2021, when the magnetic field was disturbed (the K-index reached 6). As can be seen, in the differences of the Z and F values of POS-4 and FGE, slow variations are insignificant. The difference of Y shows a daily variation up to 2 nT, which is due to the influence of temperature on the POS-4 measurements. Also, noticeable deviations of the differences from the constant level are clearly visible in the intervals when the field changed rapidly. This is due to the significant time of measuring the full field vector using POS-4 (10 seconds) and the peculiarities of calculating the components Y and Z. Similar effects are also observed in the records of the dIdD GSM-19FD magnetometer (GEM Systems).

The long-term stability of the magnetometer is controlled by the baselines values of F_0, Z_0, Y_0, i.e. by comparing the results of its measurements with the results of absolute observations at the observatory performed manually using the POS-1 scalar magnetometer and the LEMI-203 DIflux magnetometer (LC ISR, Ukraine). Figure 5 shows the average daily baseline values since October 2018. The starting date was chosen from the moment when the pillar stabilized after strengthening, the drift on the old pillar during the freezing-defrosting periods reached several nT per day. The data in figure 5 show that POS-4 gives fairly stable results in the last three years of continuous measurements. Seasonal variations with a scatter of 0.6 nT are clearly visible only in F_0, the reason is not established. Similar F_0 oscillations with a scatter of up to 2 nT are present in the dIdD magnetometer baselines and with a scatter of up to 0.8 nT in the GSM-90 scalar magnetometer baselines, all magnetometers are installed in different pavilions. We can also note the increased dispersion of Z_0 annually in the winter period, which is also manifested in Z_0 of dIdD.

4 The expedition TRANSARCTIC-2019: the works of AARI

Due to the growing development of the polar regions, geophysical interests here are of great scientific and practical interest. There are only a small number of coastal stations in the auroral zone of Russia, which are not enough to study the geophysical phenomena in this extended territory. There is only one station Spitsbergen in the polar cap area. There is a station on

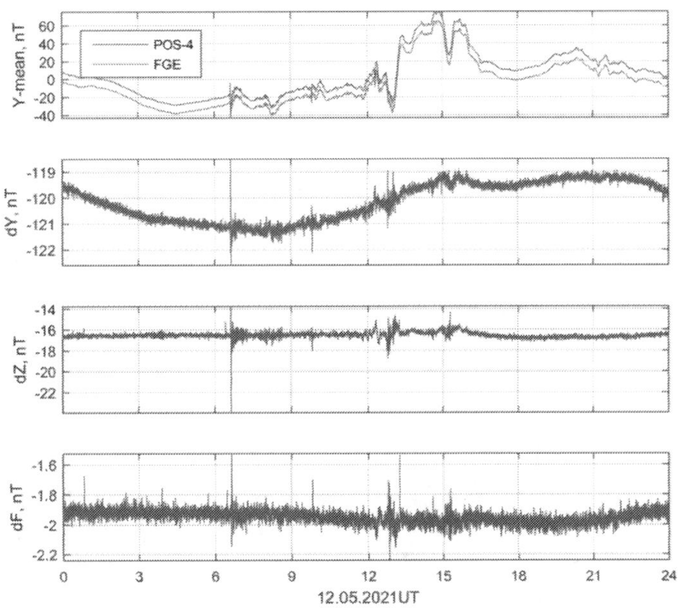

Figure 4. POS-4 and FGE-DTU data for 12.05.2021. The upper panel is values of component Y for POS-4 and FGE (the average values for the day are removed), the lower panels are the differences of POS-FGE of the components Y, Z, F.

Heys Island, which passes into the polar cap zone from the aurora zone only at certain intervals. For the polar cap, where the geophysical processes differ significantly from processes in the auroral zone, such a number of stations is not enough. Significant additional information can be provided by geophysical measurements during the drift through the Arctic Ocean a vessel frozen in ice. The first experience of joint geophysical studies on drifting ice was carried out during of the TRANSARCTIC-2019 expedition. The magnetic field measurements were included in scientific program of expedition along with other geophysical observations.

The magnetometer POS-4 was chosen to measure the magnetic field, since this device is close to the accuracy requirements of the INTERMAGNET standards, and its declared characteristics fully correspond to the conditions that can be created on the ice at a distance from the base vessel. This was confirmed by the fact that the scalar modification of such a device (POS-1) has been used by the AARI at the Cape Baranov polar station for several years without any problems.

It was also taken into account that the magnetometer is a component device, equipped with a tilt meter and allows quickly correct the orientation of the axis of the horizontal component coils during drift and possible reversal of the ice floe. That is, it is possible to measure the total field and the vertical intensity and to calculate the horizontal component even with small changes in the position of the POS-4 solenoid relative to the local vertical. The using of the modern navigation systems in conditions of drifting ice allows to measure the magnetic declination, but this expedition failed to do this. Before the start of the expedition, POS-4 passed a small test on the base AARI "Ladoga" (see figure 6a).

The magnetometer was placed in a non-magnetic heated pavilion with non-freezing design for its emergency evacuation. The pavilion was made in St. Petersburg and then de-

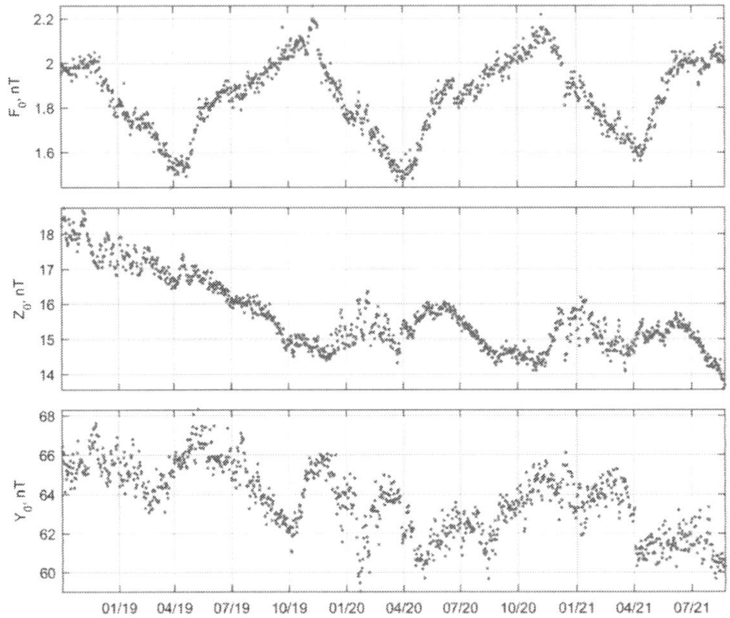

Figure 5. Daily mean baseline values of F_0, Z_0, Y_0 of the magnetometer POS-4 since October 2018.

(a) (b)

Figure 6. Magnetometer POS-4 in the TRANSARCTIC-2019 expedition: (a) the testing at the Ladoga field base of AARI; (b) a magnetometer sensor in a non-magnetic pavilion on ice.

livered by ship and helicopter to the installation site on the ice at a distance of about 400 m from the ship frozen in the ice. The magnetometer was installed in one of the corners of the pavilion on a tripod attached to the elements of the pavilion structure. The device is oriented vertically. The initial orientation of the device to the magnetic north is performed using a compass. The electronics unit and the laptop for data recording were located in the opposite corner of the pavilion (figure 6b).

In continuous operation, the measurements of the total field and vertical intensity of the magnetic field were recorded with rate of the full measurement cycle of 15 seconds. The results of the tilt meter measurements also were recorded. A Wi-Fi connection was established between the pavilion and the ship, which made it possible to constantly monitor the operation of the device, reducing the need to visit the pavilion to a minimum. During the entire time of the measurements, there were three failures of the tilt meter, which were probably caused by too large values of angular acceleration caused by the movements of the observer who was in the pavilion. These failures required a restart of the magnetometer control system. Every day, the data was sent to the Polar Geophysical Center of AARI.

Magnetic measurements were carried out in the period from 02 to 27 April 2019. At that time, the station was located at area with geomagnetic latitudes from 68° to 75°N, that is, part of the time the station was in the zone of the aurora and for some time was part of the polar cap area. During the drift of the station, the storm magnetic activity was low, but it changed noticeably in the polar cap (figure 7). Figure 7b shows that the values of the PC index, which are a numerical characteristic of the disturbance of the magnetic field in the polar cap, were not very large. Such changes in the magnetic disturbance will allow us to trace the features of the development and flow of magnetic substorms both in the area of the aurora and in the polar cap.

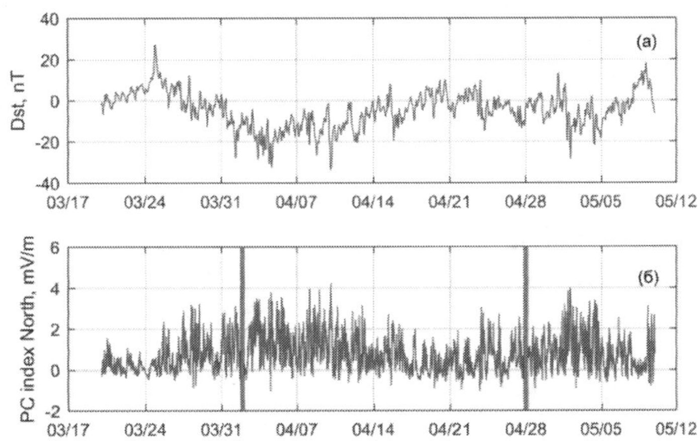

Figure 7. Global magnetic activity during TRANSARCTIC-2019: (a) Dst index (http://wdc.kugi.kyoto-u.ac.jp/dstae/index.html); (b) Polar caps index (PC North, ISGI, http://isgi.unistra.fr). Vertical lines mark the beginning and end of measurements using POS-4.

As an example, in figure 8 the initial records of the variations dZ, dH(F,Z) and dE (the eastern component, the exact orientation of the Y axis is not determined) are shown in comparison with the second data obtained at the Hornsund Observatory (HRN, Svalbard Island, Norway, the distance to the expedition is about 800 km). As can be seen, there is a consistency of some features in the horizontal components of the field, a less expressive picture for variations of the vertical component. The POS-4 record is quite clean, the most noisy is the H component, which is calculated from F and Z.

The calculation of the elements of the Earth's magnetic field using the WMM-2015 model for the 2019.25 epoch at the expedition site and at the HRN observatory (for comparison) are presented in table 1.

One of the problems that arose during the expedition was related to operational processing — daily files that are generated by the POS Manager software have a format that, when read, creates 5 duplicate values for each field element. When standard visualization and analysis tools, such as Excel spreadsheets, are used there are steps on the plots that create discomfort when viewed. An implicit problem in this case is the separation of the measurements of

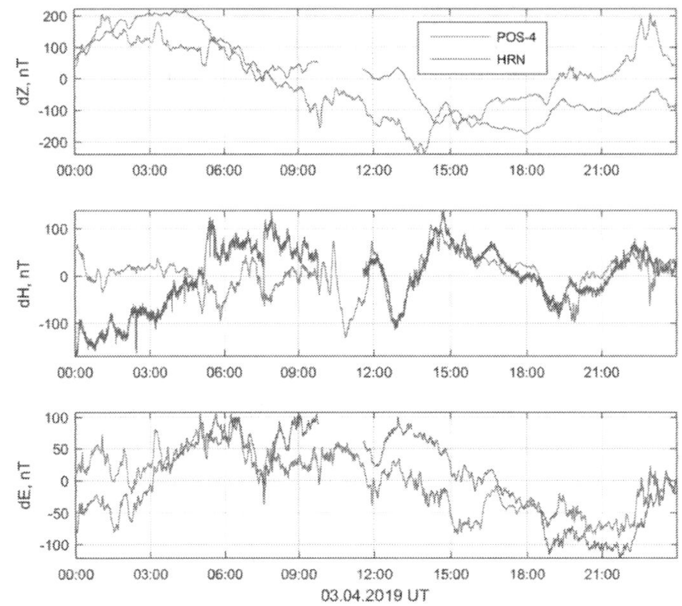

Figure 8. Comparison of the measurement results using POS-4 in the drift ice area and obtained at the HRN observatory for 03.04.2019. The average values are removed.

each of the five total field intensity values by the time in the cycle (for example, when the interval between the measurements is 3 seconds then the full cycle takes 15 seconds). The calculation of field elements (see section 1) assumes that each element is calculated by three total field intensity values that are measured simultaneously. With rapid field variations, this requirement can be significantly violated, up to the occurrence of uncertainties of the form 0/0 or $\sqrt{-1}$.

This may not be important for stationary measurements at the observatories of middle and low latitudes, but it is inconvenient for measurements in the polar regions as well as on a moving (albeit very slow) platform. It should be noted that the problem of simultaneous performance of component measurements by the method of an additional magnetic field was solved when conducting component absolute aeromagnetic surveys. For this purpose, a three-component magnetometric system was created in LO IZMIRAN together with LITMO. It consisted of three quantum magnetometers, the sensors of which were placed in three Braunbeck coil systems, of an optical sensor alignment system, and also included a navigation system, a system for recording and processing. The system was located on a specially equipped AN-12 aircraft. In the case of uniformity of the magnetic field at the site where the measurements were carried out, a fourth magnetometer could be used in the complex to simultaneously measure both the total field and the components while ensuring their mutual influence [12]. In general, there are no significant remarks on the operation of the magnetometer POS-4 in the TRANSARCTIC-2019 expedition. It proved to be a very reliable device, suitable for operation in the climatic conditions of the polar region.

Table 1. Comparison of Earth's magnetic field elements from observations and from model WMM

	data	F	Z	H	Y
Transarctic-2019	WMM	56157	55913	5224	2595
	POS-4 (03.04.2019)	56213	55975	5169	
Hornsund	WMM	54634	54069	7839	1227
	HRN (03.04.2019)	54567	53991	7905	1015

5 The Geophysical Observatory Arti, IG UB RAS, Arti, Sverdlovsk region

The Geophysical Observatory Arti of the Institute of Geophysics of the Ural Branch of the Russian Academy of Sciences is located on the outskirts of the urban village Arti, 180 km from Yekaterinburg.

The close location of the observatory to the Laboratory of Quantum Magnetometry (Ekaterinburg), the presence of a developed infrastructure and various types of magnetometers operating in according to INTERMAGNET standards, and a low background noise level makes it an excellent place to test experimental QMLab developments, calibrate and compare magnetometers, check in real measurements, see figure 9d ([13]).

Since November 2011 the POS-3 No.1 Overhauser ZT magnetometer, a prototype of the POS-4 serial magnetometer, are used at the Observatory Arti. POS-3 allows measuring the vertical and total field intensity. The main tasks for POS-3 were field measurements, including long-term measurements at the Repeat Stations when mapping the Manchazh gravimagnetic anomaly (Ural mountains) on an area of 50×50 km. Between field seasons, POS-3 is installed at the observatory for continuous measurements and various experimental observations. The general view of POS-3 on Repeat Station is shown in figure 9a. From January 2021 the observatory operates POS-3 No.5, modification ZT of the ordinary POS-4 (without horizontal coil system).

Long-term observations on Repeat Station using POS-3 made it possible to estimate the secular variation of the magnetic field on the Manchazh anomaly and to construct the magnetic maps (see figure 10a).

Joint researches of the QMLab developers and Observatory Arti magnetologists of POS-4 magnetometers allowed to detect a number of effects, both significant and less obvious in various modifications of Overhauser magnetometers. For example, an important parameter of vector POS measurements is the verticality of the axis of the solenoid in different positions relative to geographical directions and the alignment of this axis with the directions of additional fields generated by the solenoid. Regularly performed studies (figure 9b) produced the error circular diagrams (see an example in figure 10b).These diagrams allowed the developers

(a) (b) (c) (d)

Figure 9. Vector Overhauser magnetometers POS of QMLab at the Observatory Arti: (a) ZT-magnetometer POS-3 No.1 during work on Repeat Station point; (b) component Overhauser magnetometer POS-3 No.5 with a digital tilt meter on a test stand in a special verification pavilion; (c) POS-3 No.5 outdoors in a protective casing; (d) POS-4 No.3 during the intercomparison of scalar magnetometers at the observatory Arti in 2017 [13].

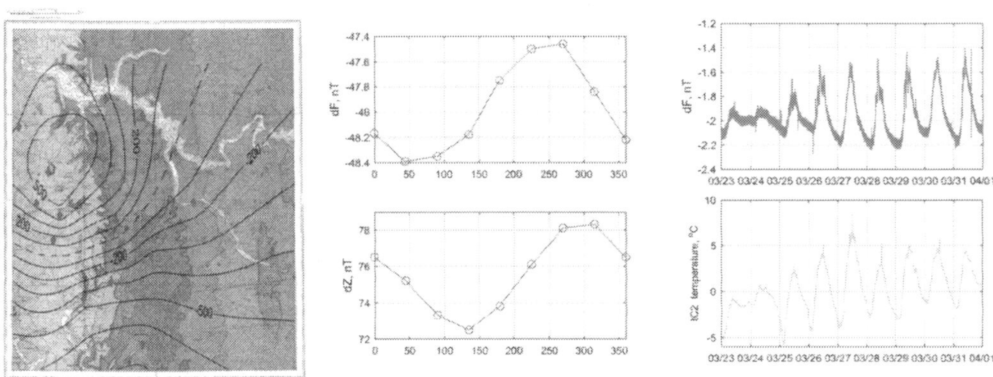

(a) (b) (c)

Figure 10. Some results of researches using POS-3 and POS-4 magnetometers: (a) the anomalous secular variation of the vertical component of Earth's magnetic field for the epoch 2019.5. The isodynamic lines are drawn through 100 nT; (b) a circular diagram for the total field and vertical intensity of POS-3 No.5; (c) presumably, the thermal effect in the titanium solenoid appeared as the difference between the F of POS-3 No.5 and POS-1 against the outdoor temperature.

to improve above mentioned POS-4 parameters by almost an order of magnitude. Currently, the systematic error when rotating the device in the azimuthal plane does not exceed 1 nT for F and 6 nT for Z.

As the magnetometer POS-4 has shown at the Observatory Paratunka, the results of measurements significantly depend on the stability of the base on which the magnetometer sensor is installed [7]. This dependence, for example, in the Canadian dIdD GSM-19FD magnetometer, is partially solved by placing the sensor and the coil system at the gimbal. In modern POS-4 models, a different approach is implemented — the slopes of the solenoid are recorded using a digital tilt meter (see figure 9b) and appropriate corrections are applied to the magnetic data.

It is noted that the POS-1 scalar sensor in a titanium solenoid is sensitive to temperature gradients. This is especially evident when working in the field with outdoor place of sensor. The sensitivity of POS-4 to the outdoor temperature in spring, when working on frozen ground, was studied at the observatory Arti. The device is mounted on an aluminum rod, sunk a meter into the frozen ground. The magnetometer is protected from direct sunlight by a plastic casing (see figure 9c). The results are shown in figure 10c.

6 Magnetometers POS-4 at the Observatory Moscow, IZMIRAN, Troitsk

IZMIRAN acquired two POS-4 Overhause vector magnetometers at the end of 2020, including POS-4 (No.5) for the Observatory in Kaliningrad, and the second (No.6) for the magnetic station in the village Vaimusha, Pinezhsky district, Arkhangelsk region. Before being sent to the station Vaimusha the magnetometer No.6 was installed at the IZMIRAN Observatory Moscow in Troitsk in the absolute pavilion (figure 11a) and tested for several months in a continuous measurement.

(a) (b) (c)

Figure 11. POS-4 magnetometer (IZMIRAN): (a) a testing of POS-4 in the absolute pavilion of the Observatory Moscow; (b) a pavilion for installing of POS-4 at the IZMIRAN station in the village Vaimusha; (c) a pillar inside the pavilion with a fluxgate magnetometer.

As an example, the figure 12 shows the comparison of the raw second values of the variations of the Y- and Z-components obtained using POS-4 and the data of the LEMI-018 fluxgate magnetometer installed at the observatory Mikhnevo at distance of 80 km from IZMIRAN. The observatory Mikhnevo was chosen for comparison due to a more favorable noise environment than in Troitsk. The middle panels show the differences between the data of two magnetometers, the lower panel shows the parameter QMC (for measurements with two opposite magnetizing fields). As can be seen in figure 12 the results of POS-4 are close to the results of the observational magnetometer, but they are more noisy during the day (primarily in the vertical component), with the exception of a small interval of 1-2 hours in the night local time. It can also be noted that the parameter QMC increases approximately every 8 hours, which indicates a regular source of noise near the measurement place.

The estimation of long-term stability was performed by comparing the hourly mean values of POS-4 and the hourly mean data obtained by the standard quartz variometer of the Observatory Moscow. Figure 13a shows the hourly values of the Y- and Z-components from June 1 to June 20, 2021, figure 13b shows the differences of these curves. Figure 13b also

https://doi.org/10.1051/epjconf/202125402018

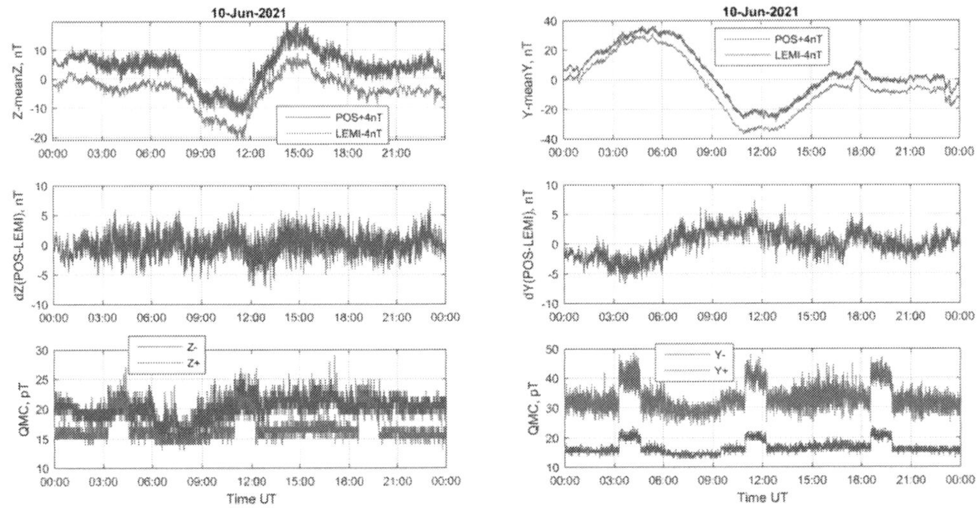

(a) (b)

Figure 12. Example of measurements by POS-4 for 10.06.2021, components Z (a) and Y (b), a full measurement cycle of 15 seconds. For comparison, the raw 1-second data obtained at the Observatory Mikhnevo using the variometer LEMI-018 is presented.

shows variations in the inclination of the solenoid in the "north-south" and "east-west" directions. As can be seen from the curves in figure 13b, there are no slow drifts in the POS-4 data, however, on some days there is a scatter in the differences, probably associated with increased geomagnetic activity. There is also no correlation between the tilts of the POS-4 solenoid and the magnetic data.

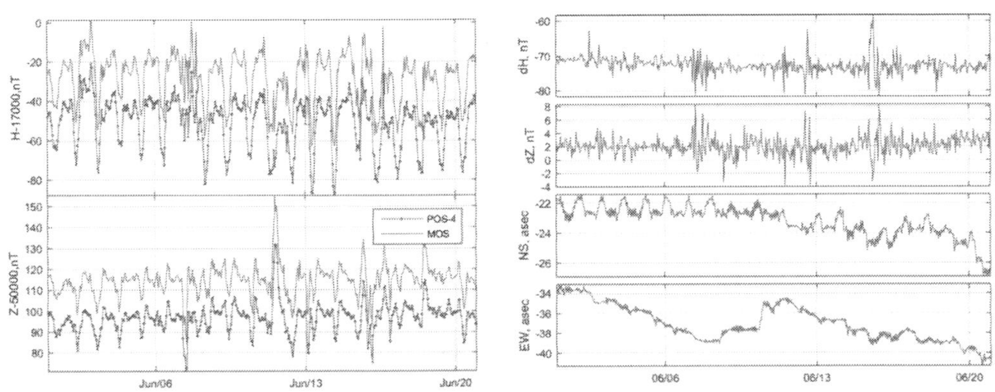

(a) (b)

Figure 13. Hourly mean data recorded by the POS-4 and by the Observatory Moscow variometer (a) and its differences and tilts of the POS-4 solenoid (b).

In August 2021, the staff of the Moscow Magnetic Observatory prepared a place for the installation of POS-4 at the IZMIRAN station in the village Vaymusha. As a result of these works, a pavilion was prepared, a pillar was installed inside. Figures 11b,c shows the pavilion itself and a pillar with a three-component fluxgate magnetometer, which was used to check the suitability of the selected place for magnetic measurements. The installation of POS-4 on a permanent place is planned for the next visit. To improve the conditions for long-term autonomous measurements, it is planned to install a heated polypropylene screen along the perimeter of the pillar with POS-4.

IZMIRAN plans to use magnetometers POS-4 for continuous and largely autonomous measurements. Long-term testing at the Moscow observatory showed that the magnetometer works quite stably, but several times there were failures (measurement stops) that required the some work of an observatory employee to manually restart the device. The cause of the failures has not been established. This moment becomes critically important for long-term autonomous measurements. There were also some problems with the software when writing files.

7 Magnetometers POS-4 at the Observatory White Sea, GC RAS

The White Sea Magnetic Observatory (WSE) was organized by the Geophysical Center of RAS on the base of the White Sea Biological Station of Moscow State University: a non-magnetic pavilion with a glass block pillar was built on a specially selected site during June 2018, and a magnetometer POS-4 was installed and continuous measurements began in August 2018. POS-4 works almost completely independently, data is transmitted online to the GC RAS (Moscow) and is available on the website http://geomag.gcras.ru/dataprod-plot.html. Photos of the pavilion and the sensor of POS-4 are shown in figure 14. The pavilion for POS-4 is not heated, i.e. the magnetometer is located in quite extreme conditions. Maintenance is performed sporadically, by employees of the GC RAS.

(a)

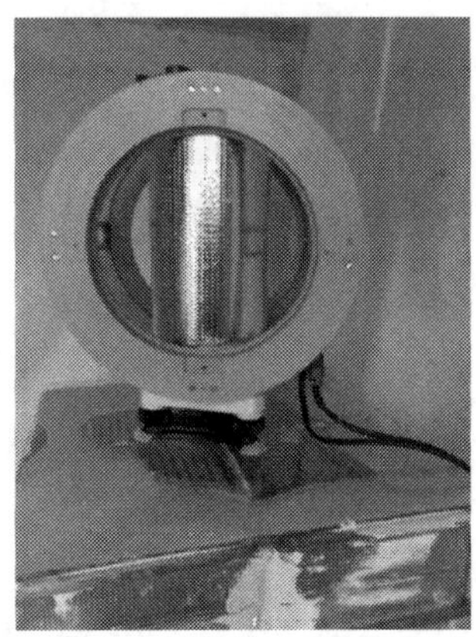

(b)

Figure 14. Magnetometer POS-4 at the White Sea Magnetic Observatory (WSE) of the GC RAS: (a) the view of the non-magnetic pavilion (under construction), (b) the POS-4 sensor on a pillar (http://geomag.gcras.ru/obs-WSE.html).

As an example, figure 15 presents the 1-second data (the full measurement cycle is 15 seconds) obtained by POS-4 at the White Sea Observatory, compared with the data of the nearest auroral observatories Abisko (ABK) and Lycksele (LYC), Sweden. As can be seen from the figure, the variations of the X-component obtained using POS-4 are structurally close to the variations at the ABK and LYC observatories, it is hardly possible to expect a coincidence of details in the magnetic field variations at spatially separated points in the auroral area. However, in the highlighted three-hour fragment shown in the lower panel of 15, it is possible to notice the synchronous manifestation of some features.

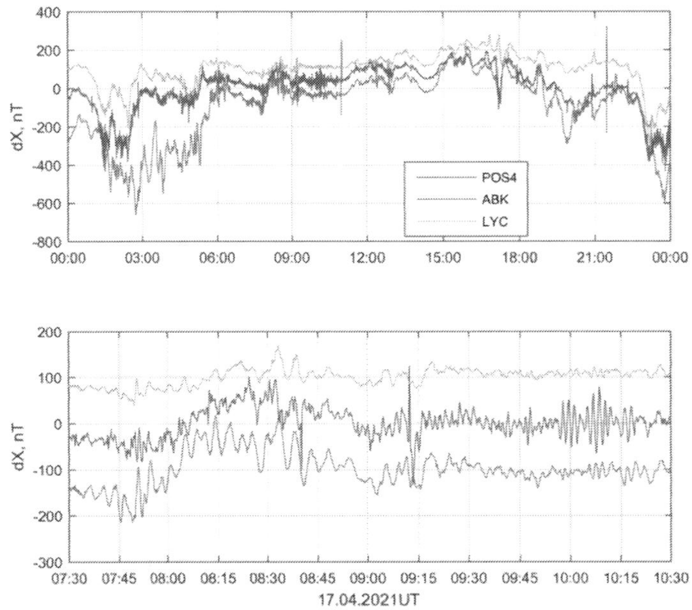

Figure 15. Comparison of the X-component variations obtained using POS-4 at the WSE observatory and at the INTERMAGNET ABK and LYC observatories (1-second values). The upper panel shows the variations for the day of 17.04.2021, the lower panel is a part of the data from 07:30 to 10:30UT, showing some details of the variations.

The three-year experience of using POS-4 at the White Sea Observatory has shown that the main problems are episodic stops of measurements (the reason is not established, but probably occurs with strong noise) and a large number of significant amplitude spikes. Some of this noise can be seen in figure 15, their amplitude reaches hundreds of nT. Presumably, the source of this noise is the power plants used at the White Sea Biological Station.

8 Discussion and conclusion

Vector Overhauser magnetometer POS-4, developed at the Quantum Magnetometry Laboratory of the Ural Federal University (Yekaterinburg), is the result of combining long-known principles and methods of measuring elements of the magnetic field vector and modern technologies, element base and microprocessor systems. Many years of experience in using POS-4 at observatories and on expeditions allowed us to identify the problem areas of the magnetometer. This made it possible for developers to make methodological, technical, design and software changes to the device that improve its characteristics.

The undoubted advantages of the POS-4 magnetometer include:

1) structural reliability, compactness, absence of moving parts and elements made of brittle materials - this makes the magnetometer suitable for field work, in which the probability of mechanical damage to the equipment significantly increases;

2) easy to install, configure, run measurements and monitor results - this becomes very important when POS-4 are used in the absence of personnel with a sufficient level of qualification;

3) POS-4 allows us to obtain the elements of the magnetic field F and Z (and H) in the absolute sense, but with the frequency of variational measurements. In fact, there is a device that performs two functions that are mandatory for INTERMAGNET observatories. A methodological question remains open with the absoluteness of the horizontal component measured using a coil system;

The main problems for POS-4, which were discovered during POS-4 operation, were:

1) a significant dependence of the measurement results on the stability of the base (the slopes of the solenoid and the turns of the coil system). The problem was manifested in long-term trends in observational conditions, for example, at the Paratunka observatory during seasonal freezing and thawing of the soil under the pillar, or in field conditions when measurements were performed from a tripod. As a solution, it was decided to use a high-precision tilt meter, the non-magnetic sensor of which could be fixed on the upper part of the magnetometer solenoid. A similar method of accounting for tilts is used, for example, in fluxgate magnetometers MAGDAS [14]. Currently, a method for compensating errors due to tilts has been developed and it is being implemented in the POS-4 software. Unfortunately, the control of rotations of the POS-4 sensor around the vertical, i.e. effects in horizontal components, remains open;

2) a methodological problem related to the choice of the orientation of the coil system fixed on the solenoid and the orientation of the scalar Overhauser sensor POS inside the solenoid. The orientation of the axis of this coil system determines which component in the horizontal plane will be determined. The orientation of the primary sensor with significant magnetizing fields in the vertical plane can increase the POS-4 intrinsic noise. Currently, the priority is to measure the eastern component and the orientation of the axis of the sensor POS perpendicular to the magnetic meridian. A similar problem, for example, is absent in the dIdD GSM-19FD, since the orientation relative to the field vector of pairs of declination and inclination coils is performed by aligning the readings in the measured channels (D+, D-) and (I+, I-);

3) there is a problem with the solenoid stand used, since its size does not provide sufficient stability on the pillar and sufficient sensitivity of the support screws when levelling. In addition, there is no possibility of precise rotation of the solenoid around the vertical axis (when studying a circular diagram or the orientation of the horizontal axis of the coil system), as is done, for example, with dIdD. Unfortunately, a possible solution by increasing the base of the stand and complicating its mechanical components may be acceptable only for the conditions of observatories and is not suitable for the field version;

4) the format used for writing data to files by the standard POS Manager program contains all the information that is obtained during measurements, including the measurement of each of the five modules is marked with its own time stamp. However, a certain problem arises for users who do not have the opportunity to develop their own software that would provide convenient work with data and minimize some methodological errors. Currently, as the experience of working with POS-4 in IZMIRAN shows, some changes have been made to the POS-4 file format, but it is not yet possible to assess how successful they are.

The tasks of automation of absolute magnetic measurements, including the creation of hybrid systems that combine the functions of absolute and variation magnetometers, remain relevant. In our opinion, the important aspects of these developments are the control of the spatial orientation of the measuring axes, noise immunity and ease of operation. The experience of using vector Overhauser magnetometers POS-4 and its ZT prototype POS-3 for several years shows that trends in modifications of these devices reflect the requests of magnetologists in the tasks of observing the Earth's magnetic field. The balance of achievements and shortcomings of the magnetometer POS-4 shown in this work gives reason to hope that in the future we can expect the appearance of a device suitable for the effective solution of a complex of scientific and applied problems, including observations under the Arctic and in tectonically active regions conditions.

Acknowledgments

The authors thank the employees of observatories and institutes who supported the operation of POS-3 and POS-4 magnetometers. The results presented in this paper rely on data collected at magnetic observatories. We thank the national institutes that support them and INTERMAGNET for promoting high standards of magnetic observatory practice (www.intermagnet.org). Work of S.Y. Khomutov was supported by budget project AAAA-A21-121011290003-0, work of P.B. Borodin, A.S. Bebnev and T.A. Cherepanova was supported by budget project AAAA-A19-119021290090-4, work of Y.G. Turbin and P.E. Baryshev was carried out as a part of the AARI planned scientific topic under the project 5.1.4 of the Targeted Scientific and Technical Program (TSTP) of Roshydromet.

References

[1] *INTERMAGNET Technical Reference Manual. Version 5.0.0. St-Louis B. (Ed.)* (INTERMAGNET Operations Committee and Executive Council, 2020) 156

[2] J.L. Rasson, A. Gonsette, Data Science Journal **10** IAGA169–IAGA173 (2011)

[3] S. Marsal, J.J. Curto, J.M. Torta, A. Gonsette, V. Favà, J. Rasson, M. Ibañez, Ò. Cid, Geosci. Instrum. Method. Data Syst. **6** 269–277 (2017)

[4] L. Hegymegi, J. Szöllősy, C. Hegymegi, Á. Domján, Geosci. Instrum. Method. Data Syst. **6** 279–284 (2017)

[5] V.I. Myukhkyurya, Geophys. Equip. (in Russian) **31** NP (1967)

[6] *dIdD Insrtuction Manual v7.0, Release 7.5* (GEM Systems, ON Canada, 2010) 50

[7] S. Khomutov, V. Sapunov, A. Denisov, D. Savelyev, I. Babakhanov, E3S Web of Conferences **11,00007** 1-5 (2016)

[8] A.Yu. Denisov, V.A. Sapunov, O.V. Dikusar, Geomagnetism and Aeronomy **39/6** 737-742 (1999)

[9] A.Y. Denisov, V.A. Sapunov, B. Rubinstein, Meas. Sci. Technol. **25/055103** 1-6 (2014)

[10] V.A. Sapunov, A.Y. Denisov, D.V. Saveliev, A.A. Soloviev, S.Y. Khomutov, P.B. Borodin, E.D. Narkhov, A.V. Sergeev, A.N. Shirokov, Magnetic Resonance in Solids. Electronic Journal **18/2** 1-9 (2016)

[11] S.Y. Khomutov, E3S Web of Conferences **20,02002** 1-18 (2017)

[12] Y.G. Turbin, V.I. Pochtarev, V.I.Yuschenko, N.V. Alekseev, E.A. Bugrov, N.G. Ptitsina, A.Y. Rotshtein, M.A. Sergeev, A.D. Cherednichok, Geomagnetism and aeronomy (in Russian) **17,2** 326-331 (1977)

[13] O.A. Kusonsky, S.Y. Khomutov, P.B. Borodin, V.A. Sapunov, D.V. Savel'ev, V.D. Savel'ev, E.D. Narkhov, L.A. Murav'ev, A.Y. Gvozdarev, I.A. Nuzhdaev, A.V. Ovcharenko, A.S. Bebnev, Conrad Observatory Journal **5** 23 (2019)

[14] MAGDAS-A Installation Manual, SERC, Kyushu Univ., Editor George N. Maeda, 35 (2005)

Application of the Riccati hereditary mathematical model to the study of the dynamics of radon accumulation in the storage chamber

Dmitryi Tverdyi[1*], *Roman* Parovik[1,2], *Evgeniy* Makarov[1,3], *Pavel* Firstov[3] and *Nazira* Alimova[4]

[1] Vitus Bering Kamchatka State University, 683032 Petropavlovsk-Kamchatskiy, Russia
[2] Institute of Cosmophysical Research and Radio Wave Propagation FEB RAS, 684034 Paratunka, Kamchatka, Russia
[3] Kamchatka Branch of the Federal State Budgetary Institution of Science of the Federal Research Center "Unified Geophysical Service of the Russian Academy of Sciences", 683006 Petropavlovsk-Kamchatskiy, Russia
[4] National University of Uzbekistan named after Mirzo Ulugbek, 100174 Tashkent, Uzbekistan

Abstract. The article proposes a mathematical model of radon accumulation in a chamber, which takes into account the hereditary properties of the environment in which radon migrates. The model equation is the fractional Riccati equation with a derivative of a fractional variable order of the Gerasimov-Caputo type, taking into account heredity, as well as taking into account nonlinearity, which is responsible for the mechanisms of radon entry into the chamber. The obtained model curves of the accumulation process are compared with real data. It is shown that the model described in the work gives a better agreement between the model and real curves of radon accumulation and can be used for a more accurate description of the processes occurring in the chamber.

The article is dedicated to the memory of the author, our colleague P.P. Firstov.

1 Introduction

Observation of the radon field in the upper soil layer is of interest both in connection with the biological activity of this gas during accumulation in basements and on the first floors of buildings, and from the point of view of developing a method for predicting strong earthquakes based on continuous monitoring of the volumetric activity of radon in the subsoil air with a high degree of detail. The radon runoff into the atmosphere is influenced by changes in meteorological values, water cut and freezing of the upper soil layer, changes in the stress-strain state of the geomedium.

According to modern concepts, variations in regional stress fields associated with the preparation of the future earthquake source lead to a change in porosity, permeability,

* Corresponding author: dimsolid95@gmail.com

temperature and pressure gradients in the upper layer of the earth's crust and, as a consequence, to a change in the rate of radon migration to the day's surface. Thus, monitoring of the parameters of the radon field at the boundary of the lithosphere - atmosphere allows us to talk about the processes occurring in the earth's crust, which can cause anomalous variations in the concentration of radon, as well as a change in the flux density of RFD (Radon Flux Density) from the soil surface [1, 2].

Since 1997, at the Petropavlovsk-Kamchatsky geodynamic test site, monitoring of the volumetric activity of radon (Rn) has been carried out in order to search for precursors of strong earthquakes [3].

In this paper, using mathematical modeling, the process of radon accumulation in a chamber with gas-discharge meters is investigated to determine RFD from the soil surface below it.

The concentration of radon inside chambers installed on the ground is the result of two main competing processes: radon influx and radon efflux. In this case, the processes are due to the presence of a radon concentration gradient in the medium and a pressure gradient. In the first case, a diffusion transfer process is formed, in the second - convective (gas transfer in a gas) or filtration (gas transfer in a porous medium).

When organizing radon monitoring points, it is necessary to know the RFD value from the surface at the selected point where the camera will be installed. In addition, areas of increased radon discharge into the atmosphere are characterized by narrowly localized zones, the search for which determines the need for a quick RFD assessment using a simple and affordable method. This method, based on the mathematical description of the accumulation of Rn in the chamber according to the approach from [4], is described in detail in [3] and was tested at the monitoring points of subsurface gases in Kamchatka [3].

In the mathematical description of the radon accumulation process in the chamber in [3], a number of assumptions were used to simplify the solution of the problem, and the accumulation process itself was assumed to be stationary when there are no abruptly changing AER (Air Exchange Rate) and RFD conditions from the surface below it. Consequently, possible factors affecting the flow rate were not taken into account, primarily changes in atmospheric pressure and temperature. In the chamber used in the work, air exchange with atmospheric air is present, but minimized. Due to weather conditions (mainly due to temperature fluctuations), the AER will vary significantly.

For a more accurate description of the accumulation process, the existing model was modified using a nonlinear function responsible for the mechanisms of Rn entry into the chamber, as well as taking into account heredity.

2 Mathematical model

Mathematical modeling of the transfer of Rn in porous soil due to the mechanisms of diffusion and convection, taking into account the hereditary properties of the medium, is considered in [5, 6]. Such mechanisms of Rn transfer in the literature are called subdiffusion and superdiffusion or anomalous diffusion [7]. In this work, we will consider a slightly different hereditary process of radon migration in the chamber.

Let us consider the following law of accumulation of volumetric activity of radon RVA (Radon Volumetric Activity) in the chamber:

$$\frac{dA}{dt} = \int_0^t K(t-\sigma)F(A,\sigma)d\sigma, \quad A(0) = A_0, \tag{1}$$

where $A(t)$ - RVA, Bq/m^3; $F(A,\sigma)$ - the function responsible for the mechanisms of entry of Rn into the chamber, in the general case, can be nonlinear; $K(t-\sigma)$ - memory function, determines the degree of influence of the function $F(A,\sigma)$ on the accumulation process; A_0 - at the initial moment of time t.

Note that there can be two limiting cases: when $K(t-\sigma)=\delta(t-\sigma),\delta(.)$ - Dirac function, in this case there is no memory effect by analogy with Markov processes, and when $K(t-\sigma)=t-H(t-\sigma)$, $H(.)$ - the Heaviside function, in this case, characterizes the presence of ideal memory. In general, memory function $K(t-\sigma)$ in equation (1) can be selected based on experimental data or properties of the medium.

Due to the fact that the rate of radon intake decreases with time, the faster, the smaller the AER of the chamber with the atmosphere, then the dynamic system (1) will be considered dissipative. Here, by dissipation, we mean that the system gradually "forgets" the mechanisms of the entry of radon into the chamber, i.e. RVA over time goes into a steady state (saturation). For this, in the dissipative system (1) the memory function $K(t-\sigma)$ can be selected in the form:

$$K(t-\sigma)=\frac{1}{\Gamma(\alpha)(t-\sigma)^{1-\alpha}},\quad 0<\alpha\leq 1, \tag{2}$$

where α - determines the intensity of dissipation, $\Gamma(.)$ - Euler's gamma function, $t\in[0,T]$ - time, T - model time.

Substituting relation (2) into equation (1), we obtain:

$$\frac{dA}{dt}=I_{0t}^{\alpha}F(A,\sigma),\quad A(0)=A_0, \tag{3}$$

where $I_{0t}^{\alpha}F(A,\sigma)=\frac{1}{\Gamma(\alpha)}\int_0^t\frac{F(A,\sigma)}{(t-\sigma)^{1-\alpha}}d\sigma$ - fractional integral of order α. If we invert equation (3) taking into account the compositional property of the fractional integral [8], we obtain the following equation:

$$\partial_{0t}^{\alpha}A(t)=F(A,\sigma),\quad A(0)=A_0. \tag{4}$$

where

$$\partial_{0t}^{\alpha}A(t)=\frac{1}{\Gamma(1-\alpha)}\int_0^t\frac{\text{[illegible]}}{(t-\sigma)^{\alpha}}d\sigma \tag{5}$$

fractional Gerasimov-Caputo derivative of order $0<\alpha<1$, [illegible] dt.

Note that if in (4) we choose the function $F(A,t)=S_D-\lambda_0A(t)$, where S_D - diffusion mechanism of penetration RVA, λ_0 - air exchange rate AER, we will get the results of work [10]. Let's choose a more general function $F(A,t)=-a(t)A^2(t)-\lambda_0A(t)+S_D(t)$, where $0<a(t)<1,S_D(t)$ - continuous on the segment $[0,T]$ functions. It should be noted that the first term in the function $F(A,t)$ responsible for slowing down the accumulation of radon in the chamber. Then we arrive at the following Cauchy problem:

$$\partial_{0t}^{\alpha}A(t)=-a(t)A^2(t)-\lambda_0A(t)+S_D(t),\quad A(0)=A_0. \tag{6}$$

The Cauchy problem (6) was solved numerically using the modified Newton method with local first-order convergence [11].

3 Simulation results

The data on the accumulation of radon in the chambers at the PRTR point (Kamchatka), the YSSR point on Sakhalin Island [9] and also GLL (Kamchatka) [3] were used as the initial data for modeling. At these points, continuous registration of radon was carried out in chambers at a depth of 1 to 3 m [3]. The chambers are a 10 liter bucket. A radon sensor is placed in this chamber through a special pipe. The minimum air exchange between the chamber and the atmosphere is also carried out through the pipe.

The simulation results were compared with the data on the accumulation of radon in the chamber at the GLL point, with an estimate of the determination coefficient between the data and the model curves.

Greatest coefficient of determination $R^2 = 0.89$ was obtained for the following values of the parameters of problem (6), the coefficients were chosen as follows: $a(t) = c(t) = S_D(t) = t(1 - A_{\max}\lambda_0)/T$, $\alpha = b = \lambda_0 = 0.25$, $T = 1314$, $A_0 = 0.484$. The simulation results are shown in the figure.

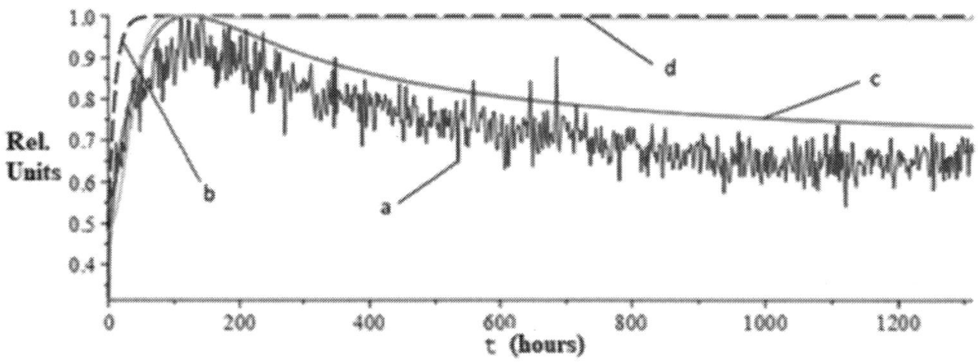

Figure 1. Curves of radon accumulation in the chamber of the GLL point and model curves. a – radon accumulation curve in the chamber at the GLL point; b – calculated curve obtained in [2]; c – calculated curve obtained by model (6) with $R^2 = 0.89$; d – calculated curve obtained by model (6) at $\alpha = 1, a = c = \lambda_0$.

In the figure, all the calculated curves were normalized to the maximum. As can be seen from the figure, the nonlinearity in model (6) makes it possible to obtain new distribution curves of RVA, much closer to the experimental ones, in comparison with the models [3, 10].

4 Conclusion

The article proposes a new mathematical model of RVA accumulation taking into account heredity. It is shown that the proposed model, taking into account the nonlinearity, can, due to additional parameters: the intensity of dissipation of the accumulation mechanism, give results that are closer to real data, in comparison with the model from [3] and the hereditary model in [10].

Acknowledgement

The work was carried out within the framework of the research project of the Vitus Bering Kamchatka State University Natural disasters of Kamchatka --- earthquakes and volcanic eruptions (monitoring forecast, study, psychological support of the population) No. AAAA-A19-119072290002-9 and subject research IKIR FEB RAS ``Physical processes in the system of near space and geospheres under solar and lithospheric influences" No AAAA-A21-121011290003-0, the work was funded by RFBR, project number 20-05-00493.

References

1. P.P. Firstov, E.O. Makarov, Seismic instruments. **51**, 4 (2015)
2. R.D. Cicerone, J.E. Ebel, J. Beitton, Tectonophysics. **476**, 3-4 (2009)
3. P.P. Firstov, E.O. Makarov, Subsoil radon dynamics in Kamchatka and strong earthquakes (KamSU named after Vitus Bering, Petropavlovsk-Kamchatsky, 2018)
4. A.V. Vasilyev, M.V. Zhukovsky, Journal of environmental radioactivity. **124** (2013)
5. R.I. Parovik B.M. Shevtsov, Math. Models and Computer Simulations. **2,** 2 (2010)
6. R.I. Parovik, Life Science Journal. **11,** 9 (2014).
7. V.V. Uchaikin, Fractional derivatives for physicists and engineers. Volume I (Springer, Berlin, 2013)
8. A.A. Kilbas, H.M. Srivastava, J.J. Trujillo, Theory and Applications of Fractional Differential Equations (Elsevier, Amsterdam, 2006)
9. E. Makarov, P. Firstov, D. Kostylev, E. Rylov, I. Dudchenko, Vestnik KRAUNC. Fiz.-Mat. Nauki., **25**, 5 (2018)
10. D.A. Tverdyi, R.I. Parovik, E.O. Makarov, P. P. Firstov, E3S Web Conf. **196** (2020)
11. R.I. Parovik, D.A. Tverdyi, Mathematical and Computational Applications. **26**, 3 (2021)

Application of spaced system of laser interferometers and tiltmeters for sharing the earthquakes precursory events

Mstislav Dubrov[1*], *Victor* Volkov[2], *Igor* Larionov[3], Jan Mrlina[4], *Vaclav* Polak[4], *Dmitriy* Aleksandrov[1]

[1] Fryazino Branch of Kotel'nikov Institute of Radioengineering and Electronics RAS, Russia
[2] Shmidt Institute of Physics of the Earth RAS, Russia
[3] Institute of Cosmophysical Research and Radio Wave Propagation FEB RAS, Russia
[4] Institute of Geophysics CAS, Czech Republic

Abstract. Our previous investigations evidently show that synchronous observation of global atmosphere and lithosphere disturbances by means of 130-1,600 km spatially distributed precise instruments is the effective method of detecting precursors of large seismic events and other dangerous natural phenomena. This study expands the spatial range of our search up to 7,000-8,000 km and allows regional and global disturbances to be shared.

1 Introduction

Solid earth movements forerunning strong earthquake can be detected at remote distances by precision geophysical instruments, which are able to measure small earth strains and tilts. Such instruments (e.g. laser strainmeter, accurate tiltmeter) being installed in accessible upper earth crust or sedimentary layers are inevitably affected by adjacent geospheres: underground water, atmosphere and ionosphere [1-4]. Abnormal instrument readings, which often observed at far distances [5] before strong earthquakes can be reasonable and qualified as earthquake precursor only if the detected affection process is global. Otherwise, it should be referred to local disturbance or instrumental obstacle.

The subject of presented investigation is the analysis of non-tidal strain-baric and tilt-baric processes recorded by system of ground-based laser interferometers within the Moscow Syneclise (Fryazino, Obninsk) [3, 6-8] and Kamchatka Peninsula (Karymshina) [8, 9] together with synchronously operating tiltmeter instruments of deep underground installations in the Bohemian Massif (Skalna, Jezeri, Pribram) [10, 11]. An applied method includes searching correlations between laser interferometers data and distant tiltmeters data before strong earthquakes. The goal is to distinguish the local, regional and global variations corresponding to abnormal pre-seismic and co-seismic processes.

Our first evaluable experience in this direction was the found statistically significant value of global coupling of strain-baric anomalies detected by means of distant Fryazino - Kamchatka instruments using the Pearson's correlation technique [12]. This kind of data processing is applied here for time intervals included the 2015-2020 strong earthquakes.

*Corresponding author: mnd139@ire216.msk.su

2 Description of strain and tilt measuring instruments and methods

Among the experimental methods and tools currently used to study geophysical processes, a special place is occupied by integrated methods and measuring systems with spatially separated instruments, united by a common platform for data collection and processing. Measurements of various deformations of solid earth, for example, such as volumetric expansion and contraction, tilt and shear deformations of the Earth's crust, carried out with their help, are fundamental research methods in geophysics, since they provide primary information for scientific analysis about the nature of the movement of the geological environment [13, 14]. Studies of the nature and spatio-temporal dynamics of seismic-deformation geophysical wave fields become especially fruitful when using spatially distributed systems of synchronously operating instruments [15-17].

The results presented in this work were obtained using instruments installed at six observation points located at a distance from each other, differing in their geological structure, seismicity and tectonic activity. In the described experiments, we used laser-interferometric strain gauges (laser strainmeters) and pendulum tiltmeter installations.

Laser strainmeters installed at the Fryazino Beam-Waveguide (underground testing site), which have been developed and modernized for many decades [3, 6-8], are located in the aseismic zone within the Moscow Syneclise (observation point FRZ in Fig.1). The crystalline basement, which lies at a depth of 1400-1600 m in the area of the testing site, is complicated by faults and is coated with a thick sedimentary cover. Two laser strainmeters with 100 m basis measuring earth surface deformation in N50°W and N30°E directions are distanced by 400 m from each other. Laser beams are enclosed in steel underground pipes at a depth of 1.5-2 m.

The nearest observation point is located at Obninsk Geophysical Observatory [18] (observation point OBN in Fig.1). Testing and experimental measurements by means of laser strainmeters was started in 2002 based on three-mirror laser interferometer-deformograph scheme. Strain records was obtained on 3-90 m basis with instrument installation at both underground flor of laboratory building and 30 m underground gallery.

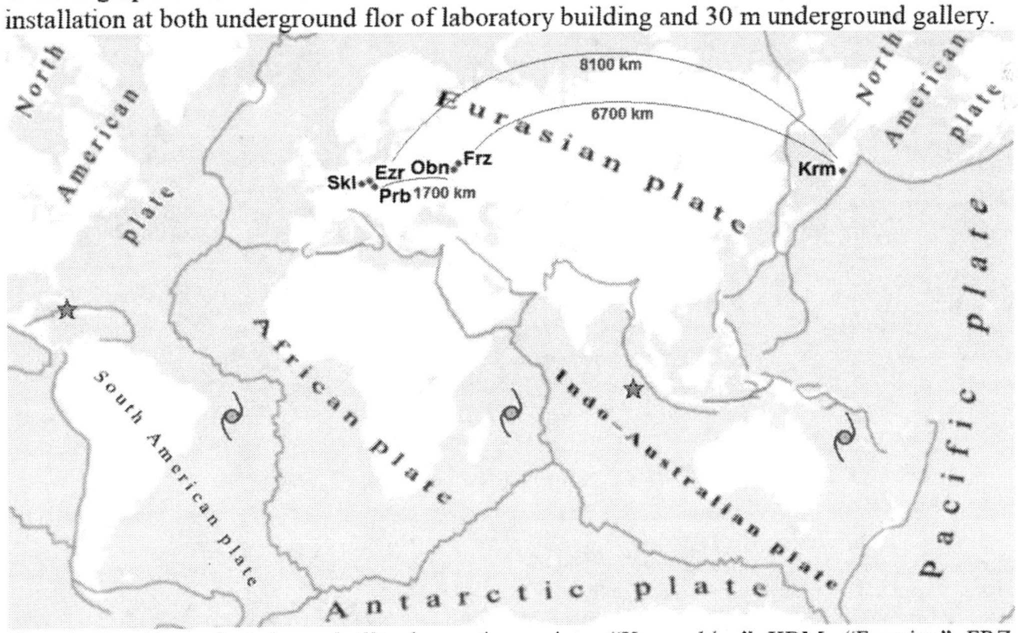

Fig. 1. Location of strain and tilt observation points: "Karymshina" KRM, "Fryazino" FRZ, "Obninsk" OBN, "Pribram" PRB, "Jezeri" JZR and "Skalna" SKL; earthquakes and tropical cyclones.

A remote laser interferometer-strainmeter [8, 9] has been operating since 2016 in the Kamchatka region, which is characterized by high seismic and volcanic activity. The 18 m measuring arm of the interferometer is oriented in the direction N50°W, which is perpendicular to the axis of the tectonic fault passing along the eastern coast of the Kamchatka Peninsula near the subduction zone (observation point "Karymshina" – KRM in Fig.1). This device is designed to conduct routine measurements of deformations of near-surface sedimentary rocks and study the relationship between the geodeformation process and its derivatives, such as the generation of geoacoustic emission, variations in the strength of the surface electric field, etc.

Laser interferometric strainmeters measure changes in length Δl of some basis with length l. Change in length Δl is expressed in fraction of laser wavelength (λ=632.8 nm) and relative linear deformation is defined as $\varepsilon=\Delta l/l$. A detailed description of used laser strainmeters and their interferometric recording system is given in [8, 9]. The highest resolution of interferometric recording servo-system of compensation type had been approved up to 0.01-0.001 nm in a 1 Hz band. The basic accuracy of the digital modulation recording system is $\lambda/2$ = 316 nm (Fig. 2 e, Fig. 3 a and Fig. 4 h) and $\lambda/4$ = 158 nm (Fig. 5 b), while appropriate tuning the interferometer modulation and acquisition software allows the resolution to be improved up to 1.0-0.1 nm [8]. Frequency-stabilized and thermally controlled He-Ne lasers are used as radiation sources in the strainmeter schemes.

Observations of earth tilts have been carried out by mean of accurate pendulum tiltmeters installed at tree tidal observatories located in the Bohemian Massif [10, 11]. Pribram Observatory (Central Bohemia, observation point PRB in Fig.1) is located in a deep abandoned mine at a depth of 90 m. Skalna Observatory is located in the West Bohemia seismically active region in an underground gallery in a granite block at a depth of 10 m (observation point SKL in Fig.1). The tiltmeter in Jezeri Observatory (North Bohemia, observation point JZR in Fig.1) is installed in a gallery about 40 m underground, which is located at a junction of Neogene clastic and fractured zone.

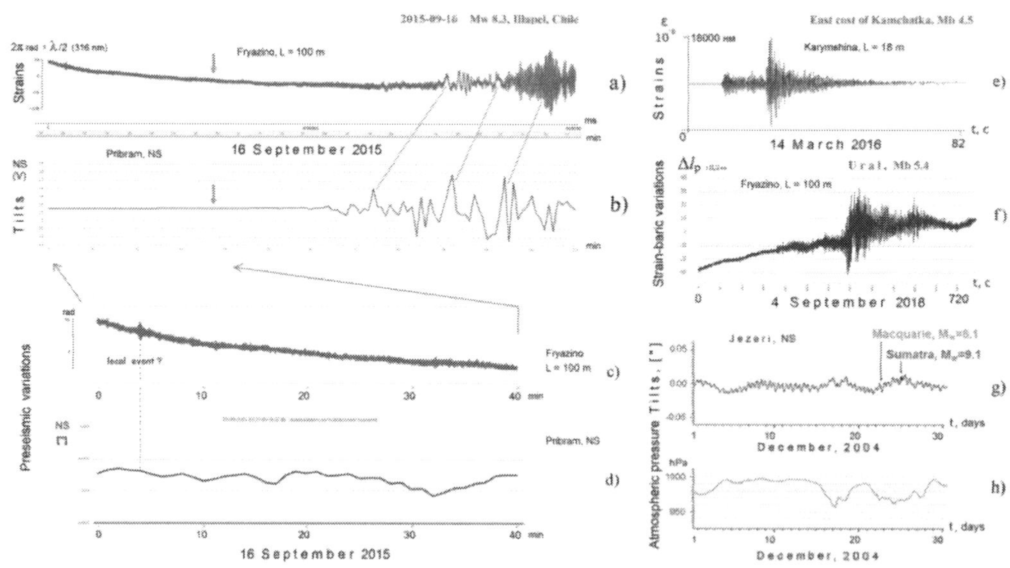

Fig. 2. The examples of strain and tilt registrations: Illapel, Chile Mw 8.3 earthquake on 16 Sep 2015 recorded at FRZ (a, c) and PRB (b, d) observational point; east coast of Kamchatka Mb 4.5 earthquake on 14 Mar 2016 recorded at KRM point (e); Ural Mb 5.4 earthquake on 4 Sep 2018 recorded at FRZ point (f); strain-baric variation at JZR point in December 2004 (g, h).

Two-component tiltmeters measuring earth tilts in NS and EW directions are describe in detail in [10,11]. The resolution of tiltmeters with vertical pendulum is estimated as 1 milliarcsecond (mas) in tidal channel and 0.1 mas in seismic-accelerometer channel. Tiltmeter with horizontal pendulum installation has resolution about 0.1 mas in tidal channel and 0.01 mas in seismic-accelerometer channel.

Selective examples of strain and tilt registrations at four observational point are shown in Fig. 2. The records are obtained by means of two 100 m laser strainmeters at observation point "Fryazino" FRZ (Fig. 2 a, c, f), 18 m strainmeter at observation point "Karymshina" KRM (Fig. 2 e), accurate pendulum tiltmeters at observation points "Pribram" PRB (Fig. 2 b, d) and "Jezeri" JZR (Fig. 2 g).

The joint strain and tilt records of the Illapel, Chile Mw 8.3 earthquake on 16 September 2015 are presented in Fig. 2 a-d. The registration at observation point FRZ is performed by 100 m equal-arm interferometer-strainmeter [6, 8], which measures the length difference between two mutually perpendicular arms and is rid of baric and other meteorology obstacles. The strain records (Fig. 2 a, c) are unfiltered wideband data obtained with 0.5 ms sampling rate and the tilt recording signals (Fig. 2 b, d) are filtered and are digitized with 1 min sampling rate. Pre-seismic strain and tilt variations 40 min before Illapel Mw 8.3 earthquake are depicted in Fig. 2 c, d. Local events and weak global disturbances can be distinguished by means of spatially distributed system of strainmeters and tiltmeters, which are sensitive to precursors of strong seismic and large atmosphere disasters [11, 19].

The records in Fig. 2 e and Fig. 2 f show earthquake registrations performed by two unequal-arm interferometer-strainmeters at observation points KRM and FRZ correspondingly. The local near east coast of Kamchatka Mb 4.5 earthquake on 14 March 2016 was recorded by means of sealed 18 m KRM strainmeter at epicenter distance of 133 km. The body and surface seismic waves are clearly distinguished, their maximum pick-to-pick amplitudes are near 18,000 nm. The remote Ural Mb 5.4 earthquake on 4 September 2018 was recorded by means of FRZ strainmeter with 100 m measuring arm enclosed into underground steel pipe, which is partially connected with outer atmosphere. In addition of standard body and surface waves, this instrument records long-period 80-100 s strain-baric oscillations, of which relative strain amplitudes reach $\varepsilon=5\cdot10^{-9}$ (absolute displacements $\Delta l\sim500$ nm on $l\sim100$ m measuring basis, Fig.2 f). In general, strain-baric interferometer output Δl_p signifies the linear function of l and P variations:

$$\Delta l_p = a \cdot \Delta l + b \cdot \Delta P, \quad (1)$$

Therefore, Δl_p describes the combined instrument response to earth strain and atmospheric pressure variations [4]. This yields a definite opportunity to observe and jointly investigate both strain and baric processes by means of a secluded laser instrument.

It is important to note that strain-baric and tilt-baric affection manifests itself in any strainmeter or tiltmeter prototype recording as namely baric earth deformation or baric errors (correction) of used measuring instrument. A case example of such interaction is depicted in Fig. 2 g, h showing the synchronous records of tilts and atmospheric pressure at observation point "Jezeri" JZR one month before the Macquarie Mw 8.1 on 23 December 2004 and Sumatra Mw 9.1 earthquakes on 26 December 2004. Strong anti-correlation (correlation coefficient $R_P\sim-0{,}9$) is evidently seen at these records: (1) joint tilt and baric variations on 18-19 December 2004, which correspond to ULF (Ultra-Low Frequency) 0.006 mHz oscillations recorded these days [7, 11]; (2) near diurnal tilt-baric variations on 21-24 December 2004 corresponding to significant ionospheric modifications distinguished on 21 and 24 December 2004 [11].

The above example shows that raising wave activities in adjacent geospheres (atmosphere and ionosphere) are probably coupled with abnormal lithosphere processes and therefore, can be attributed to the earthquake precursors.

3 Study of earthquake precursors and tropical cyclone coupling

In our recent publication [11], the atmosphere-lithosphere interaction preceding the major Macquarie Mw 8.1 and Sumatra Mw 9.1 earthquakes in December 2004 was considered as be connected to the most powerful cyclones (typhoons) activities of this season over NW Pacific and Indian Ocean in November-December 2004:

MUIFA Category 4 SSHWS, 14-25 November NW Pacific

BENTO Category 5 19 November-04 December Indian Ocean

NANMADOL Category 4 29 November-04 December NW Pacific

The hypothesis was a powerful cyclone development accompanied an earthquake preparation and caused earthquake triggering through Indo-Australian and Pacific tectonic plates swinging (see Fig 1). At the same time, precursory phenomena, including abnormal behavior of ultra-wideband (0.002 mHz-3 Hz) oscillations of the Earth, was recorded at far distances up to 1,000-10,000 km [7, 11].

The argument in favor of this statement is detecting the strain-baric and tilt-baric coupling at the regional level (Obninsk – Jezeri distance near 1,700 km) that was made provided by the comparison of laser strainmeter data from OBN with tilts recorded in JZR observational points (Fig 3).

This OBN interferometer-strainmeter prototype with 3 m measuring arm that was freely connected with outer atmosphere was installed at underground flor of laboratory building. Therefore, the strain-baric output Δl_p (Fig 3 a) is the combined instrument response to earth strain and atmospheric pressure variations: $\Delta l_p = a \cdot \Delta l + b \cdot \Delta P$.

The comparing these data with JZR tilt measurements (Fig 3 b) gave the Pearson correlation coefficient R_P=0.92 on 15-16 December 2004, i.e. in the period before the Macquarie Mw 8.1 and Sumatra Mw 9.1 earthquakes occurring. The correlation decreases down to R_P=0.85-0.67 after earthquake finishing (Fig 3, right side).

It should be note that investigation of Macquarie – Sumatra co-seismic strains in long-period domain (time intervals ~100 days before and ~100 days after earthquakes) recorded by two distant quartz extensometer yielded a two-fold increase in the correlation coefficient after these seismic events [13].

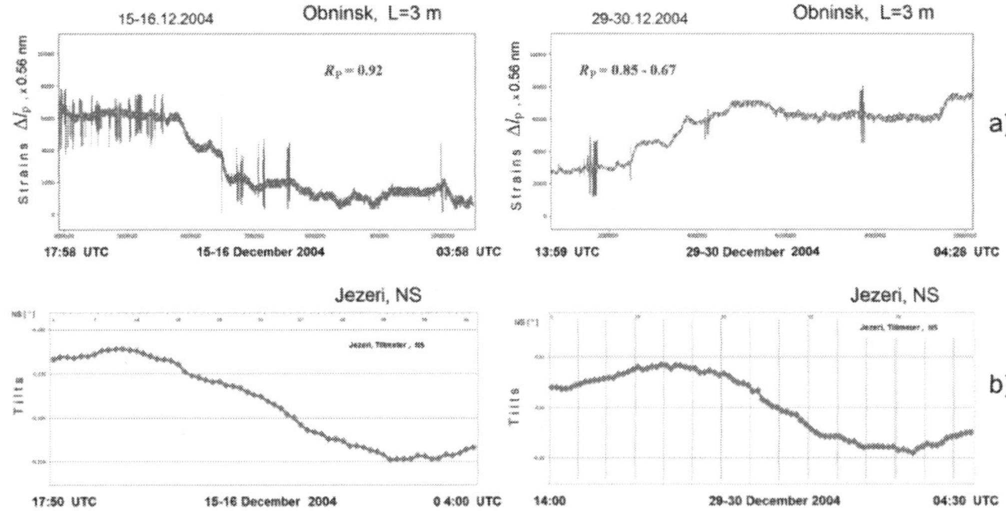

Fig. 3. Pre-seismic and post-seismic correlation of OBN strain-baric (a) and JZR tilts (b) variations in period of Macquarie Mw 8.1 and Sumatra Mw 9.1 earthquakes in December 2004.

Our study of global coupling of strain-baric variations detected by means of distant Fryazino - Kamchatka instruments was carried out for another sufficiently strong off Southwest coast of Sumatra, Indonesia earthquake on 2 March 2016. This Mw 7.8 seismic event occurred in NE border region of Indo-Australian tectonic plate (see Fig.1). Fragments of its pre-seismic and co-seismic synchronous registration at three observational points are shown in Fig.4.

The obtained results show 10-40 min oscillations in PRB pressure (a), tilts (b) and in removed linear trend (c), which are correlated with KRM strains in removed linear trend (d) and pressure variations (e) 10-12 hours before earthquake occurring. Their feature is the temporal shift (phase retardation) in KRM oscillations relative PRB oscillations for a few minutes. The retardation is most clearly seen on the synchronous FRZ (f) and KRM (h) records during the standards surface Love L and Raleigh R waves arriving. The value of observed time retardation is order of 10 min. The amplitude of these long-period oscillations, according to the registration data of two unequal-arm interferometers-strainmeters, is 3-4 times higher than the amplitude of the L- and R-waves recorded both at point FRZ (Fig. 4 f) and at point KRM (Fig. 4 h). This indicates a probable baric origin of the recorded oscillations. An approximately two-fold decrease in the periods to 5-20 minutes of such oscillations on the tiltmeter record is associated with the spatial differentiation of the Earth's surface tilts under the influence of the deformation wave front.

Earlier [12], we suggested a possible connection the observed strain-baric processes during the earthquake on 2 March 2016 (Sumatra, Mw 7.8) with the development of a powerful tropical cyclone WINSTON (Category 5 SSHWS) in South Pacific over the border region between Indo-Australian and Pacific tectonic plates (see Fig. 1). It was the ending of this cyclone activity, which finished off the north coast of Australia on 3 March 2016. If we consider this intense atmospheric disturbance as a source of recorded long-period oscillations, then it may be difficult to interpret the results obtained. Indeed, a wave with a characteristic period of 10-40 minutes (Fig. 4 f, h) arrives at the registration point "Fryazino" earlier than at the point "Karymshina". Although, the first point is 1.6 times farther to the zone of cyclonic activity 03/02/2016 than the second point: the distance of the Coral Sea - Queensland (Australia, 20°S 150°E) region from these points is 13300 km and 8100 km, respectively.

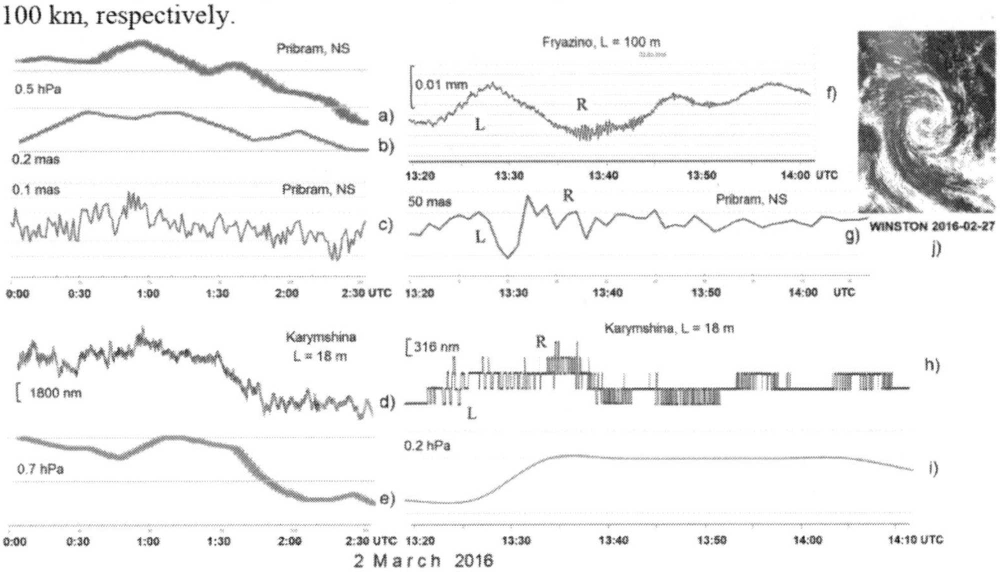

Fig. 4. Fragments of 2016 Sumatra Mw 7.8 earthquake pre-seismic and co-seismic synchronous registration at three observational point. Registration **pre-seismic**: PRB pressure (a), PRB tilts (b, c), KRM strains (d) and KRM pressure (e); **co-seismic**: FRZ strains (f), PRB tilts (g) , KRM strains (h) and KRM pressure (i); Tropical cyclone WINSTON in South Pacific – Australian region (j).

The apparent contradiction can be resolved by taking into account the spatial geometry of air flows and wave oscillations that are excited by a tropical cyclone (typhoon) in the upper atmosphere and in the ionosphere [20]. The plane of the wave front of such oscillations can be located at an angle of 40°-60° with respect to the line passing through the central region - the eye of a tropical cyclone (see the explanatory inset in Fig. 4 j). Taking into account that the vortex rotation of the cyclone air masses in the Southern Hemisphere occurs clockwise hence the wave vector of the excited waves in atmosphere and ionosphere [20] will have a significant component in the W-E direction. This model explains the fact that the wave front of the observed long-period oscillations reaches the "Karymshina" point with a time delay of 7-8 minutes relative to the "Fryazino" point, as shown in Fig. 4 f, h. An additional confirmation of the baric origin of the recorded waves is the correlation between the signal of the laser interferometer-strainmeter and variations in atmospheric pressure at the "Karymshina" observational point (Fig. 4 i).

Let us consider strain-baric and tilt-baric interaction in the period of another Mw 7.7 interplate earthquake, which happened on 28 January 2020. This earthquake occurred in the border region between South American and North American tectonic plates in Caribbean Sea, 123 km NNW of Lucea, Jamaica (see Fig 1). Synchronous registration of this sufficiently strong seismic event carried out by two laser strainmeters installed at FRZ and KRM observational points and by accurate pendulum tiltmeter at PRB observational point (Fig. 5).

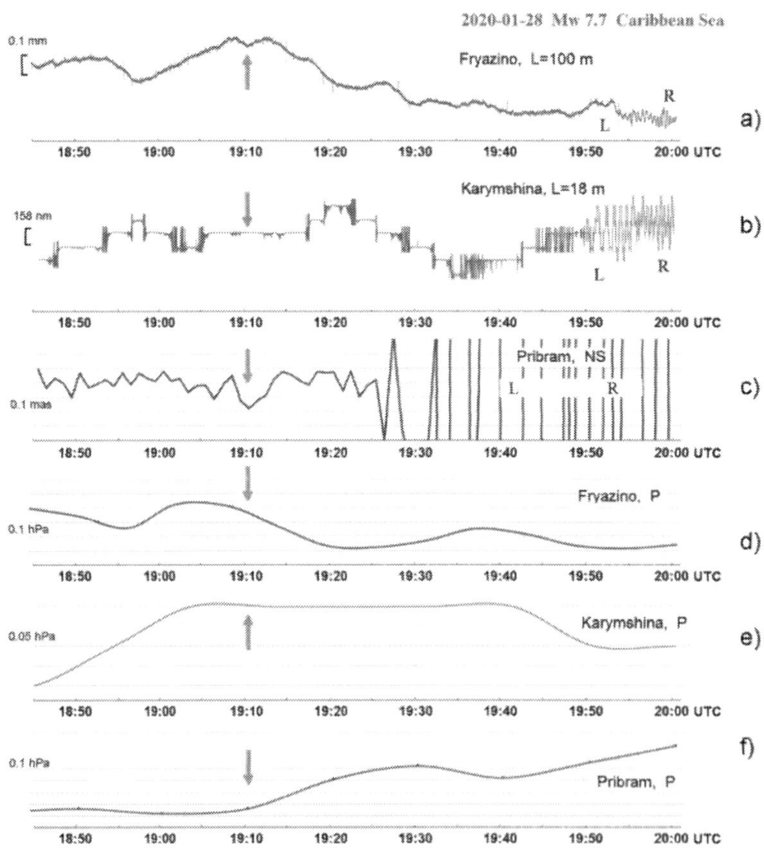

Fig. 5. Synchronous records of pre-seismic and co-seismic earth strain, tilt and atmospheric pressure variations at observation points "Fryazino" FRZ (a, d), "Karymshina" KRM (b, e) and "Pribram" PRB (c, f) in the period of Mw 7.7 earthquake in Caribbean Sea on 28 January 2020.

The analysis of the obtained data show the featured correlation of strain-baric signals before and after the earthquake occurring (t_0 = 19:10:25 UTC is pointed by vertical arrows in Fig. 5 records). Long-period 10-40 min oscillations of a replicated shape are observed on FRZ (a) and KRM (b) strain records. The feature and temporal shift (phase retardation) of KRM oscillations relative FRZ oscillations reiterate a picture of previous media interaction processes during Sumatra Mw 7.8 earthquake (Fig. 4 f, h).

The oscillation occur before earthquake starting, their amplitude exceeds the amplitude of the L- and R-waves sufficiently and the value of time retardation between KRM and FRZ phases is about of 8-10 min. It means origin of these perturbations can be wave formed variations of atmospheric pressure that is confirmed by atmospheric pressure registrations (Fig. 5 d, e, f).

The PRB tilts recording (Fig. 4 c) reveals very small 5-20 min pre-seismic oscillation amplitudes (less than 0.1-0.2 mas) while the amplitudes of the L- and R-waves are manifold higher and reach 30-50 mas. This is due to 1,700 km less epicentral distance of Caribbean Sea earthquake to PRB observational point than to FRZ point. In addition, strain-baric coefficients *b* (see relation (1) in Section 2) of used ground-based laser strainmeters are much greater than corresponding coefficient of PRB tiltmeter of deep underground installation.

In return, we can glimpse a wonderful kick in FRZ and PRB records (Fig.5 a, b) that marked by vertical arrows at t_0 = 19:10 UTC. It is just the time of the earthquake occurring and it appears on pressure records too (Fig.5 d, e, f) in form of irregular curves behavior. The shown phenomenon is not eventual as it is observed on our deformation and pressure records of some other strong earthquakes. This fact suggests that electromagnetic channel can be the permissible way of so fast transmission of seismic energy. Although other, e.g. gravitational channels cannot be excluded from the list of contenders.

The western movement of tropical cyclone activity from Indian Ocean basin into South Atlantics (www.wunderground.com/hurricane) distinguished the Caribbean Sea 2020 earthquake precursory period:

AMBALI 2019 Tropical cyclone Category 4, 3-7 December, SW Indian Ocean

CALVINIA 2019-2020 Tropical cyclone Category 1, 27 December - 1 January, SW Indian Ocean

CLAUDIA 2020 Tropical cyclone Category 1, 5-18 January, Australian region

KURUMI 2020 Subtropical storm, 21-25 January, South Atlantic

DIANE 2020 Severe tropical storm, 22 - 26 January, SW Indian Ocean

This movement ended by extremely rare event, namely subtropical cyclone development near Brazilian coast in South Atlantic. Subtropical storm KURUMI and Severe tropical storm DIANE, exciting the adjacent South American an African tectonic plates (see Fig.1) finalized an earthquake triggering process due to tropical cyclones swings: AMBALI – CALVINIA – CLAUDIA – KURUMI – DIANE, which ended by Mw 7.7 earthquake in Caribbean Sea on 28 January 2020. The two last storms KURUMI and DIANE were the sources of strain-baric and tilt-baric precursors recorded in Central and East Europe as well in Far East region.

The vivid illustration of analyzed processes is shown in Fig.6. Atmospheric pressure variations synchronously recorded at two observation points SKL and JZR spaced at 90 km allows a singularity of local cyclonic motions on the day of Caribbean Sea earthquake occurring (Fig.6 a) to be displayed. Atmospheric depression (the minima on the pressure tracks at 11:10 and 13:00 UTC) moves from SKL to JZR point with phase velocity near 49 km/h. While motion of distinct phases is about two fold faster.

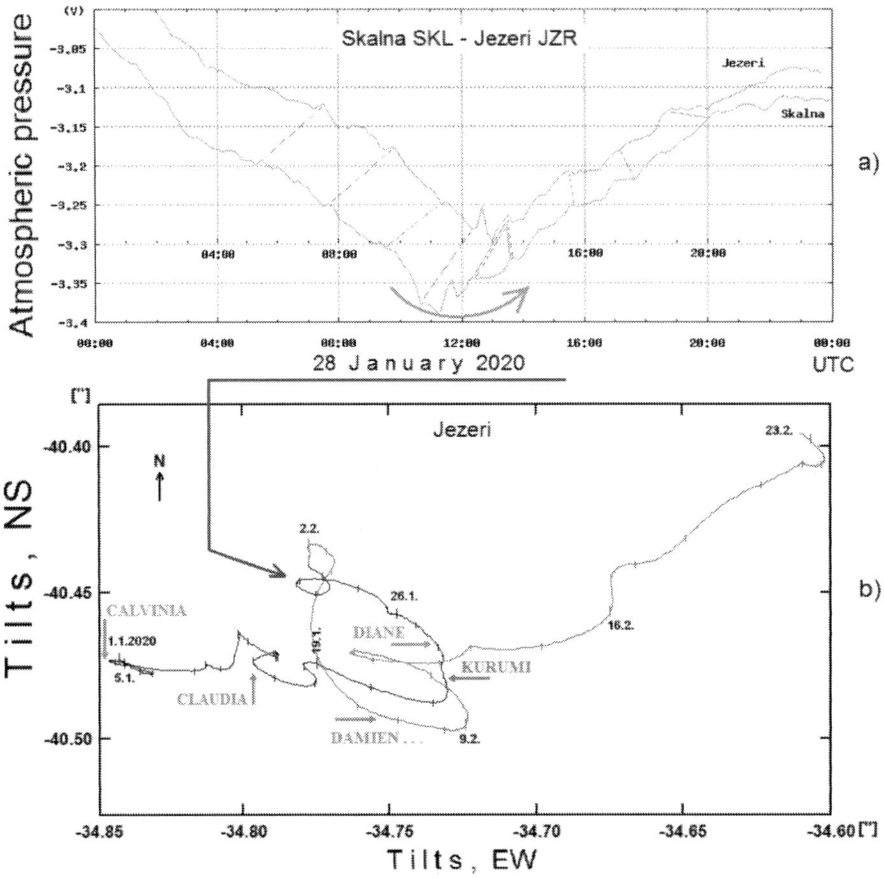

Fig. 6. Synchronous records of atmospheric pressure variations at observation points SKL and JZR on 28 January 2020 (a); vector representation of non-tidal tilts at observation point JZR during January – February 2020 together with periods of five tropical cyclones CALVINIA – DAMIEN pointed (b).

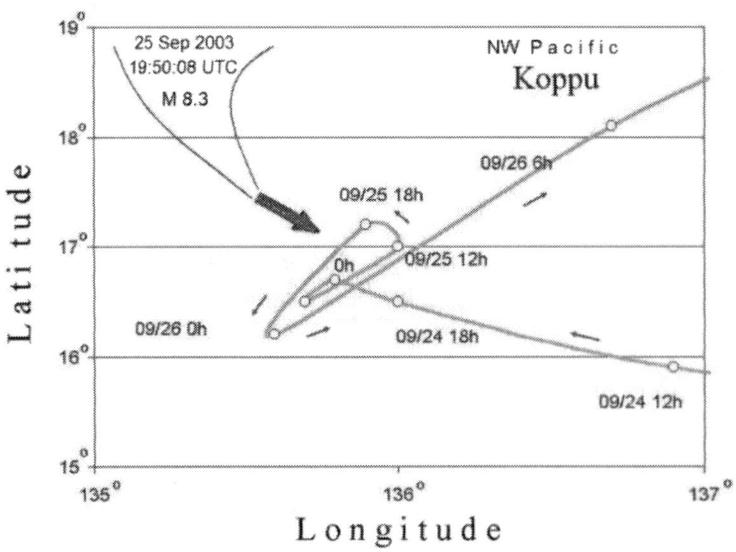

Fig. 7. Typhoon KOPPU in NW Pacific Ocean makes loops 20 hours before Hokkaido Mw 8.3 earthquake on 25 September 2003 [19].

There is time retardation in SKL phases relative JZR phases for two hours at 5:00-9:00 UTC period while SKL phase advance about of one hour is at 16:00-20:00 period (Fig.6 a). This means the curvilinear arc motion of atmospheric depression in SKL relative JZR observational point.

The loop cyclonic motions cause the loop-shaped tilt variations of the Earth's surface (Fig. 6.b). It is surprising that irregular non-tidal tilt variations are sensitive to the tropical cyclone motions, which develop more than 10,000 km apart of the point of tilts registration. At the same time, it is important that unusual tropical cyclone (hurricane, typhoon) behavior often occur before strong earthquakes.

This phenomenon we observed under studying the earthquake and hurricane coupling in various World Ocean basins. The example in Fig. 7 shows very strange typhoon KOPPU movements in NW Pacific Ocean 20 hours before Hokkaido Mw 8.3 earthquake on 25 September 2003 [19].

4 Conclusion

The novel techniques of the Earth's deformation measurements by means of strainmeter and tiltmeter systems spaced at far distances was applied for distinguish the global geophysical processes qualified as earthquake precursory phenomena. Experimental data arrays obtained at six observational points located at distances of 130-8100 km from each other have been processed and analyzed. The retarded long-period 10-40 min oscillations, which are recorded by far distanced instruments, appear due to powerful tropical cyclones excitation the tectonic plates as well as surface, atmosphere and ionosphere of the Earth. The study of strain- and tilt-baric processes preceding and accompanied three strong earthquake confirms our previous results and those together show decisively the coupling seismic events of global scale with World Ocean cyclonic activity.

Acknowledgements

The work was carried out within the framework of the state task 0030-2019-0014 and AAAA-A17-117080110043-4. Some work on recording and processing data in Czech observatory was supported by the CzechGeo/EPOS projects (No.LM-2010008 and LM-2015079).

References

1. T. Jahr, Geodesy and Geodynamics, **9**, 229-236 (2018)
2. M.N. Dubrov, O.S. Kazantseva, A.B. Manukin, V.I. Ponyatovskaya, Physics of the Solid Earth. **43**, 404–411 (2007)
3. M.N. Dubrov, L.A. Latynina, R.F. Matveev, A.V. Ponomarev, Izvestiya Phys. Sol. Earth, **34**, 983–990 (1998)
4. M.N. Dubrov, V.M. Smirnov, Geomagnetism and Aeronomy, **53**, 1, 49–59 (2013)
5. I.L. Nersesov, L.A. Latynina, Tectonophysics, **202**, 221–225 (1992)
6. Dubrov, M.N., Karmaleeva, R.M., Izv. Akad. Nauk SSSR Fiz. Zemli, 7, 81-89 (1976)
7. V.A. Volkov, M.N. Dubrov, Bull. d'Inform. des Marees Terrestres, 148, 11971–11986 (2014)
8. Dmitriy Aleksandrov, Mstislav Dubrov, Vasiliy Kravtsov and Igor Larionov, *Experience in synchronous observation of seismic-strain oscillations of the Earth by the spaced*

laser interferometers, XI International Conference "Solar-Terrestrial Relations and Physics of Earthquake Precursors", Paratunka, Kamchatka region, Russia, September 22-25, 2020, E3S Web of Conferences, **196**, 02010 (2020)

9. I. A. Larionov, Y. V. Marapulets, and B. M. Shevtsov, Solid Earth, **5**, 1293–1300 (2014)
10. B. Košt'ák, J. Mrlina, J. Stemberk, B. Chán, Journal of Geodynamics, **52**, 34–44 (2011)
11. Victor Volkov, Jan Mrlina, Mstislav Dubrov, Vladimir Smirnov, Sergey Golovachev, Vaclav Polak, Geodesy and Geodynamics, **11**, 442-454 (2020)
12. D. V. Aleksandrov, M. N. Dubrov, I. A. Larionov, Yu. V. Marapulets, and B. M. Shevtsov, Journal of Volcanology and Seismology, **13**, 3, 193–200 (2019)
13. L. Brimich, L. A. Latynina, I. Kohút, Contributions to Geophysics and Geodesy, **38**, 4, 429-440 (2008)
14. Gy. Mentes, L. Brimich, M. Bednárik, J. Bódi, *Comparison of extensometric results measured in the Vyhne Tidal Station and in the Sopronbánfalva Geodynamic Observatory*, 19th International Symposium on Geodynamics and Earth Tides, 22–26 June, 2021, Wuhan, China, Abstract book, p.33 (2021)
15. M. N. Dubrov and R. F. Matveev, Development and Study of Multicomponent Geophysical Laser Interferometer-Deformograph, J. Commun. Technol. Electron., **43**, 9, 1068-1073 (1998)
16. M. N. Dubrov and D. V. Aleksandrov, *Laser interferometer antenna array records seismo-acoustic earth strains*, 6th International Conference on Antenna Theory and Techniques, Sevastopol, 2007, Publisher: IEEE, pp.307-308 (2007)
17. P. Spudich and J. B. Fletcher, BSSA, **98**, 4, pp.1898-1914 (2008)
18. O.E. Starovoit, Vestnik VNTs, **5**, 1, 8-12 (2005)
19. M.N. Dubrov, V.A. Volkov, S.P. Golovachev, Nat. Hazards Earth Syst. Sci. Discus. **2,** 1, 935-961 (2014)
20. V.I. Zakharov, V.A. Pilipenko, V.A. Grushin, A.F. Khamidullin, Solnechno-zemnaya fizika, **5,** 2, 114-123 (2019)

Estimation earthquake occurrence probability in Kamchatka region based on seismological and complex of ionospheric precursors

Vadim Boganov[1,*] and *Aleksey* Pavlov[1,**]

[1]Institute of Cosmophysical Research and Radio Wave Propagation FEB RAS, Paratunka, Russia

Abstract. In this paper the authors present the method of estimation of a region, waiting period and probability of strong earthquakes with $K_S \geq 13.5$ ($M \geq 6.0$) in Kamchatka region based on the combination of mid-term and short-term predictive signs accompanying earthquake formation. The seismological predictive parameter ξ_P was taken as a mid-term precursor. It was calculated on the basis of the probabilistic model of seismic regime. A complex of ionospheric parameters was considered as short-term predictive signs with an earthquake waiting period of up to 5 days. It includes the K-layer, the sporadic Es layer of the r type, the critical frequency foF2, and the frequency stratification of the F2 layer. The probabilities of strong earthquakes with $K_S \geq 13.5$ ($M \geq 6.0$) that occurred over the period 2019–2021 in an expected zone, determined by the parameter ξ_P, were estimated on the basis of Bayes method provided that a complex of anomalous parameters of the ionosphere was identified.

1 Introduction

In [1], the results of ionospheric observations were presented for the purpose of short-term (3–5 days) assessment of the waiting period for Kamchatka earthquakes. At the same time, the application of the theoretical-probabilistic approach to the catalogue of Kamchatka earthquakes [2] allowed to calculate the probability of seismic events entering the seismofocal zone of the Kuril Islands and southern Kamchatka and to establish for it the growth of seismic activity that occurred during 2007–2013. This made it possible to make a medium-term forecast of major earthquakes that occurred in 2013 in this area. The task arises of developing a method for estimation the probability, area and time period of expectation of strong Kamchatka earthquakes with an energy class of $K_S \geq 13.5$ ($M \geq 6.0$) based on the joint of seismological and ionospheric prognostic signs based on the combination of seismological and ionospheric prognostic signs. This work is devoted to the solution of this problem.

2 Method of calculation of seismological precursor of earthquakes

Application of A. N. Kolmogorov's axiomatics to the catalogue of seismic events allows to consider each earthquake as an elementary event ω_i , and the whole catalogue or its separate

*e-mail: vbogd@ikir.ru
**e-mail: pavlov@ikir.ru

part as a space of elementary events Ω [2]. Each elementary event ω_i can be characterized by four random variables $\xi_i(\omega_i) = \varphi_i(\omega_i), \lambda_i(\omega_i), h_i(\omega_i), K_i(\omega_i)$, where quantities φ_i, λ_i, h_i are the coordinates of the earthquake, and K_i – energy class. (Event time $t_i(\omega_i)$, as having no mathematical expectation, is excluded from the system of random variables, but it is taken into account as a parameter that determines the occurrence of the event ω_i in the period under study ΔT). For a set of seismic events that occurred over a period of time ΔT, the probability space is a collection of three objects $\{\Omega, \tilde{F}, P\}$, where Ω – space of elementary events , $\tilde{F}$ – set of subsets of random events, P – probabilities of these events. Random events (for a specific selected period of time ΔT) can represent an arbitrary combination of a variable number of random variables (when fixing others) and form some in $\tilde{F}$ subsets A, B, C and etc.

The probability that a seismic event will fall within the specified intervals: latitude $\Delta\varphi = \varphi_i - \varphi_{i-1}$, longitude $\Delta\lambda = \lambda_j - \lambda_{j-1}$, depth $\Delta h = h_m - h_{m-1}$ and energy class $\Delta K = K_n - K_{n-1}$

$$P(\Delta\varphi_i, \Delta\lambda_j, \Delta h_m, \Delta K_n) = \int_{\varphi_{i-1}}^{\varphi_i} d\varphi \int_{\lambda_{j-1}}^{\lambda_j} d\lambda \int_{h_{m-1}}^{h_m} dh \int_{K_{n-1}}^{K_n} f(\varphi, \lambda, h, K)dK =$$
$$= F(\varphi_i, \lambda_j, h_m, K_n) - F(\varphi_{i-1}, \lambda_{j-1}, h_{m-1}, K_{n-1}) =$$
$$= P(\Delta\varphi_i)P(\Delta\lambda_j|\Delta\varphi_i)P(h_m|\Delta\varphi_i, \Delta\lambda_j)P(K_n|\Delta\varphi_i, \Delta\lambda_j, h_m) \quad (1)$$

where i, j, m and n – indices corresponding to intervals of random variables. Statistical processing of the catalog by the formula (1) makes it possible to calculate not only the average probability of occurrence of a seismic event in a given interval of geographical coordinates, depth and energy class through the corresponding conditional probabilities, but also to obtain numerical values of the step distribution function $F(\Delta\varphi, \Delta\lambda, \Delta h, \Delta K)$. As the number of events n increases and the interval Δ decreases, the relative frequency υ tends to its mathematical counterpart P, and $F(\Delta\varphi, \Delta\lambda, \Delta h, \Delta K)$ – towards sustainable continuous distribution $F(\varphi, \lambda, h, K)$.

The random event A on the basis of which the seismological precursor was calculated: «Hit of epicenters of seismic events that occurred in the selected volume V, in the specified intervals of latitude $\Delta\varphi$ and longitude $\Delta\lambda$». The catalogue of earthquakes of Kamchatka Branch of Federal Research Center "United Geophysical Service RAS" during 1962–2021 was used for the analysis.

The studied seismically active area S located along the Eastern coast of Kamchatka, including the Continental regions of Kamchatka, the Southern and Northern seismofocal zones is divided into $m = 12$ non-overlapping areas with dimensions of $S_i = 150 \times 200$ km (Fig. 1). Since seismic events with energy class $K_S^{\text{pr}} \geq 13.5$ are considered as predicted earthquakes in this paper, consequently, according to [3] the length of the gap in the focus of such earthquakes is $L \geq 11$ km. Assuming that the linear size of the earthquake preparation area is of the order of 10 rupture lengths, i.e. 110 km, the spatial cell size of 150×200 km can be chosen as the base case for calculations.

For each square S_i ($i = \overline{1, m}$) in a moving time window ΔT_k passing with step Δt, the time interval $T = 1962-2021$ calculates the probability of seismic events $P_k(S_i) = n_{ik}/N_k$, where n_{ik} – number of seismic events with energy class $K_S \geq 9.0$ and the depth of the hypocenter is less 200 km, occurred in the area of S_i for the period ΔT_k, $N_k = \sum_{k=1}^{m} n_{ik}$ – total number of earthquakes with $K_S \geq 9.0$ and with the same hypocenter depth, took place on the square $S = \sum_{i=1}^{m} S_i$ during ΔT_k. Also in each region S_i long-term (background) values of probability of seismic events hit are calculated $P_T(S_i) = n_{iT}/N_T$, where n_{iT} – number of seismic events with energy class $K_S \geq 9.0$, occured during T, $N_T = \sum_{k=1}^{m} n_{Tk}$ – total number of earthquakes with $K_S \geq 9.0$, occurred during T on the square S. In this paper, the size of the time window and its offset step are set equal, respectively $\Delta T_k = 1$ year and $\Delta t = 1$ day, a time interval at

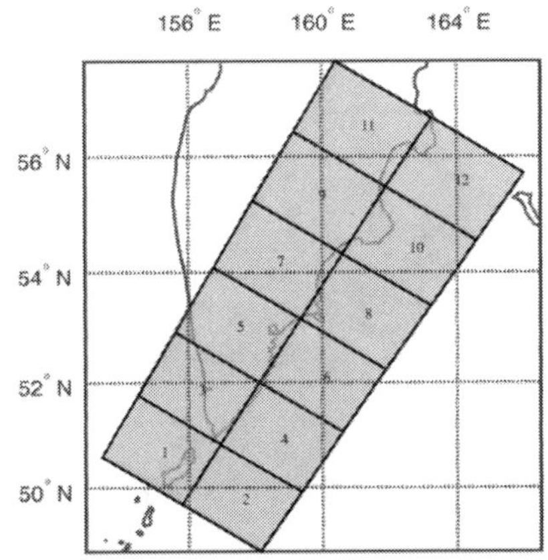

Figure 1. Division of the studied seismic area S into unit cells S_i with dimensions of 150×200 km.

which the study of the seismic regime is equal to $T = 1962-2021$. On Fig. 2, as an example, the values of probabilities of seismic events in the region S_2 are presented.

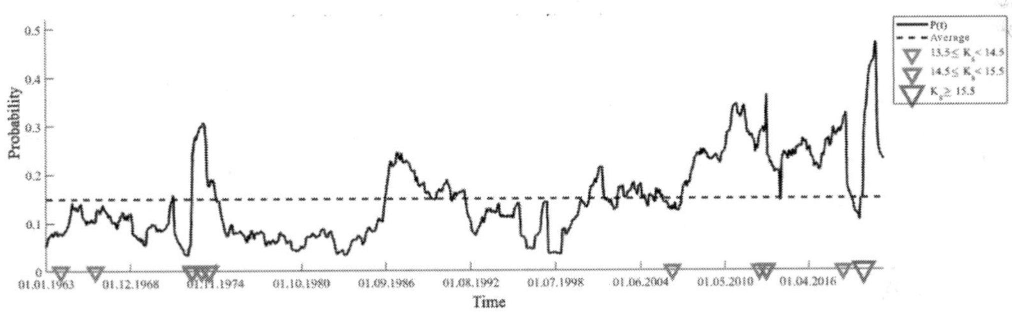

Figure 2. Time series of probability values $P_k(S_2)$. Red triangles on the time axis indicate earthquakes with $K_S \geq 13.5$ ($M \geq 6.0$).

To obtain statistically significant deviations of the current probability values $P_k(S_i)$ from their long-term (background) values $P_T(S_i)$ the parameter ξ_P, is used, determined on the basis of [4] and presented as:

$$\xi_P = \begin{cases} \frac{\Delta P_k - sign(\Delta P_k)\cdot|\sigma_{P_k}|}{|\sigma_{P_T}|}, & \text{if } |\Delta P_k| > |\sigma_{P_k}| \\ 0, & \text{if } 0 \leq |\Delta P_k| \leq |\sigma_{P_k}| \end{cases} \tag{2}$$

where P_k and $\sigma_{P_k} = \sqrt{(P_k(1-P_k)/n_k}$ current probability values and its standard deviation in the sliding time window ΔT_k; P_T and $\sigma_{P_T} = \sqrt{P_T(1-P_T)/n_T}$ long-term probability values and standard deviation $T > \Delta T_k$; n_i and n_T – the number of seismic events that occurred over time periods, respectively ΔT_k and T; $\Delta P_k = P_k - P_T$ the difference between the current P_k probability value and the long-term probability value P_T. Fig. 3 presents the timing values parameter ξ_P for region S_2.

Strong earthquakes can be preceded by periods of increased seismic activity (an increase in the number of background events), and periods of weakening of seismic activity (a decrease in the number of background events). When you search for periods of enhanced seismicity

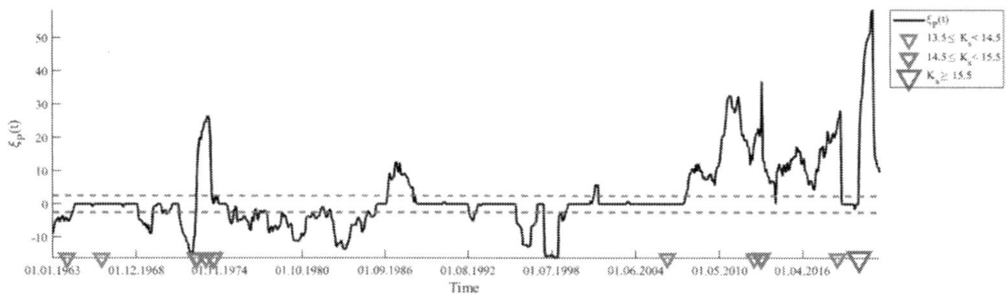

Figure 3. Temporary values of the parameter ξ_P for region S_2. Red triangles on the time axis indicate earthquakes with $K_S \geq 13.5$ ($M \geq 6.0$).

Table 1. Predictive efficiency of the parameter ξ_P.

Type of seismic anomaly	V	R	J_G	J_M
Seismic activity ($\xi_P \geq 2.5$)	0.57	0.49	1.47	0.15
Seismic calm ($\xi_P \leq -2.5$)	0.51	0.52	2.15	0.28
Seismic activity and calm	0.45	0.76	1.47	0.24

in the beginning of the anomaly in the parameter ξ_P is taken as the time of appearance of the values $\xi_P \geq \xi_P^a$, where ξ_P^a – anxiety level is selected experimentally. In the case of searching for periods of seismic calm, the beginning of the anomaly is identified by the moment of occurrence of the values $\xi_P \leq \xi_P^c$, where ξ_P^c – anxiety level is selected experimentally. Accordingly, when $\xi_P^c < \xi_P < \xi_P^a$ it is assumed that the current probability values do not significantly deviate from the background values. In this paper, the values of the criteria for the periods of activation and lull were established respectively $\xi_P^a = 2.5$ and $\xi_P^c = -2.5$.

To assess the prognostic effectiveness of a prognostic feature ξ_P following parameters are used: ratio of the number of earthquakes for which the precursor was allocated to the number of all earthquakes (reliability R), ratio of the number of precursor anomalies to the total number of isolated anomalies (validity V) [5], the effectiveness of A. A. Gusev [6] and G. M. Molchan's [7] methods.

The results of the evaluation of the predictive efficiency of parameter ξ_P for earthquakes with $K_S \geq 13.5$ are presented in Table 1 and errors diagram (Fig. 4).

As can be seen from Table 1, the share of anomalies of seismic regime activation identified by parameter ξ_P and preceding earthquakes (parameter V) was 57%, and the share of calm anomalies was 51%. The relative number of anomalies of the seismic regime of both types preceding the onset of earthquakes with $K_S \geq 13.5$, was 45%. The activation of the seismic regime was preceded by 49% of earthquakes, and calm – 52%. The J_G prediction efficiency for both types of anomalies was 1.47. Since $J_G > 1$, this predictive feature is considered informative and can be used in earthquake prediction algorithms. The J_M prediction efficiency for both types of anomalies was 0.24 and as can be seen from the error diagram (Fig. 4) the values (τ, ν) lie below the lower limit of 99% of the confidence interval (green curve), which can be interpreted as a high degree of reliability of the revealed connection between the identified anomalies of the seismic regime using the parameter ξ_P and predicted earthquakes from the considered range of the energy class.

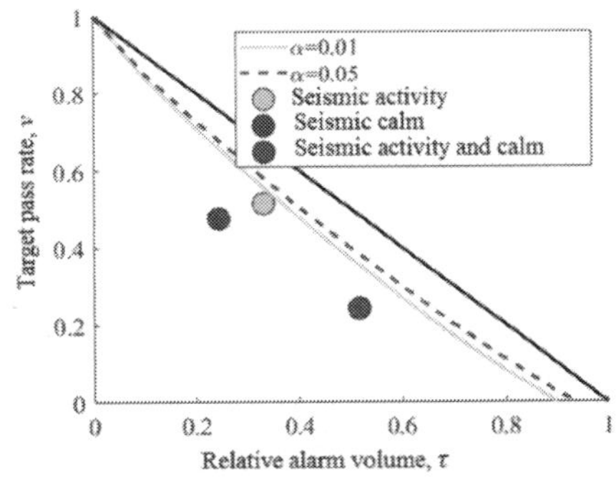

Figure 4. The diagram of errors for the parameter ξ_P in predicting earthquakes with $K_S \geq 13.5$. The lower bounds of the confidence interval of the random forecast with the significance level are marked $\alpha = 0.01$ and $\alpha = 0.05$.

Table 2. The predictive efficiency of the complex ionospheric precursors.

$n(E_A)$	$n(E)$	$n(A_E)$	$n(A)$	V	R	J_G	τ	ν	J_M
24	30	21	180	0.1	0.8	1.69	0.47	0.2	0.33

3 Method of short-term earthquake prediction based on the complex of ionospheric parameters

In this work the data of radiophysical observations made by means of vertical radiosonding are used. The automatic ionospheric station (AIS) of vertical radiosonding is located in the village of Paratunka ($\varphi = 52.97°$ N, $\lambda = 158.24°$ E). Observations are made once every 15 minutes in pulse mode at frequencies from 1 to 15 MHz.

The ionospheric precursors with an earthquake expectation period of up to 5 days were considered in article [8]. On the basis of the selected most effective ionospheric precursors, an algorithm for short-term earthquake prediction was constructed, in which a joint analysis of the ionospheric disturbances under consideration is carried out in a sliding time window of width $\Delta T = 5$ days with a step $\Delta t = 1$ day. The condition for announcing the start of the waiting period to seismic event was the execution in a time interval ΔT for at least three of the four ionospheric parameters consider the following criteria:

- K-layer formation during at least one day of ΔT interval;
- Formation of sporadic layer Es type r for at least one day interval ΔT.
- Exceeding the critical frequency foF2 layer F2 median values f_{med}, calculated for the previous 30 days, not less than 20% ($\frac{f\mathrm{oF2}-f_{\mathrm{med}}}{f_{\mathrm{med}}} \geq 0.2$) against the background of the development of the magnetospheric storm (total values of the K-index per day $\Sigma K \geq 20$);
- Stratification of the F2 layer in frequency (mode "V") for at least one day interval ΔT;

The waiting period is chosen equal to $T_{\mathrm{w}} = 5$ days. The algorithm was evaluated its predictive effectiveness in a time interval 01.01.2013–01.06.2021 for seismic events $K_S \geq 13.5$ ($M \geq 6.0$) at depths of 100 km in radius $r = 500$ km from the registration point of the ionospheric observations.

According to the results presented in Table 2, when predicting earthquakes with energy class $K_S \geq 13.5$, the reliability is 0.8 (i.e. 80% of earthquakes had a precursor), and the reliability is 0.1 (i.e. 10% of detected anomalies were realized). The efficiency value $J_G = 1.69$

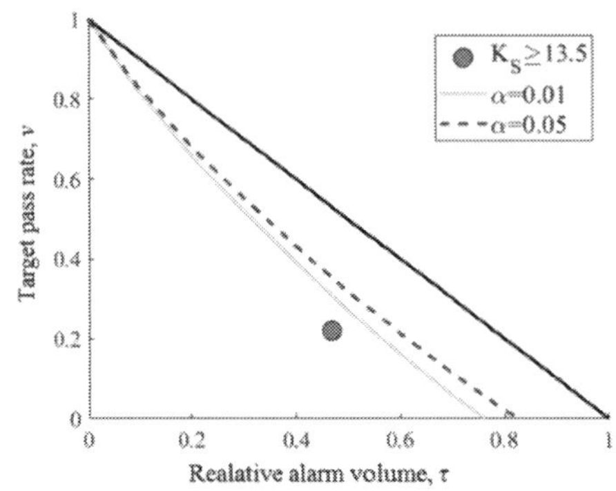

Figure 5. The diagram of errors for the algorithm of the complex ionospheric precursors for $K_S \geq 13.5$.The lower bounds of the confidence interval of the random forecast with the significance level are marked $\alpha = 0.01$ and $\alpha = 0.05$.

shows that the prognosis by this method is statistically significant and 1.69 times different from random guessing. In the error diagram (Fig. 5) the values (τ, ν) obtained for the range $K_S \geq 13.5$ lie below the lower limit of 99% of the confidence interval, which can be interpreted as a high degree of reliability of the revealed connection of the considered complex of ionospheric precursors with earthquakes of this range of energy class that occurred at distances up to 500 km from the observation point.

4 Method to estimate earthquake occurrence probability

The method to estimate the region, waiting time period and probability for earthquake with $K_S \geq K_S^{pr}$ (K_S^{pr} – is the minimum energy class of the predicted earthquakes) based on a joint analysis of the seismic parameter ξ_P and a complex of ionospheric precursors can be presented as follows:

1. Monitoring of the seismic regime in elementary cell S_i based on the seismological parameter ξ_P. Area of expectations of strong earthquake will be considered $S_{\exp}$, which includes the unit cells, which was identified anomalies of seismic regime parameter ξ_P: $S_{\exp} = \frac{1}{l}\sum_{j=1}^{l} S_j$, where $l \leq m$ – number of areas where abnormal parameter values are identified ξ_P. The average waiting period for a strong earthquake will be considered the period $\tilde{T}_w = \frac{1}{l}\sum_{j=1}^{l} T_j^w$, where l – number of areas where abnormal parameter values are identified ξ_P, $T_j^w = \tilde{T}_j^w + \sigma_T$ – the waiting period earthquake in the cell S_j, $(j = \overline{1, l})$;

2. Monitoring of the ionospheric disturbances under consideration in a sliding time window of width $\Delta T = 5$ days with a step $\Delta t = 1$ day.

3. If the appearance of at least three of the four ionospheric parameters is recorded in the time interval ΔT, then using formula 3, the conditional probability $P(D_1|EI)$ of the onset of an earthquake in the area of expectations of strong earthquake $S_{\exp}$ is calculated.

4. If the probability $P(D_1|EI)$ exceed the specified criterion, then in the area $S_{\exp}$, the waiting period T_w of a strong earthquake with a duration of 5 days is announced. The

waiting period T_w can be extended if the complex of ionospheric precursors has been registered again.

The probability of occurrence of a seismic event with $K_S \geq K_S^{pr}$ in the range S_{exp} (random event D_1), provided that anomalies of the seismic regime are identified in it (random event E), as well as a complex of ionospheric precursors (random event I), calculated based on the Bayes formula:

$$P(D_1|EI) = \frac{P(D_1) \cdot P(E|D_1) \cdot P(I|D_1)}{P(D_1) \cdot P(E|D_1) \cdot P(I|D_1) + P(D_2) \cdot P(E|D_2) \cdot P(I|D_2)} \tag{3}$$

The random event D_1 – an earthquake occurrence predicted energy class $K_S \geq K_S^{pr}$ in the range S_{exp}.

The probability of this event is defined as $P(D_1) = \frac{n_l}{N}$, where $n_l = \sum_{j=1}^{l} n_j$ – the number of earthquakes with $K_S \geq K_S^{pr}$ that occurred in the S_{exp} area during the observation period T_1, N – the number of earthquakes with $K_S \geq K_S^{pr}$ that occurred in the $S = \sum_{j=1}^{m} S_j$ area during the observation period T_1 of parameter ξ_P.

The random event D_2 – absence of earthquake with energy class $K_S \geq K_S^{pr}$ in area S_{exp}. The probability of this event D_2 is $P(D_2) = 1 - P(D_1)$.

The random event E – the occurrence of anomalous values of the parameter ξ_P.

The random event I – the occurrence of complex of the ionospheric precursors. Then the events E and I can be represented in the form: $EI = D_1EI + D_2EI$

The random event D_1EI – the occurrence of earthquake with $K_S \geq K_S^{pr}$ in area S_{exp} upon identification of an anomaly of the seismic regime (activation or calm) and the appearance of a complex of ionospheric precursors. The probability of this event is defined as $P(D_1EI) = P(D_1) \cdot P(EI|D_1) = P(D_1) \cdot P(E|D_1) \cdot P(I|D_1)$.

The probability $P(E|D_1) = \frac{n_{pr1}}{N_{S_{exp}}}$ is earthquakes occurrence probability with $K_S \geq K_S^{pr}$ in area S_{exp} when identifying seismic anomalies; n_{pr1} – number of earthquakes with $K_S \geq K_S^{pr}$, predicted when anomalies of the seismic regime appear over period T_1; $N_{S_{exp}}$ – total number of earthquakes with $K_S \geq K_S^{pr}$ in area S_{exp} over period T_1.

The probability $P(I|D_1) = \frac{n_{pr2}}{N_{S_{exp}}}$ is earthquakes occurrence probability with $K_S \geq K_S^{pr}$ in area S_{exp} when identifying the complex of ionospheric disturbances; n_{pr2}– number of earthquakes with $K_S \geq K_S^{pr}$, predicted when an ionospheric anomalies appears during the observation period T_2 (T_2 is a registration period of ionospheric parameters); $N_{S_{exp}}$ – total number of earthquakes with $K_S \geq K_S^{pr}$ in area S_{exp} over period T_2.

The random event D_2EI – absence of earthquake with energy class $K_S \geq K_S^{pr}$ in area S_{exp} when identifying anomalies of the seismic regime (activation or calm) and appears complex of ionospheric precursors. The probability of this event is defined as $P(D_2EI) = P(D_2) \cdot P(EI|D_2) = P(D_2) \cdot P(E|D_2) \cdot P(I|D_2)$.

The probability $P(E|D_2) = \frac{n_{fa1}}{N_{a1}}$ is probability of occurrence of abnormal parameter values ξ_P (false alarms) in the area S_{exp}, which did not end with $K_S \geq K_S^{pr}$ earthquakes; n_{fa1} – number of false alarms per observation period T_1; N_{a1} – the total number of cases of abnormal behavior of the parameter ξ_P in the area S_{exp} over the period T_1.

The probability $P(I|D_2) = \frac{n_{fa2}}{N_{a2}}$ is probability of occurrence of ionospheric anomalies complex, which did not end with earthquakes; n_{fa2} – number of false cases of the appearance of a complex of ionospheric parameters per observation period T_2; N_{a2} – the total number of occurrences of a complex of ionospheric parameters over the period T_2.

Table 3. List of predicted earthquakes with energy class $K_S \geq 13.5$ ($M \geq 6.0$) over period 01.01.2019–01.06.2021.

№	Date EQ	K_S	$P(D_1\|EI)$	S_{exp}	T_w
1	2019.06.25	14.3	0.69	+	2019.06.21–2019.06.26
2	2019.06.26	14.4	0.69	+	2019.06.21–2019.06.26
3	2020.01.22	14.3	0.71	+	2020.01.21–2020.02.05
4	2020.02.20	14.3	0.72	+	2020.02.11–2020.02.29
5	2020.03.25	16.8	0.73	+	2020.03.19–2020.03.28
6	2021.03.16	14.6	0.74	+	2021.02.14–2020.03.30

5 Retrospective predictive estimates of the area, period and probability of earthquake occurrence with $K_S \geq 13.5$

In the considered seismically active region, 6 earthquakes with an energy class of $K_S \geq 13.5$ ($M \geq 6.0$) occurred during the period 01.01.2019–01.06.2021. The results of the retrospective analysis according to the proposed method are presented in Table 3. Signs "+" and "-" observed earthquake respectively inside the waiting area S_{exp} and outside its borders. All six seismic events occurred during the waiting periods T_w, determined from the complex of ionospheric parameters, and hit into the waiting area S_{exp}, determined from the values of the parameter ξ_P. The conditional probability $P(D_1|EI)$ of earthquake occurrence ranged from 0.69 to 0.74.

Figure 6a shows, as an example, the waiting area for the earthquake of 25.03.2020 with $K_S = 16.8$, constructed on 19.03.2021 based on the analysis of the anomalous values of the parameter ξ_P.

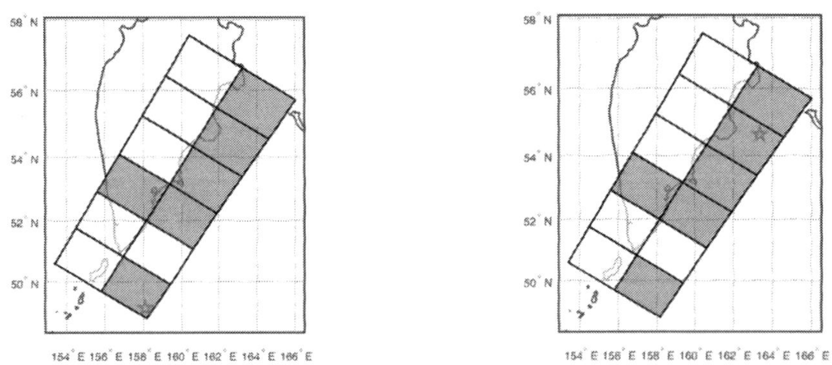

a) b)

Figure 6. Maps of expectation area S_{exp} for earthquakes: a) 25.03.2020, $K_S = 16.8$; b) 16.03.2021, $K_S = 14.6$. Red stars on the map are earthquake epicenters.

The results of evaluating the effectiveness of earthquakes prediction with $K_S \geq 13.5$ for the time period 01.01.2019–01.06.2021, when choosing as the alarm level the values of the conditional probability $P(D_1|EI) = 0.5$ and $P(D_1|EI) = 0.7$, are presented in Table 4 and on the error diagrams (Fig. 7).

The analysis of the predictive efficiency of the presented method showed that at $P(D_1|EI) \geq 0.5$ its reliability $R = 1$ (100% of earthquakes were predicted), validity $V = 0.09$ (9% of forecasts ended with earthquakes), efficiency J_G was 1.8 (with $J_G = 1$, the forecast

Table 4. The effectiveness for the method of earthquakes prediction with $K_S \geq 13.5$ for chosen levels of the conditional probabilities $P(D_1|EI)$.

Levels of conditional probability $P(D_1\|EI)$	V	R	J_G	J_M
0.5	0.09	1	1.8	0.44
0.7	0.09	0.67	1.52	0.23

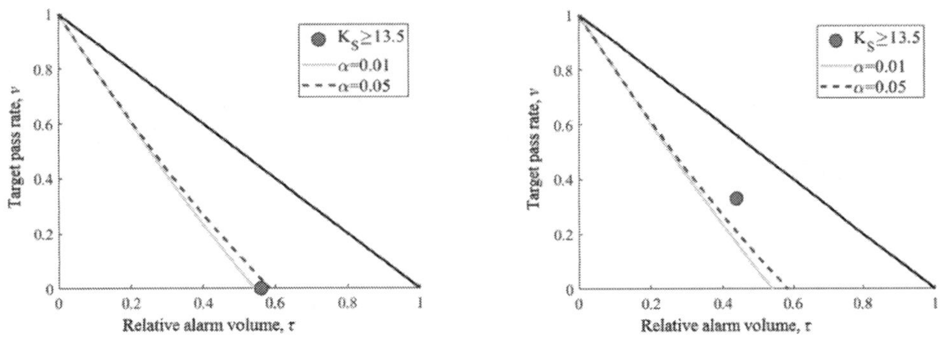

Figure 7. The diagram of errors for the method of earthquakes prediction with $K_S \geq 13.5$ ($M \geq 6.0$) based on joint analysis of seismological and complex of ionospheric precursors for chosen levels of the conditional probabilities $P(D_1|EI) = 0.5$ (a) and $P(D_1|EI) = 0.7$ (b).

is considered random). The efficiency J_M was 0.44, which indicates the presence of a connection between the considered complex of precursors and earthquakes with an energy class $K_S \geq 13.5$ ($M \geq 6.0$).

6 Conclusion

An approach is presented that combines seismological and ionospheric prognostic signs in order to assess the probability of occurrence of earthquakes with $K_S \geq 13.5$ ($M \geq 6.0$) in the Kamchatka region. A retrospective analysis was carried out using this technique for the time period 01.01.2019–01.06.2021.

The effectiveness of method for estimation the area and possible period of the occurence of seismic events with $K_S \geq 13.5$ in the joint analysis of the seismic parameter ξ_P and the complex of ionospheric precursors (at the values of the conditional probability of the onset of earthquakes $P(D_1|EI) \geq 0.5$ and $P(D_1|EI) \geq 0.7$) shows that the forecast using this method is more than 1.5 times different from a random guess.

The low validity of the V forecast may be due to the fact that, ionospheric disturbances are formed against the background of solar activity, and it is not yet possible to accurately indicate the lower threshold of the earthquake magnitude, the preparation of which is already beginning to influence the state of the ionosphere. Seismic events with an energy class $K_S \geq 13.5$ predicted by ionospheric precursors and taken into account in statistics do not include events with an energy class $K_S < 13.5$, although the ionospheric precursor has formed. Consequently, the ionospheric forecast signs that precede seismic events with $K_S < 13.5$ are false for events with $K_S \geq 13.5$, thereby lowering the forecast reliability.

The work was carried out as part of the implementation of the state task AAAA-A21-121011290003-0.

References

[1] V.V. Bogdanov, A.V. Kaisin, A.V. Pavlov, A.L. Polyukhova, C.-V. Meister, J. Physics and Chemistry of the Earth **98**, 154–160 (2017)

[2] V.V. Bogdanov, A.V. Pavlov, A.L. Polyukhova, J. Volcanol. Seismol. **4 (6)**, 412–422 (2010)

[3] Yu.V. Riznichenko, *Issledovaniy po fizike zemletrjasenij* (Nauka, Moskva, 1976)

[4] A.D. Zav'yalov, *Prognoz zemletrjasenij* (Nauka, Moskva, 2006)

[5] V.A. Saltykov, Physics of the Earth **2**, 84–96 (2017)

[6] A.A. Gusev, *Seismicity and seismic forecast, properties of the upper mantle and their relation to volcanism on Kamchatka* (Nauka, Novosibirsk, 1974) 109–119

[7] G.M. Molchan, Physics of the Earth and Planetary Interiors **61**, 84–98 (1990)

[8] V.V. Bogdanov, A.V. Pavlov, Vestnik KRAUNC. Fiz.-mat. nauki **30 (1)**, 59-78 (2020).

On the dynamics of the electromagnetic parameters of the Northern Tien Shan lithosphere

Elena Bataleva*

Research Station of the RAS in Bishkek, 720049, Kyrgyzstan. Bishkek-49

Abstract. The results of geophysical monitoring of seismically hazardous regions are of undoubted interest for studying the deep structure of the lithosphere, regional seismicity, modern geodynamics, etc. The work used experimental material, including the results of magnetotelluric monitoring and the catalogs of the KNET (Kyrgyzstan Telemetered Network), KRNET (Kyrgyz Republic Digital Network) and ISC (International Seismological Center) networks obtained in the seismically active zones of the Bishkek Geodynamic Proving Ground (Northern Tien Shan). The analysis of electromagnetic monitoring results of the Northern Tien Shan seismic generating zone was carried out in a wide frequency range. The data on the parameters of the electromagnetic field of the Earth's crust in the Northern Tien Shan are generalized and systematized. Based on the analysis of these data, the dependence of the response of seismic events in electromagnetic parameters on the distance of earthquake epicenters was studied. The most likely reason for the occurrence of anomalous changes in the electromagnetic field is the activation of deformation processes during the preparation of strong earthquakes. The results of the time-frequency analysis are presented and the features of manifestation are considered depending on the location of the hypo- and epicenters of seismic events and their magnitude. Regularities have been established in the behavior of tipper variations for remote and regional earthquakes for the first quarter of 2016 with an energy class from 6 to 10.

One of the most urgent problems of modern geodynamics is to identify the connection between variations in geophysical fields with seismic processes occurring in the Earth's crust [1-3]. The solution to this problem for the Northern Tien Shan is based on a complex of studies on the territory of the Bishkek Geodynamic Proving Ground (BGPG) (Northern Tien Shan). The geophysical monitoring system of the BGPG consists of: (i) a network of seismological observations based on a network of digital broadband telemetry seismic stations (KNET); (ii) Central Asian GPS Network; (iii) systems of geomagnetic/gravimetric observations and networks of electromagnetic monitoring of changes in the stress-strain state of the Earth's crust by the method of far-field transient electromagnetic sounding (FFES) and by the method of magnetotelluric sounding (MTS).

* e-mail: elena.bataleva@gmail.com

The BGPG MT-monitoring system includes: i) two stationary observation points (see Fig. 1) - Ak-Suu and Chon-Kurchak, where the MT-field is continuously recorded on the built-in flash memory of the measuring equipment, the recording duration depends on flash memory size and registration parameters, at the maximum polling frequency, the recording duration is about 20 days, after which the data is copied from the flash memory to the laptop, the equipment is serviced and restarted; ii) a network of profile observations at the Kentor mini-polygon (see Fig. 1, inset on the left); when carrying out specialized MT-monitoring, the measurement mode depends on the research objectives and is seasonal in nature - two sessions (spring and autumn); iii) regime points of MT-monitoring located in the most strain-sensitive zones. The time interval for recording the MT-field ranges from several hours to several days, which determines the depth of soundings.

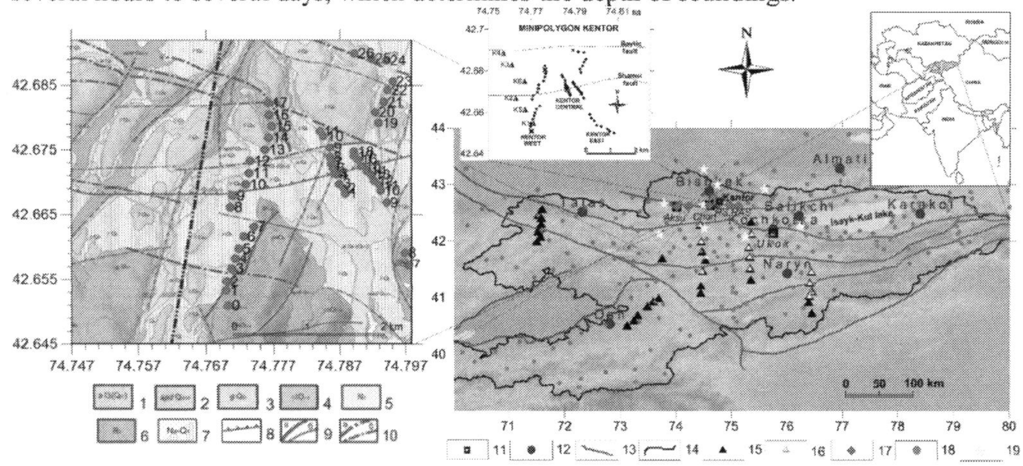

Figure 1. Location map of points of the geophysical monitoring, performed on the territory of the Central Tien Shan: 1 – modern alluvial boulder - pebble deposits; 2 – alluvial boulder - pebble deposits of the first above - flood terrace; 3 – blocky - pebble glacial deposits; 4 – Early and Middle Ordovician granodiorite; 5 – Miocene pebble - crumbly - sandy strata; 6 – Riphean complexes (undivided); 7 – Pliocene–Pleistocene alluvial boulder deposits of the Sharpyldak series; 8 – Shamsi fault; 9 – activated faults and fracture zones of the basement: a – main; б – secondary; 10 – activated faults assumed under the cover of modern sediments: a – main; b – secondary; 11 - points of stationary and profile magnetotelluric observations; 12 - settlements; 13 - main fault structures; 14 - the border of Kyrgyzstan; 15 - points of regime deep MTS; 16 - points of regime deep MTS of 2018; 17 - points of electromagnetic monitoring; 18 - points of the network of GPS observations; 19 - KNET teleseismic network sites.

At present, on the territory of the BGPG, a sufficiently large experience has been accumulated in conducting monitoring magnetotelluric (MT) observations in order to study the geodynamic processes occurring deep in the Earth in connection with the preparation of seismic events [4-8]. Reliable geoelectric models of the seismically active junction zone of the Kyrgyz ridge and the Chuy basin have been built on the basis of the performed electromagnetic methods, and interesting practical results of experimental electromagnetic research [9, 10], focused on monitoring geodynamic processes in seismically active regions, have been obtained. As a possible prognostic parameter, it is customary to use such a physical characteristic of the geological environment as electrical resistance.

At the moment, the main experimental studies on the territory of the BGPG are aimed at studying the zones of increased fracturing and fluid saturation. When examining such zones by electromagnetic methods, the following factors should be taken into account: i) distribution and orientation of cracks (random distribution of cracks will introduce the least distortion in geophysical fields); ii) the nature of the substance filling the cracks (air, water, debris and clay material, etc.). The presence of fracturing and fractured zones in rocks is one of the reasons for electrical anisotropy, i.e. different electrical conductivity in different directions. This factor makes it possible, under favorable conditions, to use the method of azimuthal magnetotelluric monitoring [4, 5] to determine the predominant direction of fractured zones and changes in their intensities at different depths.

According to the results, the strike of the dominant vertical or steeply inclined systems of cracks is established - basically this coincides with the distribution of ρk and other electromagnetic parameters in polar diagrams [5]. The increased interest in the study of fault zones is associated with their important role in the formation of the stress-strain state of the earth's crust in seismically active regions, the regime of underground fluids, as well as with the processes of mass, heat and energy exchange in the zones of their dynamic influence. These factors also determine a new stage in the development of studies of the nature of deformation processes in the Earth's crust of the BGPG territory.

The method of magnetotelluric sounding, one of the leading methods of electromagnetic monitoring of the Bishkek Geodynamic Proving Ground, allows monitoring of hidden fault structures that do not appear in the sedimentary cover, but are zones of increased fracturing and fluid permeability of the basement and are manifested by anomalies in geophysical fields. Experimental observations by the MT sounding method were carried out using the MTU series equipment from Phoenix Geophysics Ltd (Canada) [11-12]. The 2D smoothing inversion of the BGP MT data was performed using the Rodi-Mackie program. The program implements the nonlinear conjugate gradient method, which attempts to minimize the objective function, which is the sum of the normalized data residuals and model smoothness [13].

In the magnetotelluric method the relationship between the horizontal components of the electromagnetic field is expressed through a pair of linear equations,

$E_x = Z_{xx}H_x + Z_{xy}H_y$

$E_y = Z_{yx}H_x + Z_{yy}H_y$.

In matrix notation this condenses to:

$$\mathbf{E} = \mathbf{ZH}, \tag{1}$$

where **Z** is the magnetotelluric impedance tensor. The MT technique is based on the study of the impedance tensor, i.e. the stationary transfer function relating the horizontal components of the magnetic to the corresponding components of the electric field [14].

To determine the impedance tensor, it is enough to measure 4 components of the electromagnetic field - E_x, E_y, H_x, H_y. If the observation points additionally record variations in the vertical component of the magnetic field, which is a prerequisite for magnetotelluric soundings performed by the RAS Scientific Station, then you can use the Wiese method with the construction of vectors of the same name. The Wiese vectors (2) are determined by the ratio of three components of geomagnetic variations at one observation point and characterize the inhomogeneity of the distribution of electric currents induced in the Earth, and, consequently, the inhomogeneity of the distribution of electrical conductivity.

$$H_z = W_{zx}Hx + W_{zy}H_y \tag{2}$$

They are directed from the conductor perpendicular to the direction of excess currents in it. Wiese vectors (induction arrows, tippers) are constructed from the real (3) and imaginary parts (4) components of the Wiese-Parkinson matrix W:

$$\overline{\mathrm{Re}W} = \mathrm{Re}W_{zx}\overline{I_x} + \mathrm{Re}W_{zy}\overline{I_y}, \tag{3}$$

$$\overline{\mathrm{Im}W} = \mathrm{Im}W_{zx}\overline{I_x} + \mathrm{Im}W_{zy}\overline{I_y}. \tag{4}$$

The vector Re**W** is called a real vector. It characterizes the influence of currents that are in phase with the magnetic field. The Im**W** vector is called an imaginary vector, it characterizes the influence of reactive excess currents, that is, currents that differ from the phase of the horizontal magnetic field by /2. A distinctive feature of real vectors is that they are directed from zones of increased electrical conductivity to zones of low electrical conductivity. This property of the Wiese vectors makes it possible to localize geoelectric structures characterized by increased or decreased electrical conductivity. In geoelectric models, as a rule, zones of increased electrical conductivity are either zones of dynamic influence of fault structures or areas of increased fracturing. To detect and study fracture zones, both the analysis of traditional electromagnetic parameters (apparent resistivity, impedance phase, etc.) and new ones - anisotropic changes in the electrical conductivity of the geological medium monitoring studies. The geological environment, in general, and the fault structure, in particular, is subject to periodic effects of lunisolar tides, as a result of which its parameters (electrical conductivity, fluid saturation, elasticity, etc.) also periodically change (Fig. 2). Therefore, special attention in this work was paid to the analysis of variations in the magnetic tipper, their relationship with lunisolar tidal effects (Fig. 2) and seismic events (Fig. 3) recorded by various seismic networks.

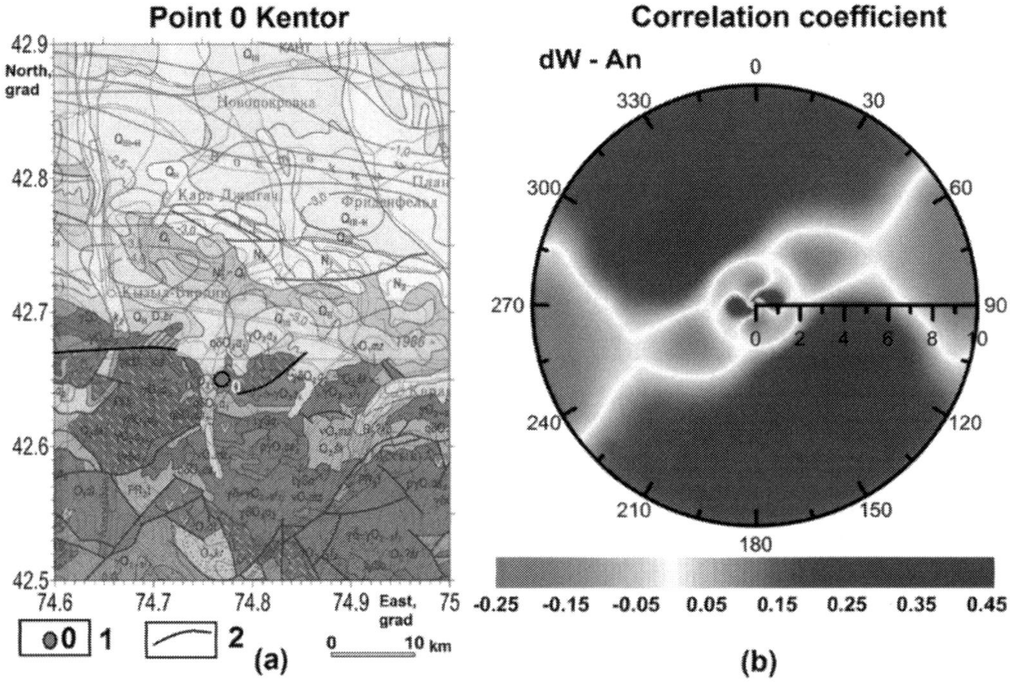

Figure 2. Correlation polar diagrams for the Kentor mini-poligon: a) - location of monitoring points and faults: 1 - local faults, 2 - MT-monitoring point; b) correlation diagrams for variations of Wiese vectors (dW) at FT-0 and: by the vertical component of lunar-solar tidal effects (An).

Taking into account that real induction arrows (Re**W**) are associated with active electric currents, and imaginary (Im**W**) - with reactive ones, we built and considered the time-frequency series of both parameters (Re**W** and Im**W**). The experimental results obtained indicate that the position of the clusters on the polar correlation diagrams (Fig. 2) is associated with the geoelectric structure of the observation point and corresponds to the orientation of the main tectonic elements in the vicinity of this point. As a result of the analysis of gravitational tidal effects and the behavior of the Wiese vector, it was found that the nature of the relationship depends on the geoelectric structure of the sounding point and is reflected in the distribution of stable clusters on the polar correlation diagrams. It was revealed that the orientation of the main tectonic faults (near the sounding point) corresponds to the position of the clusters on the polar correlation diagrams (Fig. 2).

The behavior of variations of real and imaginary induction arrows in a wide range of periods was investigated according to the data of profile magnetotelluric soundings for the Kentor geophysical monitoring minipolygon (Fig. 1) and their relationship with the seismicity distribution was analyzed (Fig. 3). Field experiments have confirmed the concept of the relationship between the stress-strain state of the medium and the change in the apparent electrical resistance through the redistribution of saline solutions between the fracture systems [15-17].

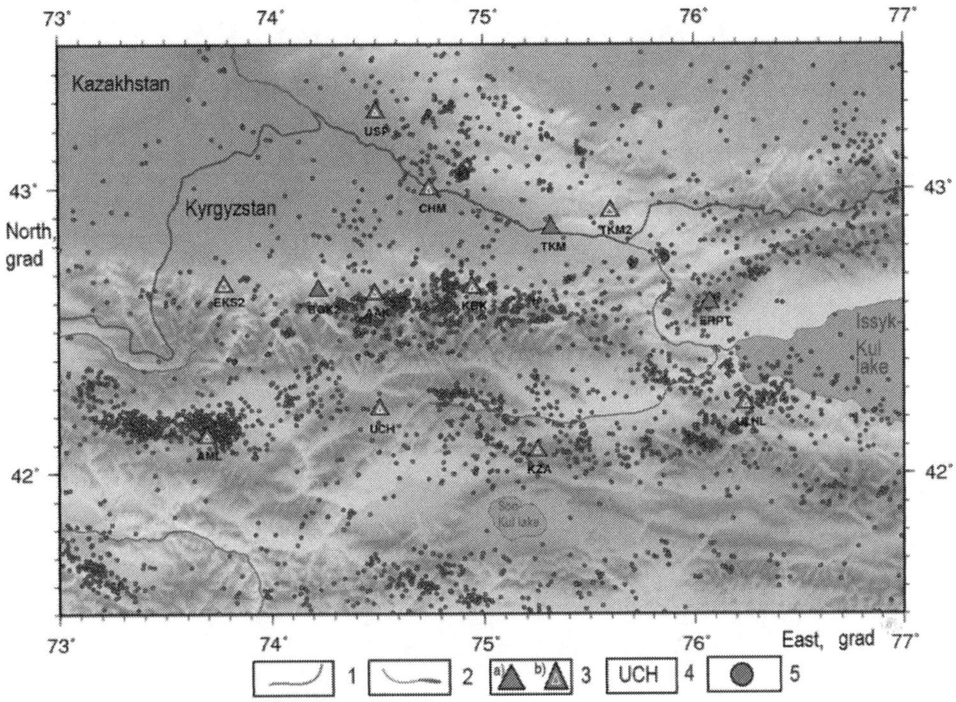

Figure 3. Distribution of earthquake epicenters:
1 – state border of the Kyrgyz Republic; 2 – rivers; 3 – KNET seismic network: a) all seismic stations as of 2000, b) actual working seismic stations as of 2021; 4 – earthquake epicenters.

Figures 4 and 5 show the time-frequency series of the Wiese vector variations for two MT-monitoring stations Ak-Suu and Chonkurchak (Fig. 1), constructed using the azimuthal MT-monitoring method [4,5]. Time-frequency series (TFS) are a form of representing the variability of the structure of the measured components of the electromagnetic field (logarithm of the sounding period) when the coordinate system is rotated by a certain angle (in degrees) [6]. We are considering the difference between the average tipper value (for January-March) and the current one for each azimuth. In both figures, there is a connection between the behavior of events of real induction arrows and the distribution of events for the first quarter of 2016. The energy class of earthquakes that occurred in the Tien Shan over a period of time and recorded by the KNET, KRNET and ISC networks was estimated from 6 to 9.5. As the results of electromagnetic observations in 2016 showed, the arrows (Re**W**) reflect changes in the stress-strain state of a gea, not only before significant seismic events. In the time-frequency series, the observed anomalous effects are quite comparable in magnitude with the same ones that were noted during the period of preparation for higher-class physical events. The confinement of earthquakes to the gradient zones of changes in ReW events is obvious, which allows us to consider the Wiese vector as an informative parameter in the study of geodynamic processes that accompany the safety and security of events.

Noteworthy is the different nature of the real induction arrows for the Ak-Suu and Chon-Kurchak stations. The stations are located in the junction zone of the Kyrgyz ridge and the Chuy basin, in direct contact from the active fault structures - Issyk-Ata and Chon-Kurchatsky, respectively. The relationship between the behavior of Re**W** and the seismicity distribution is most clearly manifested in the frequency-time series of Re**W**, built for the Chon-Kurchak station, although the noise level for it is higher than for the Ak-Suu station.

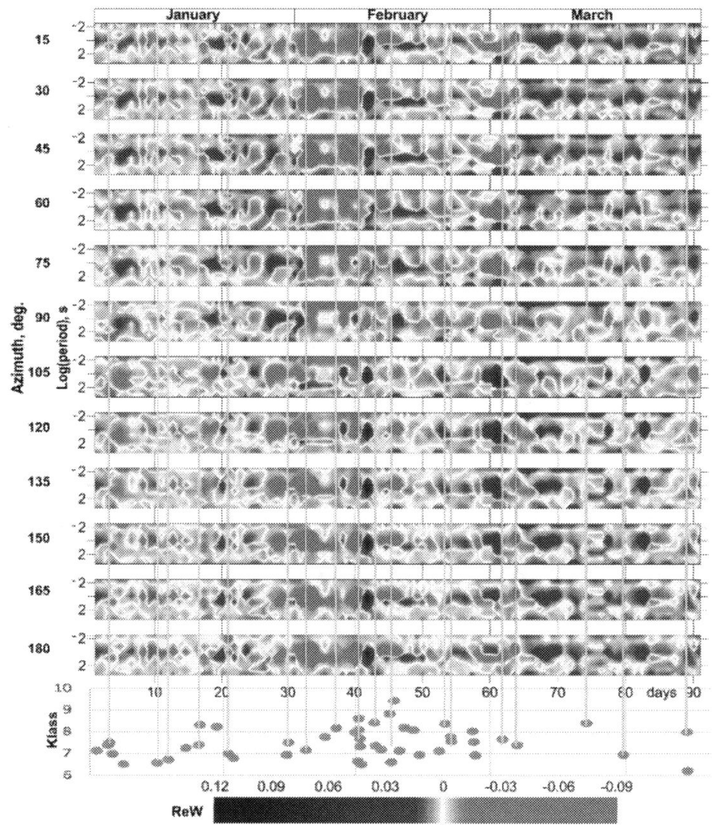

Figure 4. Time-frequency series of the azimuth MT - monitoring (variations of magnetic tipper component) at the Ak - Suu station in comparison with seismicity in January – March 2016 (ovals – seismic events).

This can be explained by the higher tensosensitivity of Chon-Kurchak and the closer location of the epicenters of seismic events. The map of the distribution of earthquake epicenters (Fig. 3) indicates that in the immediate vicinity of the Ak-Suu station, the frequency of seismic events is several times lower than for the Chon-Kurchak station. At the same time, the regularity of the reaction of the type of distant and regional earthquakes is not clearly manifested; anomalous variations can be caused by a strong distant event or a close weak one, while the amplitude of the variations can be practically the same.

Thus, an assessment of the characteristics of the stress-strain state of the geological environment and the behavior of variations in electromagnetic parameters was carried out in order to identify patterns in their relationship for seismically active regions of the Tien Shan. The paper summarizes the results of magnetotelluric monitoring on the parameters of the electromagnetic field of the Earth's crust in the Northern Tien Shan. Based on the catalogs of the KNET, KRNET and ISC networks, the dependence of the response of seismic events in electromagnetic parameters on the class and distance of earthquake epicenters is studied. The most likely reason for the occurrence of anomalous changes in the electromagnetic field is the activation of deformation processes during the preparation of strong earthquakes. The results of the time-frequency analysis are presented and the features of manifestation are considered depending on the location of the hypo- and epicenters of seismic events and their magnitude. Regularities have been established in the behavior of tipper variations for remote and regional earthquakes for the first quarter of 2016 with an energy class from 6 to 10.

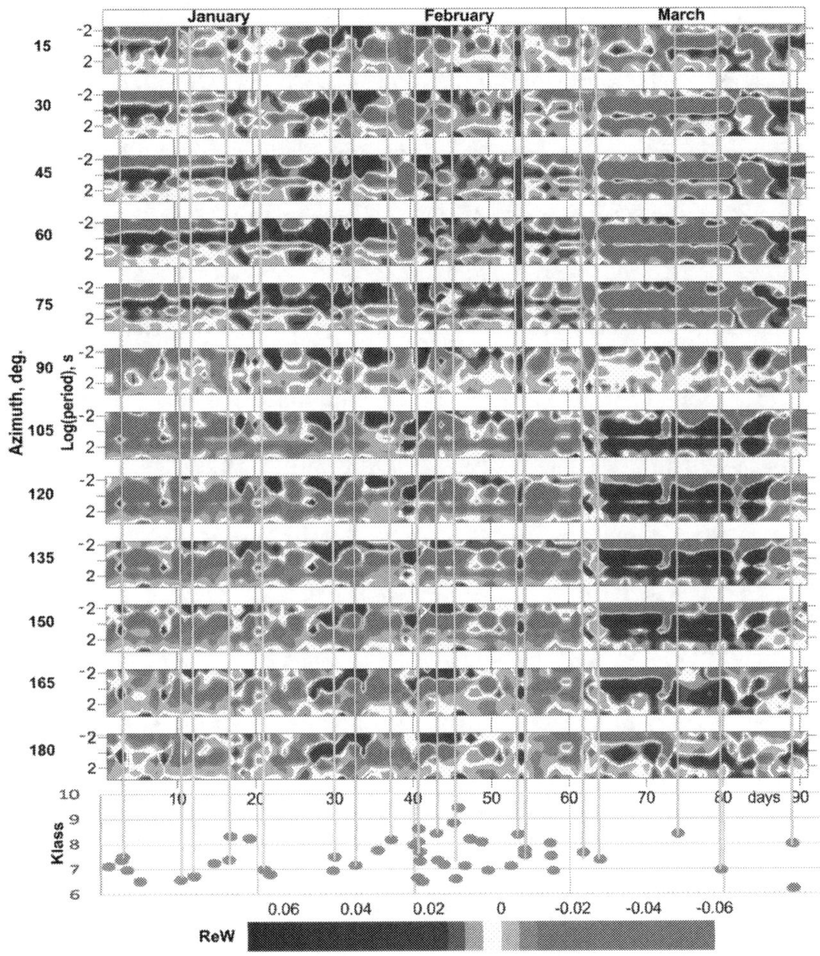

Figure 5. Time-frequency series of azimuth MT-monitoring (tipper variations) for Chon-Kurchak station in comparison with seismicity for January-March 2016 (ovals - seismic events).

The presented data clearly show that the behavior of such an electromagnetic parameter as a tipper (Wiese vector) can be considered as an indicator of geodynamic processes that accompany the preparation and implementation of seismic events.

The work was carried on the state assignment of the Research Station of the Russian Academy of Sciences in Bishkek (AAAA-A19-119020190063-2).

References

1. G.A. Sobolev, N.A. Zakrzhevskaya, J. Volcanol. Seismol. **14 (3)**, 137-144 (2020).
2. S.K. Park, M.J.S. Johnston, T.R. Madden, F.D. Morgan, H.F. Morrison, Rev. Geophys. **31 (2)**, 117-143 (1993).
3. D. Stanica, M. Stanica, Acta Geodyn. et Geomater. **4 (1)**, 99-107 (2007).
4. E.A. Bataleva, V.A. Mukhamadeeva, Geodynam. Tectonophys. **9 (2)**, 461-487 (2018) (In Russian).
5. E.A. Bataleva, V.Y. Batalev, A.K. Rybin, Izv. Phys. Solid Earth **49 (3),** 402-410 (2013).
6. E.A. Bataleva, A.K. Rybin, V.E. Matyukov, Data **4,** 99 (2019).
7. B.S. Svetov, S.D. Karinsky, Yu.I. Kuksa, V.I. Odintsov, Phys. Earth **5**, 36-46 (1997).
8. A. Rybin, V. Batalev, L. Fox, A. Ingerov, G. Schelochkov, Bulletin KRSU **6 (3)**, 87-94 (2006).
9. A.K. Rybin, V.V. Spichak, V.Yu. Batalev, E.A. Bataleva, Matyukov V.E., Russ. Geol. Geophys. **49 (5)**, 337–349 (2008).
10. V.D. Bragin, Active Electromagnetic Monitoring of the Bishkek Prognostic Test Area. PhD Thesis (Candidate of Physics and Mathematics). 135 pp, Moscow (2001) (in Russian).
11. L. Fox, Satellite-Synchronized 3-D Magnetotelluric System. U.S. Patent № 6 191 587 B1 (2001)
12. O. Ingerov, Not. Min. Inst. **162**, 15-25 (2005).
13. W. Rodi, R. Mackie, Geophys. **66**, 174–187 (2001).
14. M.N. Berdichevsky, V.I. Dmitriev Models and Methods of Magnetotellurics, 563 pp, Springer, Berlin (2008).
15. A.A. Zhamaletdinov, F.P. Mitrofanov, A.D. Tokarev, A.N. Shevtsov, Dokl. Earth Sci. **371 (2),** 403-407 (2000).
16. L. Bogomolov, V. Bragin, A. Fridman, V. Makarov, G. Sobolev, E. Polyachenko, G. Schelochkov, V. Zeigarnik, A. Zubovich, Tectonophys. **431 (1-4)**, 143-151 (2007).
17. J.P. Busby, Geoph. Prospect. **48**, 677-695 (2000).

AUTHOR INDEX

AUTHOR INDEX

EDP Sciences – Web of Conferences
17, Avenue du Hoggar
Parc d'Activité de Courtabouf
BP 112
F-91944 Les Ulis Cedex A
France

ISBN 978-1-7138-3893-7

CURRAN ASSOCIATES INC.
proceedings.com